AF699132

WITHDRAWN

ASSESSMENT OF BIODIVERSITY FOR IMPROVED FOREST PLANNING

FORESTRY SCIENCES

Volume 51

The titles published in this series are listed at the end of this volume.

Assessment of Biodiversity for Improved Forest Planning

Proceedings of the Conference on
Assessment of Biodiversity for Improved Planning,
7–11 October 1996, held in Monte Verità, Switzerland

Edited by

PETER BACHMANN

Swiss Federal Institute of Technology,
ETH Zentrum,
Zürich, Switzerland

MICHAEL KÖHL

Dresden University of Technology,
Chair of Forest Biometrics and Computer Sciences,
Tharandt, Germany

and

RISTO PÄIVINEN

European Forest Institute,
Joensuu, Finland

KLUWER ACADEMIC PUBLISHERS

DORDRECHT / BOSTON / LONDON

A C.I.P. Catalogue record for this book is available from the Library of Congress.

SD
118
.C60
1996

ISBN 0-7923-4872-9

Published by Kluwer Academic Publishers,
P.O. Box 17, 3300 AA Dordrecht, The Netherlands.

Sold and distributed in the U.S.A. and Canada
by Kluwer Academic Publishers,
101 Philip Drive, Norwell, MA 02061, U.S.A.

In all other countries, sold and distributed
by Kluwer Academic Publishers,
P.O. Box 322, 3300 AH Dordrecht, The Netherlands.

Printed on acid-free paper

All Rights Reserved
© 1998 Kluwer Academic Publishers
No part of the material protected by this copyright notice may be reproduced or
utilized in any form or by any means, electronic or mechanical,
including photocopying, recording or by any information storage and
retrieval system, without written permission from the copyright owner.

Printed in the Netherlands

FOREWORD

In some parts of the world, biological resources are being depleted faster than they can regenerate. Paradoxically, in Europe, forest resources are being under-utilised. Extinction is threatening plants and animals in some over-exploited ecosystems but in European forests, fragmentation and depleted biodiversity through intensive management are more realistic risks. Preserving forest biodiversity has internationally been stated as a common target, and thus sets limits to the use of natural resources. The countries that have ratified the environmental program "Global Biodiversity Strategy", signed in 1992 in Rio de Janeiro, and the resolutions at the Ministerial Conferences on the Protection of Forests in Europe in Strasbourg, 1990, and Helsinki, 1993, have each committed themselves to monitor nationally the state of biodiversity and to preserve the characteristic natural variation in their countries. In many parts of the world the conservation of forest biodiversity has reached a value that is equivalent or even more important than timber production.

Sustainability and long-term planning are the two terms best describing the philosophy of traditional management practices. However, the traditional planning techniques are not primarily developed to maintain sustainability of biodiversity. There is also a lack of (cost-) efficient techniques for monitoring forest biodiversity on large, representative scales as well as on the stand level. Most monitoring approaches which are currently applied quantify the productive function of forests and fail to represent the state and long term changes in biodiversity.

The gap between the international commitments and the practices in forest assessment and management is obvious. To help to close this gap the European Forest Institute (EFI) Joensuu, Finland, the Swiss Federal Institute of Technology (ETHZ), Zürich, Switzerland and the Swiss Federal Institute for Forest, Snow and Landscape Research (WSL/FNP), Birmensdorf, Switzerland, organised a conference on the Monte Verità in Ascona, Switzerland, in October 1996. The Monte Verità Conference provided a platform to exchange experience on methodology to assess and monitor the variation of ecosystems and habitats in relation to biodiversity conservation and to integrate biodiversity into regional planning of forest management and land use. During the Monte Verità Conference the state-of-the-art in the field of natural resource assessments, with special reference to forest biodiversity, was reviewed, the progress in integrating data on biodiversity in forest management planning was presented and the information needs regarding biodiversity conservation and the question to what degree assessment methods for forest biodiversity can be simplified for practical applications were discussed.

More than one hundred scientists from 22 countries met on the Monte Verità. In many parts of the world the need for conserving forest biodiversity has initiated the development of techniques for integrating biodiversity issues into regional planning, forest management and land use. Due to the complexity and the long-term dynamics of forest ecosystems as well as the heterogeneity of forests there exist no standard solutions.

During the last decades the public concern about our environment and nature conservation has continuously increased. Currently, however, the conservation of forest biodiversity has mainly to be financed from timber sales. Even if research does provide efficient monitoring tools and foresters apply management methods to conserve biodiversity, politicians are responsible for transferring the international commitments to a legal and financial basis for forest biodiversity conservation at the local level.

In organising the Monte Verità conference we were generously supported by many people. We wish to express our deep appreciation to Leena Roihuvuo, who handled most of the laborious administration and kindly offered her services to professionally compile the papers of this book. Mrs. Katja Bastianelli, Centro Stefano Franscini, provided inestimable help and professional expertise in the local organisation of the Conference. As a result of her efforts the mood of the Conference was very convivial. The Centro Stefano Franscini, the ETHZ-Centre for Seminars on the Monte Verità and the Swiss Federal Research Foundation generously sponsored this meeting. Without their financial contribution the Conference could not have been held. The papers of this volume have gone through an intensive review process. We appreciate the efforts, participation and support of all the authors and the reviewers. It is our hope that this volume may provide a significant contribution to the recognition and further development of the assessment of biodiversity for improved forest planning.

PETER BACHMANN
Professor
Swiss Federal Institute of Technology (ETHZ)
Chair of Forest Planning
Zürich, Switzerland

MICHAEL KÖHL
Professor
Dresden University of Technology
Chair of Forest Biometrics and Computer Sciences
Tharandt, Germany

RISTO PÄIVINEN
Deputy Director
European Forest Institute (EFI)
Joensuu, Finland

TABLE OF CONTENTS

International Cooperatives on Monitoring Forest Biodiversity

Information Needs, Economic and Policy Consequences of Biodiversity Oriented Forest Management

Quantification of Forest Diversity

Assessment and Monitoring Techniques for Forest Biodiversity

International Cooperatives on Monitoring Forest Biodiversity

ACHIEVING GLOBAL AND REGIONAL PERSPECTIVES ON FOREST BIODIVERSITY AND CONSERVATION

V. KAPOS AND S.F. IREMONGER
World Conservation Monitoring Centre
219 Huntingdon Road
Cambridge CB3 0DL
U.K.

Abstract

Forests are arguably the single most important repository of global biodiversity, attracting the attention of conservation planners as well as foresters. Diversity is an essential factor in maintaining forest function, so its conservation and management are important issues in forest planning.

Because species and ecosystems are no respecters of national boundaries, and international collaboration is required to ensure their preservation, conservation issues are increasingly being viewed on global and regional, as well as national scales. However, building reliable pictures of biodiversity resources at these scales is a complex task. Direct measures of species diversity over broad areas are neither possible nor meaningful. Instead, the issue may be addressed through indicators of species diversity such as ecosystem diversity, indicators of forest condition, and the identification of major centres of speciation and endemism. Even for these surrogates, problems arise in the process of scaling up from more local data sets because of the different definitions, classifications and data sources used by different agencies and groups of researchers. The broad scale data sets that are available for global and regional scale forest conservation planning include global forest cover, ecoregion distribution, protected areas coverages, endemic and important bird areas, distributions of endangered and threatened species, and information on deforestation and trade in endangered species.

Information as yet unavailable which would contribute substantially to broad scale perspectives on conservation of forest systems includes:

a) potential forest cover at global and regional scales,
b) more complete information on plant and invertebrate species diversity and distribution, and
c) measures of the relationships between people and the forest, e.g. use of forest products, and their impacts on forest ecosystems and biodiversity.

Improvements in the understanding of issues and processes in global forest biodiversity and its conservation will be derived from global assessment exercises like the Forest Resources Assessment of the FAO, increasing global compatibility of criteria and indicators for sustainable forest management, forest certification processes, and national reporting exercises, e.g. for the Convention on Biological Diversity.

P. Bachmann et al. (eds.), Assessment of Biodiversity for Improved Forest Planning, 3–13.
European Forest Institute Proceedings no. 18. © 1998 *Kluwer Academic Publishers. Printed in the Netherlands.*

1. Biodiversity and its Relation to Forests

The conservation of biodiversity has become a priority of the international community since the Convention on Biological Diversity (CBD) was drafted in Rio in 1992. Forests are the repository of much of the world's biodiversity, and therefore foresters must assume a degree of responsibility for its management and conservation. In fact, much of the forest managed for nature conservation in the tropics is held by the forestry sector (WCMC 1993).

2. The Need for a Global Perspective

For both practical and political reasons, there is an increasing need to develop global and regional perspectives on forests and their biodiversity. Ecosystems and species are no respecters of national boundaries; their protection and sound management require international collaboration and a view of the resource as an integrated whole. Global, regional and national conservation priority setting all depend on an understanding of the distribution of species and ecosystems, their protection status, and the threats to them. These parameters are all affected by being viewed in the broader context: national centres of diversity may or may not be of regional importance; sovereignty over such centres or other key ecosystems may be shared by several nations; and pressures and threats may be due to regional causes or conditions in neighbouring countries. With appropriate regional level information, a nation can evaluate the importance of its forests for their many productive and protective functions, including as refuges for rare and endangered species (particularly those unique to that country, for which the nation bears ultimate responsibility), and decide its own priorities in this context. An overview of the global state of forest systems, their composition and function, is essential for prioritising conservation efforts at the international scale.

International initiatives (such as UNCED, the CBD, the CSD, Agenda 21 and the Helsinki and Montreal Processes) are demanding a concerted effort on the part of national governments and international bodies to control the decrease in global forest cover and quality. Signatories to these initiatives are bound by law to aid and contribute to the initiatives to the best of their respective capacities. The process is experiencing some growing pains, however, in that the requirements defined by the initiatives (e.g. the Helsinki and Montreal processes on criteria and indicators of sustainable forest management) may exceed the capacities of national forest departments. This issue is being addressed by UN agencies, which are currently involved in capacity-building programmes. The improved national level information generated by these assessment and reporting exercises will facilitate the construction of regional and global biodiversity data sets.

However, assembling valid regional and global views of forest biodiversity is a problematic task. Distribution of forest types (ecosystem diversity) may be well understood for any given country, but joining national data sets is fraught with problems caused by differences in definitions, classification systems and data sources between countries. Assembling species diversity information on the global scale is problematic for all but the highest groups of vertebrates, and even these may not be fully known in some regions (witness the recent history of new primate species in Brazil). So little is known about the distribution of genetic diversity in forests, that there is little point in addressing the question of a global overview of this aspect of biodiversity. The remainder of this paper is devoted to discussing what is currently available to assemble global perspectives on forest ecosystem and species diversity.

3. Ecosystem Diversity

The greatest difficulty encountered in assembling a global view of forest ecosystem diversity is the problem of how to harmonise or make compatible data which derive from different sources and are based on different definitions and classification systems. Even the international bodies theoretically compiling global data on forest cover have had difficulty in agreeing a good, basic and globally consistent definition of what a forest is (Box 1). The use of different degrees of closure in defining forest can make an appreciable difference to the estimated total area of forest cover derived for any given country or location (Box 2), and the problems involved in harmonising different classification systems to examine the distribution of forest types and evaluate patterns of ecosystem diversity are even greater.

WCMC has recently compiled a global map of closed forest based on 70 different sources of forest cover data, which mostly date from between the early 1980s and the early 1990s (WCMC 1996b). The crudeness of its five-class classification (temperate needleleaf, temperate broadleaf and mixed, tropical moist, tropical dry and mangrove forests) reflects the difficulty of combining the many much more detailed classification systems used throughout the world.

BOX 1. Definition of Forest Cover

UNESCO (UNESCO 1973):	*Closed Forest* - trees >=5m with crowns interlocking *Woodland* - trees >=5m tall with crowns not usually touching but with canopy cover >=40%
US classification standards (FGDC 1995):	*Closed Tree Canopy* - trees with crowns interlocking, with crowns forming 60-100% cover *Open Tree Canopy* - trees with crowns not usually touching forming 10- or 25-60% cover
FAO (FAO 1993 and 1995)	*Forest (Developing Countries)* – 10% crown cover of trees and/or bamboos *Forest (Developed Countries)* – tree crown cover (stand density) of more than 20% of the area *Closed forest (tropical countries)* – tree crown cover greater than 40%

BOX 2. Sample effects of different forest definitions:

Senegal	is 40% forested using FAO's 10% tree cover definition, which includes dry woodland, but is only 2% covered by closed forest
Australia	is 5% forested according to FAO's 20% tree cover threshold, but is only 0.4% covered by dense forest (with >70% canopy cover)

Global satellite-derived forest cover data is not yet available, despite many regionally-focused programmes such as TREES and Pathfinder, which can contribute to it. A global satellite-based landcover map should be produced by the EROS Data Center (EDC) by the end of 1997. This will help to provide a consistent global view of where forests are, and some information about distribution of structurally different forests in terms of different categories of canopy closure. However, information about the distribution of forest vegetation classes is crucial to understanding the different roles of forests in carbon sequestration, hydrological cycles and other ecosystem processes, in supplying wood and non-wood forest products, and in supporting biodiversity. A global vegetation classification system is needed to ensure consistency, and it is likely that such a system will emerge from the combined efforts of FAO (ecofloristic zones), EDC and the inter-agency project on developing a General Global Nomenclature for Land Cover and Land Use involving UNEP, FAO, WCMC, ITE, ITC and WAU. Once the global classification system is resolved, it will be possible to achieve a global perspective on ecosystem diversity of remaining forests (rather than just their extent).

4. Species Diversity

Knowledge of the world's species diversity and its distribution is very incomplete. Some groups of organisms (e.g. conifers or birds; Table 1) and some parts of the world are far better known than others. The information presented in Table 1 refers only to total species diversity and does not address the question of whether individual species occur in, and/or depend on, forests. A global view of forest species diversity depends on such distinctions being made for at least some groups and is thus some way off.

On a regional scale, progress in mapping total species richness in many groups has been greater (e.g. the Mapping African Biodiversity Patterns project of the Danish Centre for Tropical Biodiversity), but as yet these efforts do not incorporate separate analysis of forest species.

However, other approaches can be used to evaluate species diversity and/or identify global priorities for its conservation. Endemic Bird Areas (EBAs; Bibby et al. 1992) and Centres of Plant Diversity (CPDs; WWF and IUCN 1994) of the world can be overlaid with forest cover data to determine forest biodiversity "hot spots".

EBAs are areas which contain at least two restricted-range bird species (those whose distribution covers less than 50,000 square kilometres). These have been reasonably accurately mapped and classified according to their major habitat type. They have also been ranked according to conservation importance. In the absence of similar data for other taxonomic groups, EBAs provide a useful measure of the importance for biodiversity of particular geographic areas.

A joint WWF/IUCN project has identified, mapped and described in detail over 200 Centres of Plant Diversity (CPD) worldwide (WWF and IUCN 1994). Far more areas than these meet the selection criteria (Box 3), but have not yet been analysed and described in detail. Unfortunately, the criteria have not been applied uniformly across the world. Moreover no size criteria have been imposed, so that CPDs vary in size from a few tens of square kilometres to over one million square kilometres. These factors seriously limit their usefulness for analysis and comparison.

In addition to showing the locations of forest biodiversity "hot spots", overlaying current forest cover with EBA and CPD distributions would allow determination of the

TABLE 1. Data availability for the major taxonomic groups.

	No. of forest occurring species	No. of endemic and /or restricted range forest species	No. of threatened forest species.
Birds	Data available for all countries. Require compilation and classifying into forest/non-forest.	Data available for all countries. Require classifying into forest/non-forest.	Data available for all countries. Require classifying into forest/non-forest.
Mammals	Data available for all countries. Require compilation and classifying into forest/non-forest.	Data available for all countries. Require classifying into forest/non-forest.	Data available for all countries. Require classifying into forest/non-forest.
Reptiles	Data incomplete.	Data available for most countries. Require classifying into forest/ non-forest.	Data incomplete.
Amphibians	Data incomplete.	Data available for most countries. Require classifying into forest/ non-forest.	Data incomplete.
Invertebrates	Data very incomplete.	Data available for a small number of groups (dragonflies, swallowtail butterflies). Require classifying into forest/ non-forest.	Data very incomplete.
Trees	Data to be completed by end of 1997.	Data to be completed by by end of 1997.	Data to be completed by end of 1997.
Other plants	Data very incomplete.	Data very incomplete.	Data very incomplete.

BOX 3. Criteria for defining Centres of Plant Diversity

- area is evidently species-rich, even though the number of species present may not be accurately known;
- area is known to contain a large number of species endemic to it.

Other characteristics also considered include:
- site contains an important genepool of plants of value or potential use to humans;
- site contains a diverse range of habitat types;
- site contains a significant proportion of species adapted to special edaphic conditions;
- site is threatened or under imminent threat of large-scale devastation.

numbers and areas of forest units which occur in EBAs and CPDs. This may be used as a tool for planning future protected areas or for managing those that already exist. However, because an area important for bird or plant conservation will not automatically be a priority for other groups organisms, the limitations of this approach must be considered in its application.

The occurrence of threatened and endangered forest species could also be mapped to identify areas where they are concentrated. This depends on the categorisation of species as to their forest-occurrence and ideally would require relatively detailed data on their distributions. Nonetheless, some advances could be gained simply from addressing presence/absence data at national or provincial scales.

A further approach to examining global patterns of forest biodiversity is the development of some sort of broad scale biodiversity index, combining information on the diversity found in different groups of organisms. Preliminary work at WCMC has generated a national biodiversity index based on species richness and endemism for vertebrates and plants, normalised and averaged across groups so that it can make use of incomplete data sets (WCMC 1994). Further development of this approach is needed so that it can address biodiversity for particular ecosystem types such as forests.

A country's forests may have high species diversity either due to a wide range of different forest types, each with its own distinct biota (e.g. the USA), or because individual forest types are highly diverse (e.g. lowland tropical moist forest). The former is generally related to the size of the country, the latter not necessarily so. Countries with very high forest diversity usually combine these two. Measures of diversity that take into account the size of the country can be derived to give an indication of the richness or importance of a country's forests per unit area.

Other approaches to identifying priority areas in terms of species diversity conservation at broad scales (Box 4) depend to a high degree on harnessing expert knowledge, but do not necessarily require detailed field survey data (WCMC 1996b).

BOX 4. Other Biodiversity Inventory Techniques for use at Broad Scales

Conservation Biodiversity Workshops (Tangley 1992)
Workshop discussions among regional field experts on species and ecosystems based on mapped info on vegetation, land use, topography, and distribution of key species generate agreed biological priorities for conservation presented in mapped form.

Conservation Needs Assessments (Alcorn 1993)
Similar to above with incorporation of socioeconomic and political perspective resulting in identified priorities that take account of social and political realities

Biodiversity Information Management System (e.g. MacKinnon 1991)
Existing habitat maps and species habitat requirements are used to model species distributions through a relational database and to monitor conservation status of species and ecosystems.

5. Habitat Condition and Management

Other information useful in establishing a global view of biodiversity is that which assesses the current condition and management of habitats. There are relatively few parameters that can serve as effective indicators of forest condition at the global scale.

5.1 FOREST FRAGMENTATION

Measures of the spatial continuity of forests and the size and shape of remaining forest patches are one means of determining something about the probable condition of existing forests and the likely state of their biodiversity. Deforestation not only removes forest cover, it also disrupts the continuity of the remaining forest and affects ecological processes within it. Remaining forest fragments may be too small to be effective habitat units for many species. The viability of their component populations may be limited by isolation from other forest areas, which restricts gene flow through pollination and dispersal. Remaining forests (both fragments and large blocks) are also affected by proximity to the forest-nonforest interface. These "edge effects" may be ecological, including influences on microclimate (e.g. Kapos 1989; Camargo and Kapos 1995) and changes in species composition (e.g. Laurance 1991; Matlack 1994), or anthropic as they relate to pressures and the probability of further disruption of the forest by human activity. Forests near edges are far more likely to be heavily exploited and disturbed by people than those in distant core areas. The smaller a forest patch, the less chance each species population has of survival in the long term.

Thus, an indication of the fragility of the forest system and its overall health as a functioning ecosystem can be gained from an examination of the size class distribution of remaining forest patches. In the case of the AVHRR data that will be the basis for the next view of current global forest cover, the smallest discernible forest patch is likely to be 3-5 km^2. However, in some countries, particularly those with low forest cover, patches much smaller than 5 km^2, and even patches less than 1 km^2 can be very important as refuges for remnant populations of forest species. Higher resolution satellite imagery could be used on a sampling basis to estimate the importance of such patches in any given country or region.

Further information about the state of remaining forests and their biodiversity can be derived from analyses of their shapes using perimeter to area ratios. High ratios indicate that more of the forest is exposed to the influence of an edge and may have been affected by human activities both inside and outside the forest. Forest units with lower ratios will have some "core" or central area that is buffered from external disturbances and therefore more likely to have intact ecosystem functionality. Average perimeter:area ratios can be calculated at national or other scales. Similar information can be obtained by examining the total "core" area of forest more than an arbitrary buffer distance from the perimeter; total "core" forest area can be calculated at national or other spatial scales. Both of these approaches have been used by FAO for some tropical countries (FAO 1993).

Both size class and core-area analyses can only be used for broad scale comparisons if the data on forest cover originate from comparable sources at equivalent resolutions. Thus, the global land cover data set being generated by EDC will be an excellent basis for such analyses, which will be even more useful in setting conservation priorities if they can be applied to different forest types.

5.2 FOREST PROTECTION

The protection status of any given forest can be a predictor of its condition; forests that have been set aside for conservation purposes are often in a better state than those designated for exploitation, though this relation is far from infallible. From the standpoint of global forest biodiversity conservation priority setting, a useful analytic technique is that of Gap analysis (Scott et al. 1993) in which the extent of existing protected areas is overlaid with maps of species and ecosystem distribution to identify gaps in the protection network. At the global scale this has been done for the five broad forest categories included in the World Forest Map (WCMC 1996a). The results of this (Box 5) show clearly that Temperate Needleleaf and Tropical Dry forests are the most poorly protected, while Mangrove is the best protected. However, to be truly meaningful for biodiversity, such an analysis needs to focus on ecosystem classifications that reflect ecosystem diversity much more directly. One such analysis has been done for the tropics using FAO's Ecofloristic Zone designations (WCMC 1995). A more detailed analysis of the Indo-Malayan Realm has been carried out (Asian Bureau for Conservation 1996) and a feasibility study for a European analysis was carried out by WCMC in 1995. Results of a global gap analysis of extant forest data from a variety of sources using a harmonised global classification will be presented at the World Forestry Congress (Iremonger et al. 1997). Similar analysis using a consistent remotely sensed global forest cover data set and improved protected area information will be yet more useful as the basis for taking steps to build a truly effective global protected areas network.

BOX 5. Global Forest Protection

Forest Type	Global Area (km^2)	Percentage Protected
Tropical Moist	11.2 million	8 %
Tropical Dry	0.8 million	5 %
Temperate Broadleaf/ Mixed	7.2 million	6 %
Temperate needleleaf	13.9 million	5 %
Mangrove	0.2 million	9 %

5.3 FOREST MANAGEMENT

Other aspects of forest management are important for the global analysis of forest biodiversity status. Forest reserves managed for watershed and soil protection, for or by indigenous peoples, and for religious or spiritual functions are all likely to be relatively effective reservoirs of forest species diversity. Forests managed for the extraction of timber and non-timber forest products are important in biodiversity conservation, depending on the type and quality of their management. The drive towards assessing sustainability of forest management according to standardised schemes of criteria and indicators (ISCI 1996) will improve our ability to monitor forest management in relation to biodiversity on a global scale (though the indicators relating directly to biodiversity have yet to be developed in most cases). The certification programme backed by the

Forest Stewardship Council (Dudley et al. 1996) also provides a basis for tracking sites that are managed sustainably, i.e. in ways not prejudicial to the native forest species, world-wide.

6. Pressures on Forests and their Biodiversity

Another approach to assessing the status of forest biodiversity is to examine the pressures on forests at a broad scale. Many of the most important pressures on forests (e.g. large scale conversion, timber harvesting, pests and diseases, invasive alien species, agricultural encroachment, airborne pollutants/acid precipitation, fuelwood extraction, poaching/illegal collection, fire, natural catastrophes) operate at such local scales that achieving any kind of global overview is very problematic.

A certain amount can be achieved using a sampling approach and extrapolating from local studies, for example on the impact of non-timber forest product extraction. However conditions vary so much from one place to another that these kind of extrapolations can be quite dangerous. Two other approaches examine pressures on forest in a more generic way and may be more appropriate to global scale analysis.

One is assessing loss of original forest cover. The absence of forest where it once existed is *de facto* evidence of pressure on forests, though this may be either current or historic. WCMC has recently compiled a preliminary estimate of original forest cover at the global scale, which can be compared with the global map of current forest cover to estimate total loss of original forest cover. Besides the difficulties caused by its intrinsically hypothetical nature, original forest cover mapping suffers from the same classification problems as assessing current forest cover and requires a globally consistent classification system before it can contribute to the identification of the forest types most at risk and critically important remnants.

The second way of assessing generalised pressures on forests is the evaluation of likely human impact or wilderness assessment. The Australian Wilderness Index (Box 6), which estimates wilderness value along a continuum, is one means of making such an assessment that might be applicable at the global scale. A global "wilderness surface" could be generated, based on road and settlement data, which could then be overlaid with current forest cover information to identify areas of key forest types exposed to both high and low degrees of human influence. Some modification will be required to take

BOX 6. The Australian National Wilderness Index (Lesslie and Maslen 1995)

The Australian Wilderness Index combines several different measures of the "naturalness" of a site:

- remoteness from access – the distance to the nearest access route in each of several grades
- remoteness from settlement – the distance to the nearest human population centre according to its grade
- aesthetic naturalness – the distance of the site from structures of modern society
- biophysical naturalness – an estimate of the intensity of disturbance of the natural vegetation often based on land use

These four measures are combined to give an overall Wilderness Index, but if adequate data are lacking (as is often the case for land use intensity) a particular component can be left out.

into account regional variations in, for example, the relative importance of different types of access such as roads and rivers. Ultimately, other factors modulating pressure such as human population density could be incorporated.

7. Conclusions

Assembling a global perspective on biodiversity is a slow and complex task. Major advances have been made in compiling global data sets on current and original forest cover, but these need to be viewed in the context of a global vegetation classification, which is still lacking. Other advances at the global scale have come from the identification of areas important for species diversity in key groups, but congruence of diversity patterns between groups has been addressed only at the regional scale, and not yet thoroughly even then.

The increasing volumes of data involved in the analysis and management of global forest biodiversity require new approaches to data management. National capacities to gather and manage data to generate useful information both for national use and to contribute to our understanding of the global picture will need to be built. Recent experiences show that the greatest challenges are organisational, not technological (UNEP/WCMC 1996). By focusing on the processes involved in creating environmental information, as opposed to concentrating on data, international efforts such as the Biodiversity Data Management Project (UNEP/WCMC 1996) are making useful advances.

In addition to contributions from national sources, international research and other efforts can also make a useful contribution to the improvement of the global biodiversity information base. Likely sources of improvement in the broad-scale information available in the near future include:

- a globally consistent land cover data set from AVHRR imagery;
- progress towards a global vegetation classification;
- analysis of the resulting spatial data sets to examine patterns of forest fragmentation and wilderness evaluation;
- FAO Forest Resources Assessment 2000, including more biodiversity-related parameters & efforts to improve global consistency;
- research to: examine the congruence of diversity patterns among groups; extend knowledge of species distributions and habitat requirements; examine effects of different forest management practices on biodiversity;
- national reporting efforts for the Convention on Biological Diversity;
- initiatives on development of criteria and indicators of sustainable forest management;
- forest management certification efforts.

Combining these and other initiatives should enable a much more coherent global view of biodiversity in forests to emerge. However it must be remembered that the crucial step in maintaining biodiversity in forests is the translation of this knowledge into improved forest management and conservation practices throughout the world.

References

Alcorn, J. (ed.) 1993. Papua New Guinea Conservation Needs Assessment. Vol. 1. Biodiversity Support Program, Washington DC, and Dept. of Environmental and Conservation. Boroko. Papua New Guinea. 216 p.

Asian Bureau for Conservation 1996. Review of the protected areas system in the Indomalayan Realm. Unpublished manuscript.

Bibby, C.J., Collar, N.J., Crosby, M.J., Heath, M.F., Imboden, C., Johnson, T.H., Long, A.J., Sattersfield, A.J. and Thirgood, S.J. 1992. Putting Biodiversity on the Map: priority areas for global conservation. ICBP. Cambridge. UK. Vi+90 p.

Camargo, J. and Kapos, V. 1995. Complex edge effects on soil moisture and microclimate in Central Amazonian forest. Journal of Tropical Ecology. 11: 205-221.

Dudley, N. Gilmour, D. and Jeanrenaud, J.P. 1996. Forests for Life. The WWF-IUCN Forest Policy Book. WWF-International. Gland. Switzerland.

FAO 1993. Forest Resources Assessment 1990 – Tropical countries. FAO Forestry Paper 112. Rome. Italy.

FAO 1995. Forest Resources Assessment 1990 – Global synthesis. FAO Forestry Paper 124. Rome. Italy.

FGDC 1995. FGDC Vegetation Classification Standards. Federal Geographic Data Committee. Reston Va. USA. Unpublished.

Iremonger, S.F., Kapos, V., Rhind, J. and Luxmoore, R.L. 1997. A global overview of forest conservation. World Forestry Congress. Antalya. Turkey.

ISCI 1996. Intergovernmental Seminar on Criteria and Indicators for Sustainable Forest Management: Background Document. Ministry of Agriculture and Forestry. ISCI. Helsinki. Finland. 131 p.

Kapos, V. 1989. Effects of isolation on the water status of forest patches in the Brazilian Amazon. Journal of Tropical Ecology. 5: 173-185.

Lesslie, R. and Maslen, M. 1995. National Wilderness Inventory Australia: Handbook of Procedures, Content and Usage. Australian Government Publishing Service. Canberra. Australia.

MacKinnon, J. 1991. National Conservation Plan for Bhutan. Annex I. Forestry Master Plan for Bhutan. ADB/RoyalGovernment of Bhutan. 120 p.

Scott, J.M., Davis, F., Csuti, B., Noss, R., Butterfield, B., Groves, C., Anderson, H., Caicco, S., D'Erchia, F., Edwards, T.C. Jr., Ulliman, J. and Wright, R.G. 1993. Gap Analysis: A Geographic Approach to the Protection of Biological Diversity. Wildlife Monographs. 123: 1-41.

Tangley, L. 1992. Computers and Conservation Priorities. Mapping Biodiversity. Lessons from the Field I. Conservation International. Washington, DC. 28 p.

UNESCO 1973. International Classification and Mapping of Vegetation. United Nations Educational, Scientific and Cultural Organization. Paris. France. 93 p.

UNEP/WCMC 1996. Framework for Information Management in the Context of the Convention on Biological Diversity. World Conservation Monitoring Centre. Cambridge. UK.

WCMC 1993. Assessing the conservation status of the world's tropical forest. Summary of the WCMC Tropical Managed Areas Assessment Project. Unpublished manuscript.

WCMC 1994. Biodiversity Data Sourcebook. Groombridge, B. (ed.). World Conservation Press. Cambridge. UK.

WCMC 1995. Biodiversity conservation in the tropics: Gaps in habitat protection and funding priorities. Compiled by Murray, M.G., Green, M.J.B., Bunting, G.C. and Paine, J.R. World Conservation Monitoring Centre. Cambridge. UK. unpublished report.

WCMC 1996a. The WWF World Forest Map. World Conservation Monitoring Centre. Cambridge. UK.

WCMC 1996b. Assessing Biodiversity Status and Sustainability. Groombridge, B. and Jenkins, M.D. (eds.). World Conservation Press. Cambridge. UK. 104 p.

WWF and IUCN 1994. Centres of Plant Diversity. A guide and strategy for their conservation. 3 volumes. IUCN Publications Unit. Cambridge. UK.

MEASURING AND MONITORING FOREST BIODIVERSITY: THE SI/MAB MODEL

FRANCISCO DALLMEIER
SI/MAB Biodiversity Program
Smithsonian Institution
1100 Jefferson Drive, SW
Suite 3123
Washington, D.C. 20560 USA

Abstract

Over the past decade, the Smithsonian Institution/Man and the Biosphere Biological Diversity Program (SI/MAB) has developed a biodiversity monitoring protocol and refined it through testing at a number of sites. The protocol has been adopted by SI/MAB colleagues and partners around the world, forming the basis for consistent measuring and monitoring at hundreds of biodiversity monitoring plots, facilitating the transfer of comparable data, and providing the framework for data analysis and dissemination to users. Program staff have trained more than 700 participants in the SI/MAB protocol and in use of BioMon (SI/MAB's computerized data management system) at annual six-week international courses and through on-site two-week courses in more than two dozen countries. Staff have also been instrumental in setting up and maintaining international biodiversity information exchange networks. The primary goals of the program – to gather, analyse, and disseminate information about biodiversity at selected forested sites throughout the world and to foster co-operation among researchers and decision-makers – are being met. The outlook for strengthening established networks and expanding the program into new areas is bright.

1. Introduction: Why Focus on Biodiversity Monitoring?

The evidence is mounting. World-wide, numerous species are going extinct, and even more that have not yet been identified are likely to be similarly threatened. Perhaps the most comprehensive data about extinctions and threats of extinction are found in the Red Lists and Red Data Books published by the World Conservation Monitoring Centre (WCMC 1992; IUCN 1995; UNEP 1995). Those data indicate that in 1994, just for the species about which enough is known to assess their status, nearly 5,400 animal species and more than 26,100 plant species were threatened; that is, they were listed as endangered, vulnerable, rare, or indeterminate. The numbers are certainly much higher, considering that there is not enough information to determine how most of the 1.75 million species described to date are faring, let alone the millions of species yet to be described.

P. Bachmann et al. (eds.), Assessment of Biodiversity for Improved Forest Planning, 15–29.
European Forest Institute Proceedings no. 18. © 1998 *Kluwer Academic Publishers. Printed in the Netherlands.*

The loss of even one species diminishes the earth's store of biological diversity, for once eliminated, a species cannot be recovered or regenerated. All possibilities the species had for bettering life are gone, including its potential to provide the basis of life-saving medicines or new or improved foodstuffs to feed a burgeoning human population. Species losses are also felt at the genetic, community, and landscape levels. When a species is gone, so is its genetic heritage that allowed the species to "adapt to local conditions and unpredictable future events..." (Rudolph 1990; see also Millar and Libby 1991 in Boone and Aplet 1994). As well, the unique roles that the species played at the community (populations of species) level and landscape (groupings of ecosystems) level are left unfulfilled.

Species are known to have survived the many alterations to habitat and landscapes caused by natural, non-cataclysmic events such as gradual soil erosion by waterways and even cataclysmic events such as earthquakes, typhoons, hurricanes, wild fires, and extended droughts. What is becoming increasingly clear is that species are less able to adapt to natural changes at local and regional scales when those changes are compounded by human-caused alterations (UNEP 1995), including large-scale deforestation for food crop cultivation; stream channel and flow alterations to gain water for drinking, irrigation, and hydropower production; and gaseous and particulate emissions from industrial enterprises that pollute the air, soil, and water and aid in changing global climate conditions.

1.1 RESEARCH: INVENTORIES AND MONITORING

It is important to be concerned about the loss of species or any other lessening of biodiversity because biodiversity is the very stuff of life. It includes the variety of life – the millions of different species and unique genes that each individual of each species carries – and the processes through which all species interact. At the least, biodiversity research should include an assessment of what the planet's huge storehouse of biodiversity contains and how it is influenced by changes to ecosystems. Such baseline data provide researchers with the information they need to determine – and to predict – probable consequences from human disturbances to the ecosystems. In turn, policy-makers and resource managers will be able to make more informed decisions about sustainable use and conservation of the earth's natural resources.

Yet, humans have only a rudimentary understanding of regional and global biodiversity and its controlling mechanisms. In developed countries, researchers have barely scratched the surface. The science is even further behind in the tropics, where areas rich in biodiversity – it is estimated that tropical forests account for approximately half of the world's species diversity (Myers 1980; Reid and Miller 1989; WCMC 1992) – are being rapidly transformed by human development into environments with low levels of biodiversity. Massive environmental transformations are being brought about through hundreds of experiments that lack adequate controls and with little recognition of the consequences.

The immense task of surveying and mapping the earth's species is largely undone, primarily because nations do not allocate enough human resources and financial support to long-term biodiversity research or to development of and training in proper monitoring methods. It may be unrealistic to assume that all species can be adequately described and assessed as to their risk of extinction, regardless of the level of commitment to accomplishing that goal. However, a well-funded, long-term, systematic program for inventorying and monitoring biodiversity through a world-wide network of representative habitats could make substantial headway.

1.1.1 Space and Time

Understanding ecosystem functions requires careful evaluation of the properties and scales of space and time. Recent research reveals a complex system of ecological interactions that determine how organisms are distributed and their abundance on local scales (Diamond and Case 1986; Gentry 1988 and 1992; Ashton 1992; Hubbell and Foster 1992; Dallmeier and Comiskey in press a, b). The evidence also indicates that ecological communities experience internal and external long-term changes with sometimes drastic turnover. This process has radical effects on the composition, structure, and diversity of the communities, resulting in imminent local extinctions and colonisations.

Such effects can only be perceived through long-term monitoring of ecological communities and recording environmental changes (Clinebell et al. 1995). But the impacts of environmental change on communities operate at various time scales, ranging from days and weeks (for example, gypsy moth defoliation in eastern North America) to a few decades (mild hurricanes and typhoon disturbances), many decades (extended drought), and centuries or millennia (catastrophic volcano activity) (Davis 1976 and 1986; Anonymous 1994) and across spatial scales that are extremely difficult to measure. It is also difficult to relate ecological processes operating at local scales to those in effect at regional or global scales.

It is important to do so because localised or small-scale events can contribute to large-scale impacts. As examples, in the Caribbean Virgin Islands the most severe drought on record (1994-95) produced a dry forest mortality of more than 13% and eliminated most seedlings and saplings (Ray et al. in press; Reilly in press; Weaver in press; Dallmeier et al. in preparation). The drought's impact on the forest was accentuated by the catastrophic effects of Hurricanes Marilyn in 1995 and Bertha in 1996. In Linden Guyana's sandy soil rain forest, bauxite mining has largely eliminated opportunities for sound regeneration of that entire ecosystem. In Bolivia's Beni region (and at many other sites around the world), annual fires intentionally set by humans as a management tool are transforming biologically rich forested areas into less productive grasslands (Comiskey et al. in press).

Sampling methodologies must be designed and implemented to measure the effects of such events in space and time and to determine their consequences for biodiversity. Reliable scientific information on habitat composition, structure, and dynamics is needed to inform decision-makers and facilitate sound management of ecosystems, particularly adaptive management strategies. The more that is known about a particular place, the greater is the range of options for promoting sustainable use and conservation of resources, a more stable local economic base, and increasing public awareness of and appreciation for the benefits provided by ecosystems at that place. As well, the greater are the possibilities that the information will prove useful at far-removed sites with similar characteristics.

2. SI/MAB'S Strategy

The Smithsonian Institution/MAB Biological Diversity Program (SI/MAB) was born of the critical need to initiate new approaches for gathering, analysing, and disseminating information about biodiversity. In 1986, UNESCO MAB and Smithsonian Institution joined to create SI/MAB with the express purpose of developing a protocol for surveying and monitoring biodiversity in a global network of forested areas under different intensities of management (SI/MAB Biodiversity News 1992 and 1994). It was envisioned that the protocol would enable users to integrate scientific research methods and analysis with strategies for getting the resulting information into the hands of a wider range of users – in particular, decision-makers and resource managers involved with issues of sustainable use and conservation of natural resources.

Over the past decade, SI/MAB staff, aided by researchers and colleagues around the world, have developed a biodiversity monitoring protocol and refined it through testing at a number of sites. The protocol has been adopted by many of our global collaborators and partners, forming the basis for consistent measuring and monitoring at more than 200 plots, facilitating the transfer of comparable data, and providing the framework for data analysis and dissemination to users.

2.1. HOW THE PROCESS WORKS

SI/MAB's biodiversity measuring and monitoring program is designed to elicit information that will assist in answering the following questions:

- How is forest biodiversity distributed regionally and globally?
- Which habitats are most important in terms of interacting assemblages of species?
- Which habitats harbour the greatest species richness and which the greatest number of exclusive species?

For conservation purposes, it is also important to know:

- How much of the existing diversity detected in the inventories is protected by conservation units?
- What new conservation units are necessary to protect the remaining habitats?
- Where are the most appropriate sites for establishing the biodiversity monitoring plot network?

The program focuses on establishment of biodiversity plots that are generally one-hectare units assembled in homologous habitats to sample forest structure and composition as well as the diversity of a site (Dallmeier 1992). This may be likened to taking a snapshot of the site that indicates what species are there, where they are located, and how abundant they are. These data form the baseline against which to assess changes in the ecosystem through repeated measurements of the same variables during the monitoring phase of the program (SI/MAB Biodiversity News 1992 and 1994). Over time, monitoring provides a series of snapshots, or a movie of events at the site-events such as the dynamics of the ecosystem and changes in vegetation that will help determine changes in interacting assemblages of other species. This allows researchers to predict the behaviour of key variables and, thus, creates an early warning system concerning changes in ecosystems and gives managers an opportunity to choose from a range of options to improve management regimes.

The number of one-hectare plots can be as large and the configuration of the sample design as random as the specific research objectives require. Data from the plots, which are set up following SI/MAB's standard protocol (Dallmeier 1992; Dallmeier et al. in press), feed other aspects of the multi-taxa surveys and monitoring. It is assumed that vegetation units as small as one hectare, when systematically sampled in combination with other taxa, can produce a great deal of information about interacting assemblages of species. A one-hectare biodiversity plot in a Caribbean dry forest on St. John Island, for example, can represent up to 65% of the vascular plants in this relatively simple forest type. A small network of such plots, if strategically located, can easily sample 90% of vascular plants on the island. (The remaining 10% may be important, and researchers should keep this in mind when designing measuring and monitoring programs.) Moving south into the more complex tropical rain forests of mainland South America requires a greater number of plots to sample most of the homologous habitats in the region. At the Beni Biosphere Reserve in Bolivia, a network of 15 to 20 one-hectare biodiversity plots is being monitored for five taxa (vascular plants, frogs, ants and ground beetles, birds, and small mammals) and is eliciting more ecological data about interacting assemblages of species than is currently known about the biodiversity in all Bolivian protected areas. In addition, such site-specific information is greatly enhanced when compared to data collected elsewhere but under similar protocols.

2.1.1. Building the Framework

The first steps of the SI/MAB protocol are to determine research objectives for the project, pose the questions that the monitoring program needs to answer, and then select components of the ecosystem/landscape for measurement. At this point, it is helpful to keep in mind that technology and resources are the bottlenecks that limit the ability to manage large amounts of biological specimens and information. In particular, lack of adequate financial and specialised human resources hampers management and processing of information.

In setting research objectives and devising monitoring questions, note that each site will have one or more specific issues that the research and monitoring should be targeted to address. Examples of such issues, arising from both natural and human causes, include regeneration following hurricanes or typhoons, habitat fragmentation, introduction of alien species, unsustainable consumption of renewable resources, and pollution. In devising a biodiversity monitoring system, it is necessary to decide if such issues apply to the study site so that researchers know whether additional multi-disciplinary research must be conducted to evaluate any changes. As examples (see SI/MAB Biodiversity News 1992 and 1994), at the biodiversity monitoring plots in Kejimkujik National Park in Nova Scotia, Canada, a major concern is the effect of acid precipitation on the forest. At the network of plots in western Africa, Cameroon, and Nigeria, a major focus is the study of medicinal plants. At Itabo and Limoy biological reserves in Paraguay, researchers are interested in the effects on natural vegetation changes resulting from the reservoir created by construction of a large hydroelectric dam. At Beni Biosphere Reserve in Bolivia, biodiversity plots are a primary component of the research project aimed at managing forest habitats for high species diversity. In China's Dinghushan Biosphere Reserve and on Hainan Island, an important objective is to understand the impacts of typhoons in the regeneration of tropical and subtropical forests.

In choosing components for monitoring, SI/MAB's protocol emphasises a multi-taxa forest monitoring approach at each site (Figure 1). Often the ecosystem/landscape relationship is divided into broad categories, including soil, flora, vertebrates, invertebrates, and climatic variables (these categories may be further divided). Such classifications assist researchers in selecting monitoring components.

Vascular plants are always measured because they represent the structure that supports other taxa. An inventory of several taxa can be completed throughout the homologous habitat, including the biodiversity plot, and the taxa that will be monitored can be selected from this group.

2.1.2. Choosing the Site

The next step is to select a site that is representative of the homologous habitat to be sampled and perhaps of the most important habitats in the larger area. Often, biodiversity plots are located in association with areas where critical management decisions apply. This is important because the findings at the plot may then be generally applicable to other homologous habitats of the area.

Frequently, site selection for the biodiversity plot requires intensive, rapid assessments and ground-truthing to determine the most appropriate location (Stohlgren 1994; Stohlgren et al. 1995a; Stohlgren et al. 1996; Dallmeier et al. in preparation). Such assessments, which normally result in economical use of limited human and financial resources, may incorporate satellite images at various scales, aerial photographs, available cartographic information, and ground-truthing that allows the identification of homologous habitats and some patterns of species diversity at the landscape level. A modified Whittaker vegetation plot is sometimes established to identify patterns of species diversity to assist in locating biodiversity plots (Stohlgren et al. 1995b).

2.1.3. Sampling Methods

Once the specific objectives are established and the site is selected, it is necessary to determine which sampling methods are best suited to achieving the inventory and monitoring objectives of the chosen taxa. Options include methodologies that result in a general overview (species lists) for short-term management or more detailed knowledge (species relative and absolute abundances) for longer term approaches.

Vegetation is central to the SI/MAB forest monitoring methodology; trees equal to or larger than one centimetre in diameter at breast height are the first to be considered as the biodiversity plots are established.

There are several reasons for this. First, flora in general have a pivotal role in the earth's terrestrial ecosystem, which contains 250,000 to 300,000 species of plants, and flora provide adequate information to interpret diversity indexes. They are also relatively easy to study because they are immobile. In addition, plants appear to be most sensitive to subtle environmental variations such as hydrological or edaphic changes. This is important given that one accepted method of extracting the maximum amount of information from sampling is to focus efforts on the most environmentally sensitive groups. Small mammals and amphibians show an intermediate level of habitat specificity. Large mammals tend to range across regions with widely disparate environmental conditions. Birds usually show relatively little local variation in species composition within similar habitat types. Trees, on the other hand, are very responsive to subtleties of local conditions, and herbaceous plants growing in the forest understory are even more sensitive.

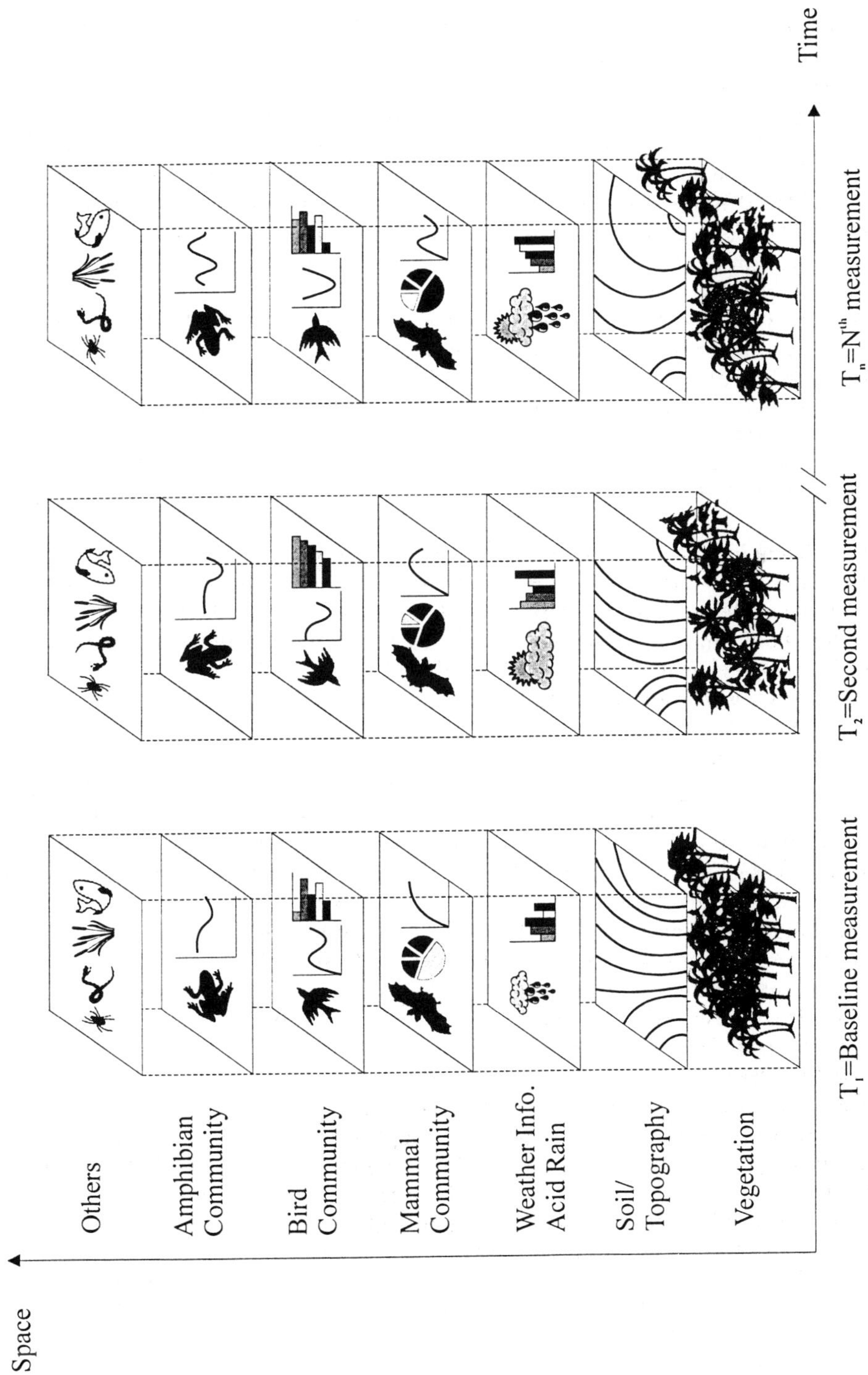

Figure 1. SI/MAB biodiversity protocol framework for multi-taxa measuring and monitoring.

When inventorying, it is convenient to choose as many species to study as human and financial resources permit. Site-specific, multi-taxa inventories at the study site should encompass well-known groups such as vascular plants, mammals, birds, and amphibians; groups that have economic importance; and flamboyant groups such as *Hymenoptera* or *Lepidoptera*. Similar criteria are used for subsequent monitoring of biodiversity. Additional layers of information can be added as needed to complement habitat information about other taxonomic groups.

Where forest biodiversity is distributed over a large area, SI/MAB emphasizes the need for careful, complete measurements of alpha diversity, or the diversity of species in a selected location, at each research site. The sites should be representative of other forest types as the objective is to expand to a larger area from that localised, representative foundation of knowledge. Often, researchers will find that for site-specific habitats, significant information about the flora and fauna already exists because of exhaustive biodiversity surveys and documentation of the alpha diversity of the habitat.

Then it is necessary to determine a pattern of local beta diversity – which species occur in which habitats and which habitats offer the most diversity. For example, of particular interest in the Amazon Basin are specialised habitats such as bamboo glades that are full of distinct, endemic species and thus are very challenging in relation to evolution.

Experience indicates that building a solid knowledge base at a single site makes it possible to expand outward by establishing transects of permanent monitoring plots running perpendicular and parallel to major environmental gradients. Parallel to the gradients, the primary environmental parameters of the transect remain very similar. These homologous habitats have the same basic structure and tree species composition as one moves away from the initial, well-documented habitat. Along the transect, it is important to look for vegetation types that do not occur in the original sample area. Often, these vegetation types are associated with distinct soil types and geological formations.

The next task is to sample in the perpendicular direction. From such work at Manu Biosphere Reserve in Peru and Beni Biosphere Reserve in Bolivia, we have found that where species richness (alpha diversity) is concerned, diversity is highest in the most mature forests (Gentry 1982; Dallmeier et al. 1996; Gerard et al. in press). However, as perpendicular sampling indicated, the forest in the region is frequently disturbed by natural events, particularly in the flood plains of the Amazon Basin where large numbers of rivers meander extensively, "chewing up" the land on one side and "spitting it out" on the other. This process allows for the development of successional forest, which occupies significant portions of the landscape. Early successional habitats have distinct species, but low diversity.

Fauna and tree diversity can be rather homogenous at sampling units in homologous habitats. For example, the mature flood-plain or upland forest of Manu and Beni reserves has a tree composition that overlaps 80% with the composition in similar sites 200 to 300 kilometres apart. Thus, even at a large scale, there is no significant change in species composition, save in distinct habitats associated with unusual geological formations, soil types, or different precipitation gradients.

2.2 SELECTING THE APPROPRIATE OPTION

Depending on site-specific objectives and the intensity of site-specific impacts, short-term or long-term options exist to assess biodiversity for improved management (Table 1). Short-term options include direct, rapid interventions or are designed to elicit enough knowledge to slow down or re-orient environmental transformations caused by human actions. That knowledge is gained through bibliographical reviews, cartographic information, and aerial or satellite photographs, combining to provide an initial overview of the area. An example of a short-term option is the attempt to protect the colony of dark-rumped petrels (*Pterodroma phaeopygia*) in the Galapagos Islands through annual extermination of introduced species of rats, cats, and pigs. The petrels are nearly extinct in Hawaii and are severely threatened in the Galapagos.

TABLE 1. Selecting the appropriate methodology for SI/MAB's forest biodiversity inventory and monitoring program (modified from Hubbell unpublished material)

Hypothesis/question	Methodology
• How is habitat diversity in a region or large-scale area measured?	Remote sensing and GIS
• What is the biogeography of selected groups of species regionally?	Cladistics
• How beta diverse is a particular site?	Transect assessment method (Foster 1993)
• How alpha diverse is the site?	.1-hectare sample transects (Keel et al. 1990)
• What is the structure, composition, diversity, and turnover of a habitat?	SI/MAB biodiversity monitoring plots
• How are forest interactions and multi-taxa monitoring assessed?	SI/MAB biodiversity monitoring plots
• What is the demography of selected species? their recruitment and dynamics?	50-hectare forest demography plots
• What are the key species of the ecosystem? (autoecology)	Detailed inventories of target groups
• What are potential species for pharmacological use	Specialised inventories
• What is the diversity of an ecosystem?	All-taxa biological inventories

In some cases, rapid assessments of biodiversity are conducted to gather additional biological information, supplementing the overview and forming the basis of recommendations for long-term options. Such options include alternative land management strategies with the goal of achieving sustainable use of resources. Nature-oriented tourism and management of areas for extraction of selected products are examples. Long-term options require in-depth understanding of the structure, composition, diversity, and dynamics at the research site.

Selection of the most appropriate methodology is based on the site-specific question and hypothesis that the inventory and monitoring will address. For example, if large areas need to be evaluated to define conservation units such as biosphere reserves, remote sensing is an essential tool. Note, however, that information obtained through remote sensing is at a coarse level and will always need ground verification. Cladistics provides information on a large scale about the relationships of species occupying similar geographical areas. In determining how diverse an area can be, different transect methodologies may be employed. SI/MAB biodiversity plots and other larger scale plots provide more specific information on the structure, composition, diversity, and turn-over of a habitat, while assessing multi-taxa interactions. Large-scale plots 50 hectares or more in size may be the most appropriate for studying population dynamics of species or the demography of selected populations. Finally, understanding a complete ecosystem will require an all-taxa biological inventory.

The characteristics of the site and larger region will also play a role in determining the extent of the inventory and monitoring program. Preservation of certain areas may require immediate steps because of imminent threats. Finally, availability of human skills, appropriate technology, time, and financial resources will largely determine how much can be accomplished.

2.2.1 Plot size

Both large- and small-scale vegetation plots are essential for strategic sampling of forest ecosystems. They complement each other, while both may be inhibited by limited technology and other needed resources. Large-scale plots, for example, can provide comprehensive understanding of a particular site. But they are not practical for most sites because they are labour intensive, they yield such large amounts of data that management of the information becomes a formidable task, and they are expensive to maintain.

Since it is not possible to set large-scale plots in many instances, it is necessary to create a system to determine how general the results from the large plots are. This can be accomplished by running extensive geographic surveys in small plots that provide essential information for calibrating species/area curves with large plots (Condit et al. in press). Standardised one-hectare plots provide a quick estimate of diversity at a site or the initial slope of the species/area curve. Within a homogeneous forest habitat, the species richness estimates from a one-hectare plot will be correlated to the total diversity in that type of forest.

The first result is an estimate of the initial slope of the species/area curve. Researchers can describe the diversity of the plot by plotting increasing cumulative area on the x axis and cumulative number of species found in that area on the y axis. Plotting log area against log cumulative number of species yields roughly a straight line, an empirical relationship that has been found in many places around the world. In diverse forest such as well-mixed forest found in most tropical areas, the curve typically will be steeper than in a species-poor forest.

This information is useful because it indicates that in sampling a large area, the total diversity of the area is related to the initial slope of diversity, assuming a homogenous habitat. In cases of environmental heterogeneity, such as a mixed forest or two or three habitats occupying the gradient from flood plain to upland, the change of species composition will be rather abrupt, with a resulting increase in diversity.

The initial slope in small-scale plots provides an index of how species rich the forest is in relation to vegetation. Comparing this information with a few large plots can establish a correlation between the initial slope and the final height of the curve and lead to a deeper understanding of how species richness varies from site to site across a broad landscape. The total number of species may not be known, but there will be a fairly tight relationship between what is measured and total diversity. This becomes an index, or surrogate, for total diversity.

There will likely be some variabilities related to how well mixed the individual tree species are. If one-species, or pure, stands exist, a one-hectare plot located in a stand of a low diversity forest will show a deviation below the regression line. If, on the other hand, the forests are extremely well mixed, then the deviation will show above the regression line.

The species-area curve is not linear on a simple number-of-species-against-area graph primarily because some species are not as common as others. Therefore, sample plots will capture portions of the rare species. As sampling moves out from the core site, the slope of the curve begins to tail off because more and more rare species will be included in the sampling unit. The process will continue indefinitely depending on the size of the area being sampled.

This process clearly indicates that if researchers are interested in how forest diversity is maintained, it is necessary to collect information about population biology, multi-taxa attributes, and how tree species interact with each other. One long-term question is whether the forests are in equilibrium or not. This matters to managers and decision-makers who must choose and manage protected areas. Small populations will obviously have greater problems with extinctions. In addition, disease and environmental factors operating at large spatial scales mean protected areas have to be big enough to accommodate the population fluctuations that will result. Biological inventories of the general ecosystem complement the data base for the site (especially for organisms such as mammals and birds that require much larger areas) and enrich knowledge of ecological interactions.

Relative to other options, SI/MAB biodiversity monitoring plots are relatively inexpensive and easy to establish while providing a substantial amount of information and a venue for comparability and compatibility of data world-wide. The methodology has proven useful in identifying major forest resource trends and suggesting potential agents of change and their associated patterns of cause and effect.

3. Training and Data Management

As noted earlier, major goals of the SI/MAB program are to maintain a global biodiversity plot network and disseminate the results to other researchers, decision-makers, and land and resource managers. This is important because the findings at the plots should inform management decisions. They also may be generally applicable in similar areas, resulting in more economical use of limited human and financial resources. Two vital elements of SI/MAB's strategy for achieving these goals are training and data management.

3.1. TRAINING

Through co-operative efforts with international, national, regional, and local governments and other institutions – among them, UNESCO MAB, UNESCO World Heritage Program, and U.S. MAB – SI/MAB has forged a program of local and regional participation in planning and carrying out each project. It is the in-country researchers who continue the work at the plots. They and local decision-makers can best describe the concerns that necessitated the project and determine the site-specific objectives. Involving inhabitants nearest to the research sites is also very helpful. In South America, for example, we have worked with members of indigenous tribes who know the landscape well. They have been instrumental in identifying species and describing traditional uses of the forests. In particular, their insights about how the forests function in relation to human-induced change can make significant contributions to the design of development projects.

To date, approximately 700 people from more than two dozen countries have participated in SI/MAB's annual, six-week international biodiversity monitoring and measuring courses and in-country training sessions. Experts in all relevant disciplines – biology, ecology, forestry, fisheries, etc. – from around the world lead the discussions at the intense, international courses. They accompany participants on field trips where the knowledge gained in the classroom is put directly to work: plots are established, baseline data are gathered, and long-term monitoring techniques are practised. These courses are held at the Smithsonian Institution research and field facility in the Shenandoah Mountains of Virginia, USA.

The in-country courses, normally consisting of two packed weeks of instruction and field work, are designed in conjunction with local, regional, and national experts and decision-makers. Field work includes the establishment of plots that become permanent, living laboratories for monitoring local forest conditions.

The courses emphasise the need for consistent, well-documented research. The SI/MAB protocol, taught and practised at the training sessions, covers all aspects of developing and carrying out a long-term biodiversity monitoring program. Through training in the protocol, we are promoting a consistent methodology for use at monitoring plots around the world so that researchers are communicating in the same "language." They can better understand the details and reports of work from other plots and determine if those results are applicable to their own work. They are also able to make more reliable comparisons among findings than when different methodologies are used.

3.2 DATA MANAGEMENT

All of this work would go for naught in the absence of an effective data management system capable of collecting, storing, and reporting reliable data in a timely fashion. SI/MAB has created BioMon, a computer-based, user-friendly system, as an integral component of the biodiversity monitoring protocol. BioMon promotes efficient collection, verification, interpretation, and dissemination of millions of data. It enables researchers within the global monitoring plot network to exchange information quickly. We are currently working with experts in related data management systems to create a link that will greatly expand the information-exchange capability of BioMon and those systems.

BioMon has proved invaluable in assisting with the task of interpreting and organising data for publication in numerous formats – from peer-reviewed journal articles and technical reports to newsletters, feature stories, and fact booklets for decision-makers. We continue to refine the system with an eye toward improving the user-friendliness of the software, strengthening the statistical analysis component, and enhancing co-ordination of the global biodiversity monitoring plot network.

4. Summary of Results

There is substantial evidence that the SI/MAB approach to measuring and monitoring biodiversity and disseminating the results of the research is achieving desired outcomes. As examples:

- The results of monitoring post-hurricane regeneration in the hard-hit dry tropical forests of U.S. Virgin Islands National Park on St. John's Island are aiding U.S. National Park Service determinations. The Park Service is studying the impacts that expanding the current road system will have on the recovering vegetation.
- In Guyana, researchers are co-operating with a bauxite mining company to assess the impacts of the mining operation on sandy-soil rain forests. Monitoring results could well be used in restoration of these highly degraded lands.
- In the Maya Biosphere Reserve that includes protected areas in Belize, Guatemala, and Mexico, the monitoring program is directed at assisting local human populations manage and use forest resources in a more sustainable manner.
- At Manu Biosphere Reserve in Peru, the 10-year assessment of biodiversity has been completed, providing the first comprehensive information about a biodiversity "hot spot."

These examples and many others are encouraging. They hold out the prospect of greater understanding of the importance of biodiversity. They also hold the promise of increased incorporation of sound, scientifically based recommendations for maintaining biodiversity in the design of development projects.

We are also working to create diverse forums for biodiversity researchers to communicate their findings and share their concerns. In 1995, SI/MAB convened an international meeting, "Biodiversity Measuring and Monitoring Symposium: The International Network of Biodiversity Monitoring Plots." The symposium attracted 314 participants from around the world, the largest group of specialists in the topic ever assembled to summarise results of forest monitoring work. The proceedings will be published in mid-1997. A second symposium will be held in the year 2000. In addition, SI/MAB staff were instrumental in developing an Internet-based communications network among 100 biosphere reserves in the Western Hemisphere. The network, informally known as EcoNetAmerica, serves as the mechanism for applying low-cost protocols and data bases to improve scientific communication and co-operation.

In conclusion, we have begun the work of creating a global network of forest biodiversity experts backed by decision-makers who understand the importance of biodiversity and its potential for bettering the human condition. Getting this combination of science and strategy right is a daunting task. SI/MAB's program was established and has evolved with that task in mind. Through careful scientific research, knowledge and understanding of the earth's storehouse of biodiversity can be increased. Through a strategy for distributing the findings of that research into the hands of policy-makers and on-site managers, the number of opportunities for successful maintenance of the biodiversity resource also increases.

References

Anonymous 1994. Scientists link CO_2 to changes in tropical forests. Global Environmental Change Report. 6: 1-3.
Ashton, P.S. 1992. The structure and dynamics of tropical rain forest in relation to tree species richness. In: Kelty, M.J., Larson, B.C. and Oliver, C.D. (eds.). The Ecology and Silviculture of Mixed-Species Forests. Kluwer. Boston. Pp. 53-64.
Boone, D. and Aplet, G. 1994. Sustaining Biodiversity in the Southern Appalachians. The Wilderness Society. Washington DC.
Clinebell, R., Phillips, O.L., Gentry, A.H., Stark, N., and Zuuring, H. 1995. Prediction of neotropical woody plant diversity from climatic and soil data. Biodiversity and Conservation. 4: 56-90.
Comiskey, J.A., Dallmeier, F. and Foster, R. in press. Forest structure and diversity in managed and unmanaged rain forest of Beni, Bolivia. In Dallmeier, F. and Comiskey, J.A. (eds.). Forest Biodiversity in North, Central and South America and the Caribbean: Research and Monitoring. Parthenon Press. London and New York.
Condit, R., Foster, R., Hubbell, S.P., Sukumar, R., Leigh, E.G., Manokaran, N. and Loo de Lao, S. in press. Assessing forest biodiversity on small plots: calibration using species-individual curves from 50-ha plots. In Dallmeier, F. and Comiskey, J.A. (eds.). Forest Biodiversity in North, Central and South America and the Caribbean: Research and Monitoring. Parthenon Press. London and New York.
Dallmeier, F. 1992. Long-term monitoring of biological diversity in tropical forest areas: methods for establishment and inventory of permanent plots. MAB Digest 11. UNESCO. Paris.
Dallmeier, F. and Comiskey, J.A. 1996. From the forest to the user: a methodology update. In Wilson, D. and Sandoval, A. (eds.). The Biodiversity of Southeastern Peru: Manu. Editorial Horizonte. Lima. Pp. 41-56.
Dallmeier, F. and Comiskey, J.A. (eds.) in press a. Forest Biodiversity in Europe, Australasia, and Africa: Research, Monitoring, and Modeling. Parthenon Press. London and New York.
Dallmeier, F. and Comiskey, J.A. (eds.) in press b. Forest Biodiversity in North, Central, and South America, and the Caribbean: Research and Monitoring, Parthenon Press. London and New York.
Dallmeier, F., Kabel, M. and Foster, R. 1996. Floristic composition, diversity, mortality and recruitment on different substrates: lowland tropical forest, Pakitza, Rio Manu, Peru. In Wilson, D. and Sandoval, A. (eds.). The Biodiversity of Southeastern Peru: Manu. Editorial Horizonte. Lima. Pp. 57-84.
Dallmeier, F., Comiskey, J.A. and Scatena, F.N. in press. Five years of forest dynamics following Hurricane Hugo in Puerto Rico's experimental forest. In Dallmeier, F. and Comiskey, J.A. (eds.). Forest Biodiversity in North, Central and South America and the Caribbean: Research and Monitoring. Parthenon Press. London and New York.
Dallmeier, F., Comiskey, J.A. and Mistry, S. in preparation. Measuring and Monitoring Forest Biological Diversity: Standard Methods for Biodiversity Monitoring Plots.
Davis, M.B. 1976. Pleistocene biogeography of temperate deciduous forests. Geosciences and Man. 13: 13-26.
Davis, M.B. 1986. Climatic instability, time lags, and community disequilibrium. In: Diamond, J.M. and Case, T.C. (eds.). Community Ecology. Harper and Row. New York. Pp. 262-284.
Davis, G.E. 1993. Design elements of monitoring programs. Environmental Monitoring and Assessment. 26: 99-105.
Davis, G.E., Faulkner, K.R. and Halvorson, W.L. 1994. Ecological monitoring in Channel Islands National Park, California. In: The Fourth California Islands Symposium. National Park Service. Washington, DC. Pp. 465-482.
Diamond, J.M. and Case, T.J. (eds.) 1986. Community Ecology. Harper and Row. New York.
Foster, R.B. 1993. Rapid quantitative sampling of vegetation without using plots. Unpublished handout at SI/MAB Biodiversity Monitoring Course.
Gentry, A.H. 1982. Patterns of neotropical plant species diversity. Evolutionary Biology. 15: 1-84.
Gentry, A.H. 1988. Changes in plant community diversity and floristic composition on environmental and geographical gradients. Annals of the Missouri Botanical Garden. 75: 1-34.
Gentry, A.H. 1992. Tropical forest biodiversity: distributional patterns and their conservational significance. Oikos. 63: 19-28.
Gerard, F., Wyatt, B., Millington, A. and Wellens, J. in press. The role of data from intense sample plots in the development of a new method for mapping tropical forest types using satellite imagery.
Hubbell, S.P. and Foster, R. 1992. Short-term population dynamics of neotropical forest: why ecological research matters to tropical conservation and management. Oikos. 63: 48-61.

IUCN 1995. IUCN Red List Categories. International Union for the Conservation of Nature and Natural Resources. Gland.
Keel, S., Gentry, A.H. and Spinzi, L. 1990. Using vegetation analysis to facilitate the selection of conservation sites in eastern Paraguay. Conservation Biology. 7(1): 66-73.
Myers, N. 1980. Conservation of Tropical Moist Forest. National Research Council. National Academy of Sciences. Washington DC.
Ray, J.R., Dallmeier, F. and Comiskey, J.A. in press. The structure of two subtropical dry forest communities on the island of St. John, U.S. Virgin Islands. In: Dallmeier, F. and Comiskey, J.A. (eds.). Forest Biodiversity in North, Central and South America and the Caribbean: Research and Monitoring. Parthenon Press. London and New York.
Reid, W.V. and Miller, K.R. 1989. Keeping Options Alive. World Resources Institute. Washington DC.
Reilly, A.E. in press. Hurricane Hugo: winds of change....or not? Forest dynamics on St. John. U.S. Virgin Islands. In: Dallmeier, F. and Comiskey, J.A. (eds.). Forest Biodiversity in North, Central and South America and the Caribbean: Research and Monitoring. Parthenon Press. London and New York.
SI/MAB Biodiversity News 1992 No. 2 Winter and 1994 No. 4 Winter/Spring. Smithsonian Institution, SI/MAB Program, Washington DC.
Stohlgren, T.J. 1994. Planning long-term vegetation studies at landscape scales. In: Steele, J.H. and Powell, T.M. (eds.). Ecological Time Series. Chapman and Hall. New York. Pp. 209-241.
Stohlgren, T.J., Binkley, D., Veblen, T.T. and Baker, W.L. 1995a. Attributes of reliable long-term landscape-scale studies: malpractice insurance for landscape ecologists. Environmental Monitoring and Assessment. 36: 1-25.
Stohlgren, T.J., Falkner, M.B., and Schell, L.D. 1995b. A modified-Whittaker nested vegetation sampling method. Vegetation. 117: 113-121.
Stohlgren, T.J., Kalkhan, M.A., Chong, G.W. and Schell, L.D. 1996. Rapid assessment of plant diversity patterns: a methodology for landscape, Environmental Monitoring and Assessment.
UNEP 1995. Global Biodiversity Assessment. United Nations Environment Programme. Cambridge University Press. Cambridge.
Weaver, P. in press. The effects of environmental gradients on hurricane impact in Cinnamon Bay Watershed, St. John, U.S. Virgin Islands. In Dallmeier, F. and Comiskey, J.A. (eds.). Forest Biodiversity in North, Central and South America and the Caribbean: Research and Monitoring. Parthenon Press. London and New York.
WCMC (World Conservation Monitoring Centre) 1992. Global Biodiversity: Status of the Earth's Living Resources. Chapman and Hall. London.

NEEDS AND PROSPECTS FOR INTERNATIONAL CO-OPERATION IN ASSESSING FOREST BIODIVERSITY: AN OVERVIEW FROM WWF

NIGEL DUDLEY
Equilibrium Consultants
23 Bath Buildings
Montpelier
Bristol BS6 5PT
UK

JEAN-PAUL JEANRENAUD
WWF International
Avenue du Mont Blanc
CH-1196 Gland
Switzerland

Abstract

Several current attempts to develop indicators of forest biodiversity at national and international level are briefly summarised. Some efforts have also been made to identify global indicators of biodiversity "hotspots", for example through the use of endemism and richness in bird populations. However, there has in general been much less co-operation in developing indicators of biodiversity at the stand level, with the result that many systems are being developed, using a range of criteria, techniques and degrees of accuracy. Thus there is a need for international co-operation, perhaps under the auspices of the Convention on Biological Diversity, to address the following issues:

- agreement on *definitions* relevant to biodiversity measurement at the stand level;
- identification of indicators operating at various *levels* of detail;
- development of *rapid field assessment methods* for biodiversity;
- research on *indigenous knowledge* of likely indicators in particular habitats;
- investigation of the *cultural framework* of biodiversity indicators;
- proposals for *capacity building* for assessment in developing countries.

1. Introduction: Setting Indicators at the National Level

Since the Earth Summit in 1992, there have been a number of attempts to develop criteria and indicators of forest condition. Most of these are measurements at a national or even international level, and thus inevitably rely on very broad methods of assessment. Some important examples include:

P. Bachmann et al. (eds.), Assessment of Biodiversity for Improved Forest Planning, 31–41.
European Forest Institute Proceedings no. 18. © 1998 *Kluwer Academic Publishers. Printed in the Netherlands.*

- Helsinki Process for European countries;
- Montreal Process for other temperate and boreal countries;
- Tarapoto Process (or Proposal) for Latin American countries;
- Dry Zone Africa Process; and
- FAO/UNECE Forest Resource Assessment.

Most of these examples have made an effort to develop indicators of biodiversity as part of their objectives. Interim results are summarised in Table 1 below.

Similar initiatives are being discussed in Asia and Central America. In addition, there have been some attempts to identify biodiversity "hotspots" on an international level, through the use of generalised indicators. Some of the best of these are the use of endemic and localised bird populations by Birdlife International (Bibby et al. 1992) and the study co-ordinated by IUCN and the Smithsonian Institute on Centres of Plant Diversity (Davis et al. 1995).

2. Setting Indicators of Biodiversity at the Forest Stand Level

International co-operation has concentrated mainly on assessment of biodiversity at the national or international level. Under these conditions indicators are inevitably very general and can act only as fairly coarse filters.

While there has been considerable study of indicators at the forest or stand level, such work has generally been less co-ordinated, and as a result has varied enormously in technique, degree of accuracy, and even general philosophy about biodiversity. Instead of setting out to define indicators, most research has started by determining *impacts* of changes, such as changes in populations of species following disturbance, and this information has then sometimes been used to develop indicators.

A number of researchers have attempted to define the importance to biodiversity of a particular forest stand in terms of naturalness (Anderson 1991) or authenticity (Dudley 1996). Some have drawn on general information about population biology (Hayward 1991) and vegetative change (Leak 1992).

Many other researchers have used indicator species as means of identifying old-growth or semi-natural forests, and in some cases the degree of naturalness. Birds are perhaps the most common indicators, with those dependent on old-growth characteristics such as dead wood (Raphael and White 1984) being particularly popular (Schreiber and de Calesta 1992), as in the case of woodpeckers in Scandinavia (Angelstam and Mikusinki 1994; Virkkala et al. 1993) and North America (Bull and Johnson 1995; Lennartz 1988) and large birds of prey (Tjernberg 1986). Changes in bird communities under even-aged management have been recorded in several areas including Florida (Repenning and Labisky 1985) and the UK (Ford et al. 1979), as have the impacts of forest fragmentation (Robbins 1980; Rolstad and Wegge 1987) and other forms of land management (Virkkala 1987). Birds have also been used as indicators in the tropics (Johns 1996), including comparisons between plantations and natural forests (Carloson 1986).

More recently, lower plant forms such as fungi (Bader et al. 1995), lichens (Söderström 1988) and bryophytes (Gustafsson and Hallingbäck 1988) have assumed an increasing importance, as their dependence on certain forest conditions has been noted and identified. In Northern Sweden, detailed indicators of forest richness have been

TABLE 1. International attempts to set indicators of biodiversity

Initiative	Details
Helsinki Process	An initiative of the Finnish and Portuguese governments, launched in June 1993 at the *2nd Ministerial Conference on the Protection of Forests* in Helsinki, which agreed on a General Declaration and four Forestry Resolutions. The Helsinki process has now drawn up indicators of good forest management, including some relating to biodiversity, such as (i) changes in area of natural and ancient semi-natural forest types; (ii) changes in number and % of threatened forest species; (iii) changes in the proportion of stands managed for conservation; (iv) changes in the proportion of mixed stands; (v) proportion of annual area of natural regeneration.
Montreal Process	Launched in October 1993 at a *Seminar of Experts on Sustainable Development of Boreal and Temperate Forests*. It has drawn up criteria and indicators of sustainable forest management with 10 non-European temperate and boreal countries. Biodiversity indicators include: (i) extent of area by forest type, age class and successional stage; (ii) extent of area by forest type under protection,, including in IUCN protected area categories; (iii) fragmentation of forest types; (iv) status and number of forest dependent species; (v) population levels of representative species.
Tarapoto Process	Launched by the Amazon Cooperation Treaty at a *Regional Workshop to Define Criteria and Indicators of Sustainability of the Amazon Forest* in Peru in February 1995. Biodiversity indicators include: (i) area by forest type in categories of protected areas; (ii) impact of activities in other sectors on forest conservation; (iii) rates of conversion of forest to other uses; (iv) measures for *in situ* conservation of endangered forest species; (iv) measurement for the conservation of genetic resources; (v) rate of natural regeneration, species conservation and survival.
Dry-Zone Africa Process	Launched at a *UNEP/FAO Expert Meeting on Criteria and Indicators for Sustainable Management in Dry-Zone Africa* at Nairobi in November 1995. Biodiversity indicators include: (i) area by type of vegetation; (ii) extent of protected areas; (iii) fragmentation of forests; (iv) area annually cleared of forests containing endemic species; (v) number of forest dependent species at risk; (vi) average number of provenances.
FAO/UNECE Forest Resource Assessment	Development of the Forest Resource Assessment 2000 includes an attempt to widen the scope of the assessment to cover aspects of biodiversity. An expert meeting in Kotka, Finland, in June 1996 identified new criteria including "naturalness" of forests.

devised using fungi and lichens (Karström et al. 1993). In the UK, flowering plants have been used both as an indicator of forest age (Peterken 1974; Peterken and Game 1984) and also to describe changes under different management regimes (Kirby 1990), leading to detailed inventory work (Spencer and Kirby 1992).

In theory virtually any living group can be used to indicate forest condition, and among the other groups used as indicators are mammals (Friend 1982; Prost and Rakstad 1987; Scrivner and Smith 1984; Thomas 1988), amphibians (Bury 1983; Welsh 1990); arthropods (Huhta 1979; Lasebikan 1975), microarthropods (Seasted and Crossley 1981), insects (Holloway et al. 1992; Khen et al. 1992), and stream invertebrates (Newbold et al. 1980). Use of top predators has assumed particular importance in parts of North America (DellaSala et al. 1995).

However, relatively little work has been done to make generalised assessments of the likely indicator values of groups and species. In practice, choice of indicators often reflects the particular interests of the specialists involved, rather than an attempt to provide an overall assessment. Exceptions are some work on identifying useful assemblages (Kremen 1992) and drawing up hierarchies of indicators (Noss 1990) and identifying indicators for larger areas such as has been attempted for Alaska (Sidle and Suring 1986).

Biodiversity indicators at stand level are increasingly being used to help identify conservation areas and to develop forest management strategies. Greater international co-ordination of aims and methods is therefore urgently needed. In the remainder of this paper, some general suggestions are given for international co-operation in developing biodiversity indicators at the forest stand level.

3. International Co-operation in Developing Biodiversity Indicators at the Forest Stand Level

Perhaps the most suitable forum for developing stand-level forest biodiversity assessments would be the Convention on Biological Diversity, either through its own Subsidiary Body on Scientific, Technical and Technological Advice (SBSTTA) or through a specially appointed sub-group. WWF has identified six main issues that any such body should focus upon.

3.1 AGREEMENT ON AIMS AND DEFINITIONS

Any international co-operation must start with basic agreements about what is being assessed. Three issues that require discussion are:

- Some measure of the *quality* of biodiversity from an ecological standpoint. Assessment of biodiversity requires more than a simple estimation of total number of species, and should include such issues as endemism, fragility of species, authenticity of species' distribution (i.e. the natural range of a species), etc.
- Agreement about whether assessment is in *pristine* or *disturbed* habitats, and analysis of the different techniques required in each case.
- Assessment of biodiversity *resilience* to disturbance, outside factors, etc.

3.2 AGREEMENT ABOUT THE LEVEL OF ANY SURVEYS

Biodiversity can be measured at a number of different levels, depending on the time, resources and expertise available. For simple surveys to determine forest management practices, conservation strategies and areas requiring further study, very broad-scale survey techniques are probably adequate. At the other end of the scale, detailed surveys of local provenances and genetic variation may be needed for studies of genetic biodiversity. Five general levels of assessment can be identified:

i) **National survey**: a survey to monitor such issues as compliance with international agreements, provide data for global surveys etc. This relies on very general data, often in the form of questionnaires from international agencies, national-level studies etc.

Methods include:
- division of forests into broad classes, such as *natural forest*, *semi-natural forest* and *plantation*;
- use of age-class to identify old forest; and
- use of Red Lists of endangered species.

ii) **Landscape survey**: an initial survey to identify areas worthy of further study. This uses techniques that are either very simple to apply and thus can be used over a wide area in a fairly short time, or more specialised methods such as interpretation of satellite imagery that cover large areas in a single examination.

Methods include:
- use of satellite images to identify unfragmented, primary or old-growth forest;
- use of aerial and/or video surveys of forest cover, with the option of including computer simulated mapping from video images; and
- ground-level surveys of canopy structure on or from elevated land to identify old-growth characteristics in forest or particular indicator tree species.

iii) **Structural survey – forest authenticity**: a ground-level survey of structural characteristics of forests, to identify the *potential* for authentic population levels and composition of biodiversity.

Methods include:
- identification of underlying geology and soil type;
- survey of species mix, age composition and stand structure; and
- identification and classification of snags and down logs.

Forest structures can identify forests that have a natural pattern and process, and can indicate areas that have the potential for high biodiversity by virtue of having a wide range of suitable habitats.

However, structural characteristics do not offer proof that a forest has high biodiversity. A forest can look natural but if, for example, it has been cleared at some time in the past and later regrown, or if certain components have been removed at some period (such as people collecting dead wood for fuel) then the biodiversity may have been substantially altered. To gain more knowledge about whether the potential for high biodiversity is realised, more detailed surveys are needed, such as those using indicators.

iv) **Indicators survey**: a survey using specialised indicator species to identify rare habitats and microhabitat continuity. Indicators must be chosen carefully to "capture" as large a proportion of biodiversity as possible. Ideally, a "portfolio" of indicators will be used to identify different aspects of ecological authenticity.

For example, some fungi grow on down logs while others are confined to snags; using examples of both will give a more complete picture of forest function than one or the other. However, this method is only an *indication* of biodiversity rather than a complete survey and a balance must be struck between detail and limitations of time and resources. Indicators also need to be continually re-assessed to check whether they are supplying valid information. Examples of indicators include:

- lichens and fungi associated with different ages and conditions of trees;
- vascular plants associated with ancient woodland; and
- birds associated with old-growth forest.

v) **Genetic survey**: detailed stand survey of genetic variation within a stand to identify local varieties, different provenances, etc. Knowledge of genetic surveys is still developing, possible methods include:

- historical surveys to identify dates of planting and management, and sources of seed;
- identification of local variation in species where this can be identified by visible characteristics (such as some land snails, vascular plants and birds);
- detailed studies of genetic variation in particular species.

These five levels are summarised in Table 2 below. Note that the five levels also have a decreasing level of generality, i.e., generally speaking:

- techniques used for landscape surveys will be applicable in many areas;
- elements for structural surveys will be applicable over biomes;
- indicator species will be confined to regions (although in some cases these will be large areas); and
- studies of local provenances will rely on very site-specific data.

In many cases, assessment at the simplest level will be enough to make informed management choices. However, the option to develop successively more sophisticated surveys offers important advantages where more difficult management decisions have to be made, or where rare or particularly important ecosystems are under consideration.

TABLE 2. Different levels of biodiversity survey

Level of survey	Examples of techniques	Results
National	Surveys, use of Red Lists etc	Broad picture of forest biodiversity status
Landscape	Satellite images, aerial survey, landscape survey	Identification of sites worthy of greater study
Structure	Soils, forest structure, presence of dead wood	Potential for authentic biodiversity
Indicator species	Species indicating forest types or microhabitats	Identification of authentic forests and micro-habitat continuity
Genetic	Detailed studies of genetic variation	Biodiversity at genetic level, presence of local provenances

3.3 DEVELOPMENT OF RAPID FIELD ASSESSMENT METHODS FOR BIODIVERSITY

Good indicators should supply the maximum amount of information for the minimum amount of work. This means that they should:

- be relatively easy to identify and easy to locate in the field;
- give a maximum amount of information about overall conditions; and
- give information at the appropriate level.

Care should be taken to avoid using more complex survey methods than are necessary. For example, use of rare birds that need particular types of dead wood for nesting may be overly complex as a means of identifying forest types; it would be quicker and easier to survey the dead wood. In general, the greater the degree of specialisation, the better the likely result of using an indicator. On the other hand, species that are so specialised that they are likely to be extremely rare are of little use. An ideal indicator species is one that will be fairly common *only* when a certain set of ecological conditions are present.

3.4 RESEARCH ON INDIGENOUS KNOWLEDGE OF LIKELY INDICATORS IN PARTICULAR HABITATS

Indigenous people and others living in close proximity and involvement with a particular forest system may have a much greater knowledge of key elements in biodiversity than an incoming scientist. This knowledge may focus on particular elements of use to human communities, such as:

- medicinal herbs;
- food plants; and
- conditions necessary for game animals, etc.

Many indigenous or local people have a high level of understanding about the ecological conditions needed by literally hundreds of different plant species. This knowledge can sometimes be useful in the identification of richness of secondary forest fragments, as is the case for example in the *tembawangs* or fruit gardens developed by the Dayak people in parts of Borneo, which are now the nearest equivalent to natural forest in areas that have been heavily cut over, such as parts of West Kalimantan (Michon and de Foresta 1995).

Systems for involving local people in forest biodiversity assessment have been successfully developed in some places, and this knowledge should be applied more generally.

3.5 INVESTIGATION OF THE CULTURAL FRAMEWORK OF BIODIVERSITY INDICATORS

Quite apart from their biological value, indicators will only be effective if users feel both confident to use the various indicators and are convinced of their value.

Indicators that can only be used by a few specialists have limited value, as do those that may cause antipathy in certain users. People trained in forestry may be happiest with structural characteristics, while biologists and amateur naturalists may get positive pleasure from the time and effort needed to track down obscure species.

Indicators also need to be carefully explained and put into context. The use of the spotted owl in the western United States has helped further polarise an already bitter debate by (wrongly) suggesting that people are offered a choice between the life of a small bird and a logging community. The dangers of over-emphasising indicators, rather than the whole ecological and cultural values of the forest, should also be noted.

3.6 PROPOSALS FOR CAPACITY BUILDING FOR BIODIVERSITY ASSESSMENT IN DEVELOPING COUNTRIES

In addition to practical and organisational aspects, some overall strategy is needed with respect to building capacity for biodiversity assessment in developing countries.

4. Conclusions: Putting Biodiversity Assessment in Context

Use of indicators of biodiversity at that stand level could be of critical importance in identifying forests worthy of conservation and in measuring the success of sustainable forest management.

However, biodiversity indicators are only one part of a measurement of overall forest quality, and any assessment needs to be placed in this wider context. WWF has developed criteria for measuring forest quality, which also consider a number of environmental, economic and social issues, as follows. Four broad categories of quality criteria have been proposed:

Authenticity

- composition of trees, etc.
- spatial variation of trees with respect to age, size, proportion of dead timber, etc.
- continuity of forest over time
- accommodation of natural disturbance patterns and regeneration strategies
- management practices which mimic natural ecological processes

Forest health

- health of trees
- health of other forest flora and of fauna
- robustness to changing environmental conditions

Environmental benefits

- biodiversity conservation
- soil and watershed protection
- impacts on other natural or semi-natural habitats
- climatic stabilisation

Other social and economic benefits:

- timber products
- non-timber products
- employment, indirect employment and subsistence activities

- recreational value
- forest as homeland for people
- historical values
- educational values
- cultural and aesthetic values
- spiritual and religious values
- local distinctiveness

WWF is currently undertaking a project to produce guidelines for measuring forest quality, in all its aspects, at the stand level.

5. Acknowledgments

Thanks are due to many people for their comments and ideas during the development of the thinking presented in this paper. We would like to thank, in particular, Anders Lindhe of WWF Sweden for comments about different levels of survey and the participants in the COST E4 Forest Reserves Research Network Meeting in Fontainebleau, France, September 12-14 1996 for discussion of these issues.

References

Anderson, J.E. 1991. A conceptual framework for evaluating and quantifying naturalness. Conservation Biology. 5: 347-352.

Angelstam, P. and Mikusinki, G. 1994. Woodpecker assemblages in natural and managed boreal and hemi-boreal forest – a review. Annales Zoological Fennica. 31: 157-172.

Bader, P., Jansson, S. and Johnson, B.G. 1995. Wood inhabiting fungi and substratum decline in selectively logged boreal spruce forests. Conservation Biology. 72: 355-362.

Bibby, C.J. et al. 1992. Putting Biodiversity on the Map: Priority Areas for Global Conservation. Birdlife International. Cambridge.

Bull, E.L. and Johnson, E.A. 1995. Habitat use and management of pileated woodpeckers in northeast Oregon. Journal of Wildlife Management. 57: 335-345.

Bury, R.B. 1983. Differences in amphibian populations in logged and old-growth redwood forests. Northwest Science. 57: 167-178.

Carloson, A. 1986. A comparison of birds inhabiting pine plantations and indigenous forest patches in a tropical mountain area. Biological Conservation. 35(3).

Davis, B.D., Heywood, V.H. and Hamilton, A.C. (eds.) 1994. Centre for Plant Biodiversity: A Guide and Strategy for their Conservation. Volume 1. WWF and IUCN. Gland.

Davis, B.D., Heywood, V.H. and Hamilton, A.C. (eds.) 1995. Centre for Plant Biodiversity: A Guide and Strategy for their Conservation. Volume 2. WWF and IUCN. Gland.

DellaSala, D., Olson, D.M. and Crane, S.L. 1995. Ecosystem management and biodiversity conservation: applications to inland Pacific Northwest Forests. In: Baumgartner, D. and Everett, R. (eds.). Proceedings of a Workshop on Ecosystem Management in Western Interior Forests. Washington State University Cooperative Extension Unit. Pullman. WA.

Dudley, N. 1996. Authenticity as a means of measuring forest quality. Biodiversity Letters: 3: 6-9.

Ford E.D., Malcolm, D.C. and Atterson, J. (eds.) 1979. The Ecology of Even-Aged Forest Plantations. Institute of Terrestrial Ecology. Cambridge.

Friend, G.R. 1982. Mammal population in exotic pine plantations and indigenous eucalypt forests in Gippsland, Victoria. Australian Forestry. 45(1).

Gustafsson, L. and Hallingbäck, T. 1988. Bryophyte flora and vegetation of managed and virgin coniferous forests in southwest Sweden. Biological Conservation. 44: 283-300.

Hayward, G.D. 1991. Using population biology to define old-growth forests. Wildlife Society Bulletin. 19: 111-116.

Holloway, J.D., Kirk-Spriggs, A.H. and Khen, C.V. 1992. The response of some rain forest insect groups to logging and conversion to plantation. Philosophical Transactions of the Royal Society of London. Series B 335: 425-436.

Huhta, V. 1979. Evaluation of different similarity indices as measures of succession in arthropod communities of the forest floor after clear cutting. Oecologia. 41: 11-23.

Johns, A.D. 1996. Bird population persistence in Sabahan logging concessions. Conservation Biology. 75(1): 3-10.

Karström, M., Lindahl, K., Olsson, G.A. and Williamson, M. 1993. Indikatorarter för identifiering av naturskogar i Norrbotten. Naturvårdsverket. Stockholm.

Khen, C.V., Speight, M.R. and Holloway, J.D. 1992. Comparison of biodiversity between rain forests and plantations in Sabah using insects as indicators. In: Adams, K.L. and Miller, F.R. (eds.). Proceedings of the Oxford Conference on Tropical Forests: Wise Management. 30 March-1 April 1992. Oxford Forestry Institute. Oxford.

Kirby, K.J. 1990. Changes in the ground flora of a broad leaved wood within a clear fell, group fell and a coppiced block. Forestry. 63: 242-249.

Kremen, C. 1992. Assessing the indicator properties of species assemblages for natural areas monitoring. Ecological Applications. 4: 768-785.

Lasebikan, B.A. 1975. The effect of clearing on the arthropod population of a Nigerian rainforest. Biotropica. 7: 84-89.

Leak, W.B. 1992. Vegetative change as an index of forest environmental impact. Journal of Forestry. 90(9): 32-35.

Lennartz, M.R. 1988. The red-cockaded woodpecker: old-growth species in a second-growth landscape. Natural Areas Journal. 8: 160-165.

Michon, G. and de Foresta, H 1995. The Indonesian agro-forest model. In Halladay, P. and Gilmour, D.A. (eds.). Conserving Biodiversity Outside Protected Areas: The role of traditional agro-ecosystems. IUCN in collaboration with AMA Andalucia and Centro de Investigación F González-Bernáldez. Gland. Switzerland.

Newbold, J.D., Erman, D.C. and Roby, K.R. 1980. Effects of logging on macroinvertebrates in streams with and without buffer strips. Canadian Journal of Fisheries and Aquatic Sciences. 37: 1076-1085.

Noss, R.F. 1990. Indicators for monitoring biodiversity: A hierarchical approach. Conservation Biology. 4: 355-364.

Peterken, G.F. 1974. A method for assessing woodland flora for conservation using indicator species. Biological Conservation. 6: 239-245.

Peterken, G.F. and Game, M. 1984. Historical factors affecting the distribution of vascular plant species in the woodlands of central Lincolnshire. Journal of Ecology. 72: 155-182.

Prost, J.R. and Rakstad, D.S. 1987. Small mammal communities in three aspen stand-age classes. Canadian Field Naturalist. 101: 363-368.

Raphael, M.G. and White, M. 1984. Use of snags by cavity-nesting birds in the Sierra Nevada. Wildlife Monographs. 86: 1-66.

Repenning, R.W. and Labisky, R.F. 1985. Effects of even-aged timber management on bird communities of the longleaf pine forest in northern Florida. Journal of Wildlife Management. 49: 1088-1098.

Robbins, C. 1980. Effect of forest fragmentation on breeding bird populations in the piedmont of the Mid-Atlantic region. Atlantic Naturalist. 33: 31-36.

Rolstad, J. and Wegge, P. 1987. Distribution and size of capercaillie leks in relationship to old forest fragmentation. Oecologia. 72: 389-394.

Schreiber, B. and de Calesta, D.S. 1992. The relationship between cavity-nesting birds and snags on clearcuts in western Oregon. Forest Ecology and Management. 50: 299-316.

Scrivner, J.H. and Smith, H.D. 1984. Relative abundance of small mammals in four successional stages of spruce-fir forests in Idaho. Northwest Science. 58: 171-176.

Seasted, T.R. and Crossley Jr., D.A. 1981. Microarthropod response following cable logging and clear-cutting in the southern Appalachians. Ecology. 62: 126-135.

Sidle, W.B. and Suring, L.H. 1986. Management Indicator Species for the National Forest Lands of Alaska. Technical Publication R10-2. US Department of Agriculture Forest Service. Alaska Region. Juneau. Alaska.

Söderström, L. 1988. The occurrence of epixylic bryophyte and lichen species in a natural and a managed forest stand in northwest Sweden. Biological Conservation. 45: 169-178.

Spencer, J.W. and Kirby, K.J. 1992. An inventory of ancient woodland for England and Wales. Biological Conservation. 62: 77-93.

Thomas, D.W. 1988. The distribution of bats in different ages of Douglas-fir forests. Journal of Wildlife Management. 52: 619-626.

Tjernberg, M. 1986. The Golden Eagle and Forestry. Swedish University of Agricultural Science. Department of Wildlife Ecology. Report number 12. Uppsala.

Virkkala, R. 1987. Effects of forest management on birds breeding in northern Finland. Annales Zoologica Fennica. 24: 281-294.

Virkkala, R., Alanko, T., Laine, T. and Tiainen, J. 1993. Population contraction of the white-backed woodpecker *Dendrocopus peucotus* in Finland as a consequence of habitat alterations. Biological Conservation. 66: 47-53.

Welsh, H.H. 1990. Relictual amphibians and old-growth forests. Conservation Biology. 4(3): 309-319.

UNEP'S GLOBAL BIODIVERSITY ASSESSMENT

SIPI JAAKKOLA
UNEP, Regional Office for Europe
15 chemin des Anémones
1219 Châtelaine/Geneva
Switzerland

Abstract

UNEP completed the Global Biodiversity Assessment (GBA) in 1995. Underlying this endeavour was an attempt to mobilise the global scientific community to analyse the present state-of-the-art knowledge and understanding of biodiversity and the nature of our interactions with it. A total of over 400 experts from over 50 countries were formally involved. The GBA was an independent, critical, peer-reviewed scientific analysis of the current issues, theories and views regarding the main aspects of biodiversity. The assessment did not concern itself with the assessment of the state of national or regional biodiversity. Its perspective was global with a focus on general concepts and principles. It drew attention to possible policy implications of its major findings and to existing gaps in knowledge and capacity. The paper discusses the key results of the Global Biodiversity Assessment and its implications to related forestry assessments.

1. Introduction

United Nations Environment Programme (UNEP) which is the host organisation of the Convention on Biological Diversity emphasises that for environmental policy-making, an essential element is the collection and dissemination of knowledge generated by scientific research. In this regard the scientific community has been and shall continue to be UNEP's key partner in carrying out our mandate. It was in this spirit that UNEP commissioned the Global Biodiversity Assessment (GBA) project. Underlying this endeavour was an attempt to mobilise the global scientific community to analyse the present knowledge and understanding of biodiversity and the nature of our interactions with it.

Unlike the global agreements on Climate Change and Ozone Depletion, no formal scientific assessment was carried out prior to the final negotiations of the Convention on Biological Diversity. The Parties to the Convention clearly recognise the lack of knowledge regarding biodiversity, and the urgent need to develop our knowledge base in this area. The Assessment does not concern itself with country-level or regional biodiversity. Its perspective is global with a focus on general concepts and principles. It does not present any policy recommendations, although it draws attention to possible policy implications of its major findings and to existing gaps in knowledge and capacity.

P. Bachmann et al. (eds.), Assessment of Biodiversity for Improved Forest Planning, 43–50.
European Forest Institute Proceedings no. 18. © 1998 *Kluwer Academic Publishers. Printed in the Netherlands.*

Although the GBA provides an analysis of a wide range of biological and social science issues pertaining to biodiversity, its range is by no means exhaustive. Issues such as fair and equitable sharing of benefits, financial mechanisms and technology transfer have not been treated as extensively as the more scientific issues.

UNEP believes that the GBA provides a compendium of knowledge for the benefit of those involved in the implementation of the Convention on Biological Diversity. The Assessment will also provide a significant conceptual input in the implementation of the relevant chapters of Agenda 21 and some initiatives put forth by the Commission on Sustainable Development (CSD). Clearly the aim of the Assessment was not to present a consensus document. It is, however, an important step in building scientific consensus and creating the foundation for implementing political consensus.

This paper introduces the Assessment, discusses its main findings and highlights its implications in forestry making relevant linkages where and when appropriate. No originality is claimed since the bulk of this material is published in the two monographs listed under References.

2. The Global Biodiversity Assessment

The Global Biodiversity Assessment (GBA) is an independent, critical, peer-reviewed analysis of the current issues, theories and views regarding the main aspects of biodiversity. The GBA was endorsed by the Global Environmental Facility in 1992. In May 1993, UNEP approved the GBA project, and held the initial meeting of the its Steering Group to establish the policies for preparing the Assessment and to approve a draft list of contents. The major issues associated with biodiversity were addressed by establishing 13 Sections:

- Introduction
- Characterisation of biodiversity
- Magnitude and distribution of biodiversity
- Generation, maintenance and loss of biodiversity
- Biodiversity and ecosystems function – basic principles
- Biodiversity and ecosystems function – biome analysis
- Inventory and monitoring of biodiversity
- The resources base for biodiversity assessments
- Data and information management and communication
- Biotechnology
- Human influences on biodiversity
- Economic values of biodiversity
- Measures for conservation of biodiversity and sustainable use of its components.

Each Section was headed by a team of expert scientists, with Co-ordinators and Lead Authors in order to achieve broad geographical coverage and balanced representation of developing and developed countries views. A total of over 400 experts from over 50 countries were formally involved. Teams met in workshops to plan and write each Section. The draft text was submitted to extensive peer review: governments and organisations were asked to help nominate reviewers. More than 1,100 scientific experts from over 80 nations were solicited for comments, resulting in more than 500 written reviews. The revised GBA Sections and a draft Summary for Policy-Makers were then subjected to a comprehensive review workshop held in Panama in June 1995. The GBA was launched in Jakarta, November 1995.

3. How Much Biodiversity is There and Where Is It Found

The current distribution and magnitude of biological diversity can be viewed at three levels: the diversity of ecological communities, species diversity, and genetic diversity.

3.1 INADEQUATE KNOWLEDGE OF THE DIVERSITY OF ECOLOGICAL COMMUNITIES

An ecosystem is a community of organisms and their environment which functions as an integrated unit. Forests are ecosystems. So are rotting logs, ponds, rivers, range-lands, whole mountain ranges, and indeed the planet itself. The species composition, structure and function within them changes continually over time.

Ecosystems are frequently delineated along boundaries that correspond to variations in the physical environment – such as soil types, climate and elevation, and variation in faunal of floristic composition. Various classification systems have been devised, each of which defines bio-regions or ecosystems somewhat differently. Since ecosystems are often defined with a specific purpose in mind, there is no single measure of ecological community diversity which can be uniformly applied.

3.2 THE EXTENT AND DISTRIBUTION OF SPECIES' DIVERSITY

Recent estimated of the total number of species range from 7 to 20 million, out of which we believe a good working estimate is about 13 to 14 million. Only about 1.75 million species have been described scientifically, of which slightly under a fifth are plants or vertebrates. Even for the 1.75 million species described there is no comprehensive listing. In many cases the area from which species were officially described has been so dramatically altered that it is no longer possible to re-find the species there. Also, many of the earlier descriptions of species failed to record site characteristics and habitat conditions at the collection sites.

Estimates of species diversity are relatively good for plants and vertebrates although many remain to be described: in Brazil alone, five new species of monkey were described during the last decade. Less well characterised groups of organisms include bacteria, arthropods, fungi and nematodes, while species that reside on the deep-sea floor and below the ground are especially poorly known. Scientists know very little about most of the properties of most of the 1.75 million species that have been described, such as their reproductive biology, demography, the chemicals they contain, their ecological requirements and the roles they play in ecosystems.

Domesticated species represent a tiny fraction of Earth's biota. Of the estimated 320,000 vascular plant species, about 25% have edible properties, but only about 3,000 species are regularly exploited for food. Additional 25,000-50,000 plant species are used in traditional medicine. Only about 30 of the estimated 50,000 vertebrate species have been domesticated. An additional 200 species of fish, molluscs, crustaceans, frogs, turtles and aquatic plants are grown for food and other products, and an increasing number of fungi and other micro-organisms are also eaten or used in fermentation processes, in industrial processes, or for drugs. Additional efforts in taxonomy can aid in the identification of more economic uses.

4. The Downward Spiral

During the last few millennia species have been made extinct as a result of human activities. Prehistoric colonisation of islands in the Pacific and Indian ocean some 1,000 to 2,000 years ago may have lead to the extinction of as many as a quarter of the world's bird species. Since 1600, 484 animal and 654 plant species are recorded as having gone extinct although this is almost certainly an underestimate, especially as regards tropical regions. Based on an assumed average life span of 5 to 10 million years for organisms with adequate fossil records, the extinction rate for these groups has been estimated to be fifty to hundred times the average expected natural rate. In addition, there has been widespread loss of populations and genetic resources.

Because of the world-wide loss or conversion of habitats that has already taken place, tens of thousands of species are already committed to extinction. It is not possible to take preventive action to save all of them. Projections of impending extinctions due to habitat loss can be made using the empirical relationship between the number of species and the area of a habitat, derived from island biogeography.

For some groups of vertebrates and plants, between 5 to 20% of the identified species are already listed as being threatened with extinction in the foreseeable future. These estimates depend strongly on predictions of future rates of forest loss which may increase or decrease, and on the effects of fragmentation. They will also be modified by the effects of conservation action such as the protection of areas of high diversity.

Even if species do not become extinct, many of them will lose distinct populations or suffer severe loss of genetic variability through habitat loss or fragmentation. Natural and agricultural systems are being degraded through soil erosion, introduction of exotic species, fragmentation and pollution; the effects of these most recent human-induced changes will not emerge for some time yet – suggesting that a substantial increase in the level of future extinction of species or populations as a result of human activity is already inevitable.

5. The Speed and Scale of Changes in Biodiversity

5.1 THE IMPACT OF HUMAN ACTIVITIES ON ECOSYSTEMS

The net effect of human activities may possibly be a greater overall diversity in the types of ecosystems and landscapes around the world, some of which are extremely important to societal well-being. Human activities are directly responsible for creating agroecosystems and cultural landscapes, for example. However, these increases in the diversity of ecological systems have come at the expense of impoverishment of a great number of natural communities, and the reduction of at least some ecosystem services.

Although total area may or may not have changed for major vegetation types, the plant species composition within them has often been dramatically altered by practices such as various forms of logging, grazing, introduction of exotics and fragmentation. As an example, northern temperate forests have changed dramatically in North America and Europe as a consequence of intensive logging, exploitation, and replanting, often with introduced species, over the last several hundred years. In terms of total area, forest and woodland communities have decreased globally by about 15% since pre-agricultural times. Northern temperate forest are currently relative stable in total area. In the early to mid-1980s, humid tropical forests were losing approximately 10 million hectares, or just under 1% per year globally, although there is good evidence that rates of forest conversion in parts of Central and South America have declined since. Dry tropical forests may have lost even more area. Since 1700 there has been a five-fold increase in cropland, and since 1800 about a 24-fold increase in irrigated crop land. The total area of grassland has remained roughly constant over the past 300 years, with loss through conversion to cropland being balanced by gain through deforestation.

Coastal and lowland areas, wetlands, native grasslands and many types of forests and woodlands have been particularly influenced or destroyed. For example, the dry tropical forest of the Pacific coast of central America once covered 550,000 square kilometres; now, less than 2% is intact. The United States have lost 54% of its wetlands; Thailand has lost more than 50% of its mangroves, and the Atlantic forest of Brazil has been reduced to approximately 10% of its original 1 million square kilometres. Of a total area of 600,000 square kilometres of coral reefs, about 10% have already been eroded beyond recovery.

5.2 METHODS OF ESTIMATION

Two quantitative methods have been used to estimate pending rates of extinction. The first uses careful demographic analysis in selected species, and one assessment has projected that 50% of species in selected subsets of mammals, birds and reptiles are likely to go extinct within 100 to 1,000 years. Because of the stringent requirements for data, this approach has only been applied to a small number of very well known species.

The second approach uses the observation that the total number of species found in an area is related to the size of that area in a fairly simply way, quantified by a species-area curve. Several estimated have been made of the potential loss of species due to tropical deforestation by using this empirical relationship and making assumptions about the future rate of loss of habitat. Published estimates of species that will become extinct or committed to extinction in tropical forests over approximately the next quarter-century range from 2% to 25% in the groups examined. This would be equivalent to 1,000 to 10,000 times the expected background rate. If recent rates of loss of closed tropical forest (about 1% globally per year) were to continue for the next 30 years, the equilibrium number of species in the forest, as calculated by species-area techniques, would be reduced by approximately 5 to 10%. These potential extinctions would not be immediate: it could take decades or even centuries to reach the new equilibrium number of species.

Estimates are likely to be most accurate for groups such as beetles, birds, plants and mammals, for which empirical data are easily available, but they may apply equally well to less-studied groups. These estimates to not take into account the potential effects of fragmentation, increases or reductions in the rate of deforestation, or the potential effects of mitigative measures that might be taken in the interim that could retard or even prevent extinction. Such measures include the setting aside of protected areas which include "hot-spots" or concentrations of species' diversity.

More qualitative methods of assessing the probability of species' long-term survival use of categories of IUCN – the World Conservation Union. Threatened species are those that are thought to be at significant risk of extinction in the foreseeable future because of random or deterministic environmental factors, or by virtue of their rarity. Estimates for regions are constructed by taking into account those species for which enough is known about their population size, trends, and potential threats, and extrapolating estimates of habitat loss. In 1994, the minimum estimate for numbers of globally threatened animal and plant species was about 5400 animals and 26000 plants. About 11% of bird species, 18% of mammals, 5% of fish, and 11% of plants were catalogued as threatened. An analysis of the change in species' categorisations over time implies that 5% of bird and mammal species will be extinct within 200-300 years although these changes in status reflect not only biological factors but the state of knowledge of the species involved and the rate of data entry. However, for the vast majority of the 1.75 million described species of the many millions of undescribed species, no assessment of status has been made.

6. The Underlying Causes

The primary causes underlying the loss of biodiversity are demographic, economic, institutional and technological factors, including:

- increasing demand for biological resources due to increasing population and economic development;
- failure of people to consider the long-term consequences of their actions, often due to a basic lack of knowledge;
- failure of people to appreciate the consequences of using inappropriate technology;
- failure of economic markets to recognise the true value of biodiversity
- failure of economic markets to apply the global values of biodiversity at local level;
- institutional failure to regulate the use of biological resources resulting from the growth in urbanisation, changes in property rights, and shifting cultural attitudes;
- failure of government policies to address the over-use of biological resources; and
- increasing human migration, travel, and international trade.

The specific mechanisms by which the underlying casual factors result in the reduction and loss of populations, the extinction of species and the transformation and degradation of ecological communities manifest themselves in the loss, fragmentation, and degradation of habitats; the conversion of natural habitats to other uses; overexploitation of wild resources; the introduction of non-native species; the pollution of soil, water, and atmosphere; and more recently, signs of long-term climate change.

In continental terrestrial ecosystems, the most important mechanisms is the loss, fragmentation and degradation of habitat. On islands, species introductions and habitat loss have been equally important mechanisms. In oceans, overharvesting and pollution are the most important factors. All of these mechanisms have been major influences on species and populations in freshwater ecosystems.

7. The Role of Research, Monitoring and Inventory

Enhanced research, inventory and monitoring are necessary for the promotion of responsible policy-making and management. Research into the uses and applications of biodiversity and its components is important, so is further research into ways in which biodiversity contributes to the provision of ecological services. Monitoring and inventories are needed so that newly discovered life-forms can be properly documented and that the status of individual species and particularly ecosystems can be gauged over time.

Monitoring programmes should be designed at a scale relevant to the ecological processes being monitored and the pertinent conservation or management questions being asked. Often, landscape-scale monitoring is necessary to evaluate the aggregate impact on species or ecosystems of local activities or events. Similarly, both a national and a global scales are needed to understand and effectively manage such species as migratory birds. Monitoring programmes need to be evaluated over time in order to refine techniques and to test new technologies.

8. Conclusions

One of the main conclusions of the Assessment was: unless action is taken now to protect biodiversity, we will lose forever the opportunity of reaping its full potential benefit to humankind. Priority actions need to focus on improving the knowledge base, correcting past failures in policy and ensuring that conservation and sustainable use of the planet's resources and the equitable sharing of benefits are made an integral part of all socio-economic development.

8.1 MANAGING BIODIVERSITY

The conservation and sustainable use of biodiversity needs to become an integral component of economic development by correcting policy and market failures. This will require much greater levels of co-operation and co-ordination than has previously been seen in traditional sectoral approaches to the management of natural resources.

A balanced mix of incentives and disincentives, working alongside conservation laws, market adjustments, and traditional regulatory techniques, is needed in managing biodiversity at the national level. Institutional and legal frameworks are needed to ensure that conservation and sustainable use of natural resources are integrated successfully into the wide range of social, cultural and economic context in which actions must be taken.

The adoption of more ecologically based management systems, which take into account the effects on biodiversity of extracting goods and using ecological services promises a way of balancing human socio-economic and long-term ecological considerations.

8.2. CONSERVING BIODIVERSITY

A wide variety of measures can be used to conserve biodiversity, including both in situ and ex situ methods. Effective in situ approaches include legal protection of endangered species, the preparation and implementation of species management or recovery plans, and the establishment of protected areas to conserve individual measures, such as the preservation or establishment of safe corridors through areas of intensive human use. Such measures must also return some economic benefit to local human populations if access and other restrictions are to be respected.

Ex situ conservation centres such as arboreta, aquaria, botanic gardens, seed banks, clonal collections, microbial culture collections, field gene banks, forest nurseries, propagation units, tissue and cell cultures, zoological gardens, and museums, can help to conserve stocks of both wild and domesticated animals, plants, fungi and micro-organisms, but are less able to maintain their populations. Serious gaps exist even in the ex situ coverage of those species that are known to be of direct economic importance, particularly in the tropics. Indeed the only species that have been sampled in any depth are some of the major crop plants, certain pathogens of humans and crops, and "model organisms" used in scientific research.

Restoration and rehabilitation of habitats, which depend on the availability of material and its multiplication ex situ, will come to play increasingly important role in re-establishing degraded and damaged ecosystems.

8.3 SUSTAINABLE USE OF BIODIVERSITY

Sustainable use of biodiversity is a key component of sustainable social and economic development. Management systems must take this into account explicitly, recognising that social and economic measures may be just as important as technical considerations. Flexibility of management is needed so as to be able to respond to changing social, biological and physical environments, while still maintaining essential ecosystem functions. Appropriate incentives and the enforcement of management decisions and policies must be ensured. Finding the appropriate balance depends on the particular cultural, legal, economic, ownership tenure and biological circumstances in each country.

8.4 EQUITABLE SHARING OF BENEFITS

An equitable sharing of income and assets is an important component of a strategy for conservation of biodiversity. In particular, an equitable distribution of the benefits of biodiversity conservation is a prerequisite for creating the incentives needed to maintain the Earth's biological wealth. Local benefit sharing has the effect of lowering the opportunity cost of forgoing conservation to commercial or other issues such as arable agriculture, pasture or industry.

References

United Nations Environment Programme 1995. Global Biodiversity Assessment. Cambridge University Press. Cambridge.

United Nations Environment Programme 1995. Global Biodiversity Assessment: Summary for Policy-Makers. Cambridge University Press. Cambridge.

Information Needs, Economic and Policy Consequences of Biodiversity Oriented Forest Management

BIODIVERSITY PROTECTION AND FOREST MANAGEMENT – SOME ECONOMIC AND POLICY ASPECTS

BIRGER SOLBERG
Norwegian Forest Research Institute
P.O. Box 5044
N-1432 Aas
Norway

Abstract

Three aspects regarding biodiversity protection and forest management are addressed in the paper: the role of economics, policy interventions, and main research challenges. It is argued that in most cases economics as well as policy interventions are necessary in securing optimal biodiversity protection. Among the most important research challenges is to get more empirical facts about how alternative forest management practices effect biodiversity, how risk and uncertainty should be included in a consistent way, and which policy means to introduce given the urgency of policy interventions.

1. Introduction

The threat to biological diversity is causing high general concern, not least in the forest sector. A large quantity of literature already exists about ecological impacts of species losses, but considerably less is written on economic and policy aspects. General and rather thorough overviews are given by Barbier et al. (1995), Brown et al. (1993), McNeely (1988, 1993), Myers (1993), Pearce and Moran (1994), Perrings and Opschoor (1994) and Randall (1991). In addition, works like Adamowiez et al. (1996), Hoen and Veisten (1994), Linddahl and Naskali (1994), Veisten and Hoen (1994), Veisten et al. (1993), address more specifically analyses related to forestry.

There are many issues which are of interest to discuss in this paper, but given the topics of the other presentations of the proceedings, I have found it relevant to concentrate on the following three questions:

i) What can economics contribute with in the context of biodiversity protection and forest management?
ii) Are forest policy interventions necessary, and does biodiversity as a policy goal have special consequences for forest policies and forest policy analysis?
iii) What are the main research challenges in this field?

The concept of forest biodiversity is not easy to define, and it is measured in many ways. In this paper I will by forest biodiversity simply mean diversity of life in forests. For

P. Bachmann et al. (eds.), Assessment of Biodiversity for Improved Forest Planning, 53–62.
European Forest Institute Proceedings no. 18. © 1998 *Kluwer Academic Publishers. Printed in the Netherlands.*

practical reasons I will divide it into two levels – species diversity (including genetic diversity) and ecosystem diversity. The latter could be viewed as a vector of species diversity in a spatial context.

2. Contribution of Economics

There are several reasons why economics is of high interest in the context of forest management and biodiversity protection, and in my opinion the following three are among the most important ones: (a) understanding the problem; (b) estimating the costs, benefits and distributional impacts of biodiversity protection; and (c) structuring the forest management planning problem.

a) Understand the problem: What are the driving economic forces causing biodiversity loss?

We may distinguish between proximate causes and fundamental causes of biodiversity loss. Proximate causes are the popular explanations of biodiversity loss like logging, agricultural clearance of forested land and pollution. Fundamental causes are those underlying the proximate causes – they are basically socio-economic and institutional, and most of them are caused by the following factors: population growth, economic growth, technological change; and institutional and policy arrangements.

Economic theory and analysis can contribute to a deeper understanding of all these factors, in particular of economic growth, but also of institutional aspects and technological change. These fundamental factors are also interrelated – it is hard to analyse them in isolation from each other.

As outlined in e.g. Solberg (1997), economic theory is of high interest in explaining the behaviour of various types of agents in different types of markets. This is important, not least to understand how various policy measures (in forestry as well as in other sectors) influence the use of forest land.

Related to this, economic theory also gives a basis for analysing what type of information about marginal costs and benefits can be derived from the observed market prices.

b) Quantifying the costs, economic benefits and distributional impacts

Economics makes it possible to estimate the costs and benefits of biodiversity protection measures and their distribution among regions and interest groups. There are two types of costs: the first are the benefits forgone by increased protection, for example the value of the roundwood not being harvested because harvesting is restricted in the forest area due to biodiversity considerations. These costs are in principle easy to calculate – in practice one has to use rather complicated models, for example partial equilibrium models, to include that harvest may increase on the margin in other places.

The second type of costs are the direct costs of protection – the transaction costs. Both these costs represent the use of resources (labour, energy, capital, natural resources) which could have been used in other activities – and are opportunity costs. The magnitude of these marginal costs has to be compared with the marginal benefits.

During the past 10-20 years, a rapidly increasing number of studies have been made on the valuation of the benefits of biodiversity protection. New methods have been developed and existing methods have been considerably improved. Barbier et al. (1995: 844-857), Braden and Kolstad (1991), Hausman (1993) and Kriström (1996) give excellent overviews of the present state-of-the-art regarding methods and methodological problems.

As far as the quantification of economic benefits is concerned, Pearce and Moran (1994) and Barbier et al. (1995) give an overview of studies related to forests, but most of the quoted studies there refer to tropical countries and have limited relevance for Europe. The proceedings of the 'International symposium on the non-market benefits of forestry', arranged in Edinburgh in June 1996, will give an overview of the state-of-the-art in European forestry. One may summarise the experience so far of the valuation of biodiversity related to forestry as follows:

- The methods have various weak points, e.g., the hypothetical and not real nature of payments, aggregation bias, payment vehicle bias, starting point bias and validity.
- Nevertheless, the methods give interesting results and, if properly done, good indications of the strength of attitudes for various types of biodiversity protection measures.

One of the interesting recent developments in this field is to perform experiments where people are first hypothetically asked about their willingness to pay, and then later are being asked to pay for the issue, e.g. in a real fund organized for biodiversity protection. A recent study from Norway (Navrud and Veisten 1996) found that the actual payment was from about ten to forty percents of the initially stated hypothetical willingness to pay, depending on methods used.

The costs and benefits allow to estimate the distributional impacts of biodiversity protection measures for regions and groups of individuals. This is important for policy making and for judging the chances of success of the measures.

c) Structuring the forest management planning problem

Economics and economic theory can be very useful in combination with bio-economical planning models, for example in the many large-scale forest management optimisation models that exist. By applying an objective function in such models of maximising net present value, shadow prices of high value for interpreting the costs of constraints due to biodiversity considerations can be calculated. The consistency in the comparisons can be secured by carrying out a sensitivity analysis with the optimisation model. In particular when combined with a GIS-framework, this kind of studies are of high interest in forest management decisions.

Some people have strong aversions against bringing in economics at all in considerations of biodiversity protection, arguing that it is quite wrong to allow economic values to play any role in determining what species should or should not be saved. It is argued that it is immoral to use economic values on species, as the species that are threatened have a right of their own to survive.

One can put at least two arguments against this reasoning. The first is that the economic view is itself a moral view – as it is based on the utilitarian assumption that it is people of the present generation which are the decision makers. It is not possible to get the next generation's views – what is possible is just to focus on how the present generation values the welfare of the next generations. The second main argument is that due to economic growth and population growth in particular, the pressure on species loss is getting stronger and stronger. The loss of species is high today, and it seems unrealistic to believe that all species that are threatened can be saved. We will most likely have to make choices as to which areas and which species to protect, simply because of the lack of resources to save all that could be saved if unlimited resources were available. If all threatened species are equally important, the logical criterion would be to get as many of those species saved per unit of cost. A cost efficiency criterion like this immediately brings in economics.

3. Policy Aspects

3.1 ARE FOREST POLICIES NECESSARY?

Forest policies have during the past decade shown a growing concern about biodiversity (Tikkanen and Solberg 1995). It is relevant to investigate under which conditions policy means are of interest. As discussed i.a. in Solberg (1986), there are at least the following possible justifications for policy interventions:

i) Existing markets do not function properly, so that the prevailing timber prices (and other prices) do not reflect the true marginal costs and benefits for the society. Main reasons for this could be:
   - Monopolistic or monopsonic timber markets, and relatively strong market regulations.
   - A regulated forest property market (for example forest properties can not be sold without public concessions).
   - A capital market with significant differences between real interest rates after tax on loans and on savings (partly due to the tax systems).
   - Inappropriate subsidies, fees and taxes creating major differences between micro- and macro-economic costs and benefits (for example inappropriate forest revenue systems and too high subsidies for agriculture or forest industries).
   - The factors (ii) - (vi) below.

ii) The time perspectives are different between micro-agents (e.g. forest owners) and the society at large.

iii) Some markets do not exist, i.e. some prices are lacking. This could be forest goods and services like recreation, wildlife/hunting, sports fishing, nature conservation. The representation of the future generations, i.e. the realistic future market could be included here or in the above mentioned factor "time perspective". In several cases biodiversity is a global good (not only a national concern) – and this global market is not functioning today.

iv) Distributional effects. The society at large would in many cases have to consider distributional effects like regional and personal income.

v) Risk and uncertainty. The society at large can spread the risk over many projects and many agents/inhabitants thus facing a different risk environment than a single agent. (Among other things this will depend upon the possibilities and costs of insurance, and degrees of irreversibility).

vi) Conflicts between different interest groups. The optimal sharing of the production surplus is not evident – there are different political opinions on how the resources of the society should be distributed. One may say that all the above mentioned factors (i) - (v) are valid assuming that there is one decision maker (policy maker) or a consensus among various policy makers, whereas the conflict/interest group argument is based on non-consensus, and comes in addition to the factors (i) - (v).

In principle, only one of the factors above is sufficient to justify policy interventions. In the present state of forest biodiversity, most of the factors are valid in most countries.

3.2 DOES FOREST BIODIVERSITY CONSERVATION LEAD TO SPECIAL POLICY IMPLICATIONS?

One may point at several special implications for forest policy analysis and policy making when setting biodiversity protection as a policy goal – I will emphasise the following four:

1. Uncertainty and risks are a major challenge. In addition to the "normal" uncertainty factors in all kinds of policy analyses, the uncertainty aspects become more important because biodiversity protection automatically raises the question of possible irreversible outputs. The risk of irreversibility implies a special value for avoiding such an output – an option value – which in many circumstances might be high. As outlined by e.g. Zinkhan (1995), forest managers have considerable options over time to abandon, delay, expand, or contract certain practices in response to changing operating, market and political situations. The concept and measures of biodiversity are not generally agreed upon.

   There is not enough information available on the effects - both short and long term – of the decrease in biodiversity and of the influence of forest management on biodiversity. Most scientists would probably argue that biodiversity loss is characterised more by uncertainty than by risk (i.e. we have very little empirical basis for probability distributions of the likely events). This implies particular problems regarding what are optimal decisions – cf. for example Brown (1983) and Katzner (1989, 1995).

2. When setting biodiversity production as a policy goal, the conflict and distributional importance increases as a consequence of the fact that the number of goal elements increases. An additional reason is that this new goal element is burdened with more uncertainty than most of the other goal elements – which in itself gives more space for different opinions and larger differences in the preferences of the various stakeholders.

   The distributional aspects are closely linked to conflicts. Conflicts may arise because at least some stakeholders may feel that they have to bear too much of the costs or are getting too little of the benefits over time.

A particular type of a distributional problem is how much consideration should be given to the future generations. For biodiversity this question is crucial, but the next generation with their preferences is not yet here. One of the most difficult aspects will be to handle intra-generational (i.e. within a generation) equity in relation to the inter-generational equity aspects. One may say that inter-generational equity, or justice between generations, is the main principle behind the notion of sustainability. However, for the poorest of the present generation who often live in very bad ecological conditions it is not easy to understand that resources that could help to survive today must be saved for further generations. There is no clear answer to what is the optimal income distribution (income in a wide sense including ecological conditions). If we for example follow Rawl's principle of justice (Rawl 1971), it is the welfare of the poorest segments of the population which should be compared between generations. The risks/uncertainties and conflicts are not decreased by this.

In many cases the risks related to biodiversity differ considerably between stakeholders. As such, risks become an important distributional effect, which might be of high relevance in the policy analysis.

3. Safe minimum standards (so-called SMS) have been discussed as an alternative to economic analysis (Bishop and Ready 1991, Bishop 1993). In my opinion SMS do not substitute economic analysis, because when deciding on a SMS the marginal costs (and benefits) of the standard most often have to be analysed. Economic reasoning is then unavoidable. SMS is more relevant as a complement to economics, as it secures under "normal" economic impact situations a minimum ecological standard. If a certain SMS should imply very high costs, then certain institutional procedures could be initiated to discuss the standard.

4. The complexity of forest policy analysis increases as the uncertainty and the conflict issues become more important. All stages in the policy process will be influenced by this. It is important to take the time dimension into account at all stages. Theories of dynamic analysis and behaviour under uncertainty and conflicts will have more importance. The possibilities for policy analysis to arrive at "absolute" and "sure" results will be reduced (if they ever have existed) – more modesty and qualitative statements are likely to be the results of the analysis.

5. A few main consequences for policy-making of including biodiversity protection as forest policy goal can be drawn based on the discussion above:
   - Uncertainty aspects have to be realised from the very beginning of the policy process. It has to be stated as explicitly as possible for the biodiversity issues at stake. For some of those it may be a rather good documentation of dose/response effects etc., whereas for others we may have almost no information at all, i.e. classical uncertainty.
   - Likewise, the time dimension should always be considered. Particularly important is the aspect of learning over time, and corresponding possibilities for monitoring uncertain impacts and adjusting the policy means over time according to any new knowledge obtained.

- It may be highly advantageous to use policy means which are flexible and promote keeping options. In order to protect biodiversity in some circumstances it may be a wise policy e.g. to rent forest land for some time to prevent a harvest operation which may entail a high risk of biodiversity loss, and thus buy time for learning from ongoing research (or invest in new research), thus avoiding irreversibility.
- Conflict considerations and distributional effects have to be included in most cases. It is important to establish processes/procedures which ensure that the main stakeholders' interests are taken into account in an appropriate way, and that the information and data available is transparent.
- Finally, because the time and risk dimensions are so important, it may in many situations be a considerable advantage to view and practice policy analysis in a monitoring aspect as a tool for adjusting the course over time. As such, policy analysis will be a policy mean in itself, and thus as mentioned by e.g. Wallace et al. (1995), will itself have to be regularly evaluated.

4. Main Research Challenges

There are several research challenges to address – in my opinion the following are among the important ones:

1. Probably the most vital issue related to management and biodiversity protection is to get more empirical facts about how biodiversity (at various levels and definitions) and for various ecosystems is influenced by alternative forest management programs. By a forest management program I mean a clearly defined set of use over time of the control variables available. Control variables are variables which can be (or are) decided by forest management, like:
   - Controlled burning (frequency and size of area burned)
   - Natural regeneration (intensity and size of area)
   - Planting including supplementary planting
   - Tending (intensity and area size)
   - Fertilisation
   - Harvesting (including clearfelling)
   - Protection area (size)
   - Area distribution of the various forest management alternatives over time

   Today we have too little empirical knowledge about this; while waiting for more information we have to take decisions in the situation of classic uncertainty.

   One should use dynamic forest models to map the trade-off frontiers between timber production and biodiversity and combine this with economic data. Such models do exist today for boreal forests – the weakest point at present is the empirical knowledge about the changes in species composition as a function of changes in forest management control variables – or more specifically the partial derivatives:

$$\frac{\partial s_{i,j,t}}{\partial u_{k,j,,t}} \qquad \forall_{i,j,k,t} \tag{1}$$

where

$s_{i,j,t}$ is nos. of species in area j at time t
$u_{k,j,t}$ is control variable k used in area t at time t
i is species type, i ε (1, I)
j is area, j ε (1, J)
t is time period, t ε (1, T)
k is type of control variable, k ε (1, K)

Note that the level of detail of $s_{i,j,t}$ should be decided by the definition of biodiversity used.

It has been advocated, due to the lack of accurate knowledge, to mimic the natural forest succession – i.e. how a natural forest without human intervention will develop. This may sound good, but it does not solve the main challenge related to sustainable forest use: how a marginal deviation by human intervention will influence biodiversity.

2. The research related to forest management and biodiversity should be more co-ordinated, concentrated and multidiciplinary than today. During the last 3 years I have met several researchers in this field in Europe. My impression is that although very good work is being done, much more could be gained by pooling resources with other research groups working on the same ecosystem but on other species. There is a high potential for improvement to be gained by increased co-ordination and concentration of efforts.

3. With more basic information about the relationship between loss of species and forest management practices, we can construct more reliable production frontiers balancing biodiversity and timber production, and combine these with economics.

   In the meantime we have to live with the present uncertainty and risk. But decisions have to be taken and it is important to take those decisions in a consistent way. Here, theory on risk, and more important, uncertainty have to be utilised more. More research is needed on how to include risk/uncertainty in a consistent way.

4. As far as forest monitoring is concerned, it is important to concentrate the monitoring efforts on the most important species, i.e. from the biodiversity point of view on those that are threatened by forest management practices.

   In addition, it is important to monitor not only the ecosystems, but also the policy means used and how they influence the forest ecosystem. For research it is important to document the costs as well as the diversity gain of the various forest policies proposed.

5. We have to differentiate between geographical area of different sizes, as it is a totally different matter if we want to save threatened species at a county level, national level, regional level or global level. The smaller the geographical area, the

bigger the costs. In my opinion the most important level when talking about extinction of species is the global level – that is the irreversible level from the genetic point of view, and should be given the highest attention.

6. Policy measures to protect forest biodiversity have to be implemented with the time dimension in mind. Many ecosystems that are highly valuable from the biodiversity point of view are currently under heavy pressure now and it is necessary to act quickly to save them. An important challenge is to find appropriate policy measures for this.

   One needs, at national as well as international levels, priority lists of threatened forest ecosystems based on at least the following criteria: ecological value, cost of protection, and corresponding chances for successful protection. The compilation of such lists calls for multidisciplinary research efforts. It is also unavoidable that several subjective assumptions have to be made in the lack of sufficient information, but if the assumptions are explicitly specified, such priority lists are likely to clarify the situation for decision-making.

5. Concluding Remarks

The pressure caused by the competing uses of land and the growing demand for services from forests call for careful intersectoral decision making in land use questions, particularly concerning forest management and biodiversity protection. Economic analysis will have to play an important role in such decisions, combined with ecological and risk/uncertainty information and assumptions.

There are several reasons why policy interventions most often are needed to protect biodiversity. Biodiversity as a policy goal usually leads to increased conflicts and uncertainty compared to a situation where biodiversity is not considered. The question of the optimal protection of biodiversity is today probably the most important challenge for forestry, and should be adequately addressed.

References

Adamowitcz, W.L., Boxall, P.C., Luckert, M.K., Phillips, W.E. and White, W.A. (eds.) 1996. Forestry, economics and the environment. CAB International. UK.

Barbier, E.B. et al. 1995. The economic value of biodiversity. In UNEP: Global Biodiversity Assessment. Cambridge University Press. Cambridge. Pp. 827-912.

Bishop, R.C. 1993. Economic efficiency, sustainability, and biodiversity. Ambio. 22(2-3): 69-73.

Bishop, R.C. and Ready, R.C. 1991. Endangered species and the safe minimum standard. American Journal of Agricultural Economics. 73: 309-312.

Braden, J.B. and Kolstad, C.D. (eds.) 1991. Measuring the demand for environmental quality. North-Holland. Amsterdam. 370 p.

Brown, K., Pearch, D.W., Perrings, C. and Swanson, T. 1993. Economics and the Conservation of Global Biological Diversity. Working Paper No. 2. Global Environment Facility. Washington DC.

Brown, S.P. 1983. A note on environmental risk. Journal of Environmental Economics and Management. 6: 282-286.

Hausman, J. (ed.) 1993. Contingent Valuation: A critical assessment. North Holland. Amsterdam.

Hoen, H.F. and Veisten, K. 1994. A survey among users of Oslomarka: Attitudes towards the state and management of the forest area. Skogforsk. Report No. 6/94. Agricultural University of Norway. Department of Forest Sciences. (In Norwegian – English summary).

Katzner, D.W. 1989. The Shackle-Vickers approach to decision-making in ignorance. Journal of Post Keynesian Economics. 12: 237-259.

Katzner, D.W. 1995. Time, Ignorance and Uncertainty in Economic Models. University of Michigan Press.

Kriström, B. 1996. Contingent valuation. Work report 233. University of Agriculture. Department of Forest Economics. Umeå.

Lindal, M. and Naskali, A. (eds.) 1994. Valuing biodiversity. Proceedings from a workshop. Scandinavian Forest Economics No. 34. Rovaniemi.

McNeely, J.A. 1988. Economics and biological diversity: Developing and using economic incentives to conserve biological diversity. IUCN. Gland. Switzerland. 200 p.

McNeely, J. 1993. Economic incentives for conserving biodiversity – lessons for Africa. Ambio. 22(2-3): 144-150.

Myers, N. 1993. Biodiversity and the precautionary principle. Ambio. 22(2-3): 4-79.

Navrud, S. and Veisten, K. 1997. Validity of non-use values in contingent valuation: An empirical test with actual payments. Draft paper. Agricultural University of Norway. Department of Forest Sciences. Ås.

Perrings, C. and Opschoor, J.B. 1994. The loss of biological diversity: some policy implications. Environmental and Resource Economics. 4: 1-12.

Pearce, D. and Moran, D. 1994. The economic value of biodiversity. Earthscan. London.

Randall, A. 1991. Total and nonuse values. In: Braden, J.B. and Kolstad, C.D. (eds.) 1991. Measuring the demand for environmental quality. North-Holland. Amsterdam. Pp. 303-321.

Rawls, J. 1971. A theory of justice. Harvard University Press. Cambridge.

Solberg, B. 1986. Forest sector models as methodological tools in forest policy analysis. In: Tikkanen, I. (ed.). Silva Fennica. (20)4: 419-427.

Solberg, B. 1997. The role of economics in forest policy analysis. In: Saastamoinen, O. and Harjo, A. (eds.). Proceedings of the conference "Economic and legal aspects of forest policy in the Scandinavian countries and Russia". Research Notes xx. Faculty of Forestry. University of Joensuu. Finland.

Tikkanen, I. and Solberg, B. 1995. Evolution of Forest Policy Science in Finland and Norway. In: Solberg, B. and Pelli, P. (eds.). Forest policy analysis – methodological and empirical aspects. EFI Proceedings No. 2. European Forest Institute, Joensuu, Finland. Pp. 69-90.

Veisten, K. and Hoen, H.F. 1994. Attitudes towards and willingness to pay for changes in Norwegian forest management practices. Skogforsk. Report No. 4/94. Department of Forest Sciences. Agricultural University of Norway. Ås. (In Norwegian – English summary).

Veisten, K., Hoen, H.F., Navrud, S. and Strand, J. 1993. Valuing biodiversity in Norwegian forests: A contingent valuation study with multiple bias testing. Memorandum No. 7, 1993. Department of Economics. University of Oslo. Norway.

Wallace, M.G., Cortner, H. and Burke, S. 1995. Recommendations for improving policy evaluation in natural resources. Paper for XX IUFRO World Congress. Tampere.

Zinkhan, F.C. 1995. Forest economics – the management of options and values. J. Forestry 99(1): 25-29.

INFORMATION NEEDS FOR BIODIVERSITY ASSESSMENT DERIVED FROM INTERNATIONAL FORESTRY DISCUSSIONS

JARI PARVIAINEN
The Finnish Forest Research Institute
Joensuu Research Station
Box 68
FIN-80101 Joensuu, Finland

RISTO PÄIVINEN
European Forest Institute
Torikatu 34
FIN-80100 Joensuu, Finland

Abstract

International agreements, initiatives for developing criteria and indicators for sustainable forest management and the public discussion on certification commit all countries to preserve biodiversity and to develop a more balanced management of forest environment. To achieve these goals, most representative natural forests have been totally protected and production forests managed with more nature oriented principles. According to the historical forest use, human impact, forest ownership structure and natural forest development cycle, different forestry practices can be distinguished.

Research on stand level biodiversity aims at developing silvicultural guidelines. Studies on stand structure and dynamic processes play a key role in the description of biodiversity indicators. Further research on stand level is needed in defining the key indicators/parameters for biodiversity with standards (e.g. dead wood component), differences in biodiversity indicators between natural forests and managed forests (naturalness of forests), influence of disturbances on biodiversity indicators, and economic comparisons between biodiversity oriented and wood production oriented silviculture.

At landscape level the most urgent need is to develop forest inventory methods and to monitor the forest ecosystem as a whole. This requires an efficient use of GIS and other techniques to visualise the different forest protection and management units. The forest ecosystem inventory should include, in addition to the traditional forest stand characteristics, assessment of the protection areas (representativeness, geographical location, site classes, age structure, size), location of forest areas with regard to other land use alternatives, key biotopes and other protective zones and watershed management areas. Further development is also needed in the definitions of terms (e.g. forest protection classes, origin of forests) and in the harmonisation of classifications.

P. Bachmann et al. (eds.), Assessment of Biodiversity for Improved Forest Planning, 63–70.
European Forest Institute Proceedings no. 18. © 1998 *Kluwer Academic Publishers. Printed in the Netherlands.*

1. Introduction

The public discussion on criteria and indicators for sustainability or ecolabelling demands wood production oriented forest management to be replaced by a more balanced management of forest environment including the preservation of biodiversity. These targets are tried to be met through 1) total protection of the foremost regionally representative natural forests and 2) by managing production forests with more nature-oriented principles.

Natural or semi-natural old forests are important for preserving the original biodiversity, as they provide a habitat for certain specialised forest related species. The only means to preserve the flora and fauna typical for natural forests is by establishing a sufficiently dense and representative network of conservation areas. Improvement in biodiversity depends, however, not only on conservation measures but also on how production forests are managed and how well valuable and unique key biotopes are preserved within these forests. Thus, there are no strict scientific criteria for deciding how much forests should be reserved for conservation only (see Schuck et al. 1994).

The basic approach for developing international general parameters for biodiversity in forests is to proceed to the management level in which forestry is practised in individual countries. According to the traditional forest use, human impact, forest ownership structure and natural forest development cycle, different forestry practices can be distinguished as shown in the following (see also Oldeman et al. 1994):

- utilisation of untouched boreal/temperate forests by large scale forestry practices (e.g. Canada, Russian Siberia)
- utilisation of untouched tropical rain forests by small scale selective cuttings (e.g. in Amazonian forests, Indonesia)
- utilisation of modified natural or semi-natural boreal/temperate forests by small scale family forestry practices (e.g. Finland, Norway, partly Sweden)
- utilisation of modified semi-natural temperate forests by small scale family forestry practices (e.g.. Denmark, Germany, France, Austria)
- intensive plantation forestry in the tropics, temperate or boreal zone (e.g. South-Brazil, Chile, New Zealand)

By defining the criteria or certificates for biodiversity oriented silviculture at national or management unit level these differences should be recognised and taken into consideration.

2. Ongoing Processes

Since the Rio 1992 Earth Summit, governments, environmental groups, the business community and consumers have agreed that there is a need to develop a common understanding of the definition and characteristics of sustainable forest management. Four governmental efforts are underway to reach consensus on indicators of sustainable forestry (Harmonization of... 1995; ISCI Background Document 1996).

Through the Helsinki process, European countries have developed and reached a pan-European, binding consensus (Geneva, June 24, 1994 and Antalya Statement, January 28, 1995) on criteria and indicators. In the first Ministerial Conference on the Protection of

Forests in Europe, the principles of the preservation of genetic biodiversity at the European level were agreed upon. The Helsinki process aimed at identifying the important characteristics of sustainable forest management and defined criteria (6) and indicators (27) to serve as a basis of measuring the progress towards that goal. Through progress assessments, European countries intend to be able to demonstrate that at the national level individual countries are managing their forest lands in a sustainable way. The first interim report on the progress was produced at the end of April 1995, co-ordinated by the Liaison Unit in Helsinki (Interim Report 1995). The final report on the questionnaire on sustainability in European countries has now been completed and published in late 1996.

The Montreal process is a parallel effort to the Helsinki process, including non-European temperate and boreal forest countries: Australia, Canada, Chile, China, Japan, Mexico, New Zealand, Republic of Korea, Russia and The United States. In February 1995 an agreement was reached on a number of international level criteria (6) and indicators (67) (Sustaining the World's...1995, Santiago-Agreement). Their purpose is to provide a method to describe and develop a broader list of criteria and indicators for national processes than those agreed upon at the international level.

The Tarapoto Proposal was launched in February 1995 by the Amazonian Cooperation Treaty (ACT). It consists of 8 Amazonian Basin states (Bolivia, Brazil, Columbia, Ecuador, Guyana, Peru, Surinam and Venezuela) which share the special interests of this region. The proposal comprises 12 criteria and 77 associated indicators, grouped into three categories (national, management-unit and global level.

The Dry-Zone Africa Initiative was initiated in February 1995 by FAO and UNEP. It embraces the sub-Saharan dry-zone African countries. 7 criteria and 47 indicators for monitoring sustainable forest management at the level were developed.

The International Tropical Timber Organization (ITTO) adopted the year 2000 as its target date for sustainable management of all tropical forests. ITTO has developed guidelines for the sustainable management of natural tropical forests as early as 1990 for planted tropical forests, conservation of biodiversity in tropical production forests and tropical forest management, but they are not identical to the lists of either the Helsinki or the Montreal Process.

Perhaps the most wide-ranging definition work of non-governmental organisations is the undertake by the Forest Stewardship Council, FSC, set up by the WWF (the World Wide Fund for Nature International). The parameters of sustainability adopted by FSC are in agreement with those brought forward in connection with the follow-up work of the European Ministerial Conference. Differences arise from the difference in emphasis given to certain aspects. Conservation bodies attach primary importance to the evaluation of the sustainability of forest management from the perspective of conservation and ecology.

3. Criteria and Indicators

The term used in this connection is 'criteria of sustainability' which describe the various dimensions of sustainability. The purpose is to use indicators in measuring changes in the criteria of sustainability, i.e. in relation to time. As a matter of fact, indicators are instruments, parameters compiled from statistics or from data obtained by conducting various measurements in the forest. The information provided by indicators is then evaluated in connection with societal or international debates, and they form the basis for the setting of policies to implement sustainability.

In addition to quantitative indicators, there is a need for descriptive parameters with which to assess the realisation of the measures taken (e.g. in legislation). Development work in this area has started. Clarification is particularly needed as regards the socio-economic aspects of sustainability.

The characteristic features of sustainability in European forestry have been defined as the following six criteria (the Helsinki process, Ministerial Conference 1993):

- forest resources and the carbon cycle
- the health and vitality of forests
- multiple functions of the forests
- the biological diversity of forest ecosystems
- maintenance of the protective functions of forests (soil and water system)
- maintenance of socio-economic functions

Each criteria consists of a number of indicators (in the group 'forest resources', for instance, trends in the development of the forested area and the centrally important characteristic stand features of the forest) and their development is monitored in relation to time. In all, at the European level there are at the moment 27 agreed indicators. Taken together, the set of criteria and indicators suggests an implicit definition of the conservation and sustainable management of forest ecosystems at the country level. It is recognised that no single criterion or indicator is alone an indication of sustainability.

4. Forest Biodiversity in the Helsinki and Montreal Processes

The largest number of indicators that has been discussed during the Helsinki and Montreal processes concern the criteria of forest biodiversity. The definition of biodiversity in a broad sense, as described by the Helsinki process, is the following:

Recalling the definition of biological diversity agreed upon in the Convention on Biological Diversity: viz. *'Biological diversity means the variability among living organisms from all sources including,* inter alia, *terrestrial, marine and other aquatic ecosystems and the ecological complexes of which they are part; this includes diversity within species, between species and of ecosystems'.*

For the present, consensus on forest biodiversity has been reached only with regard to the factors of most central importance. In the matter of ensuring the preservation of forest biodiversity in production forests, the proportion of mixed woods and genetic biodiversity are the only factors agreed upon. Discussion have been held in issues such as fragmentation of forests, how to define naturalness, age distribution, the proportion of decaying wood in forests, and the methods to be applied in forest regeneration and timber harvesting. A uniform indicator platform applicable in different forest vegetation zones is still lacking.

The criteria 4 "Maintenance, conservation and appropriate enhancement of biological diversity in forest ecosystems" of the Helsinki-process contains the following indicators with separate definitions:

4.1 Changes in the area of:
   a. natural and ancient semi-natural forest types
   b. strictly protected forest reserves
   c. forests protected by special management regime
4.2. Changes in the number and percentage of threatened species in relation to the total number of forest species (using reference lists e.g. IUCN, Council of Europe or the EU Habitat Directive)
4.3. Changes in the proportions of stands managed for the conservation and utilisation of forest genetic resources (gene reserve forests, seed collection stands, etc.); differentiation between indigenous and introduced species
4.4 Changes in the proportions of mixed stands of 2-3 tree species
4.5. In relation to the total area regenerated, proportions of annual area of natural regeneration

To compare with the Montreal process, biodiversity is defined by indicators as follows:

Criterion 1:
Conservation of biological diversity. Biological diversity includes the elements of the diversity of ecosystems, the diversity between species, and genetic diversity in species.

Indicators:
Ecosystem diversity

a. Extent of area by forest type relative to total forest area (a)[1]
b. Extent of area by forest type and by age class or successional stage (b)
c. Extent of area by forest type in protected area categories as defined by IUCN[2] or other classification systems (a)
d. Extent of areas by forest type in protected areas defined by age class or successional stage (b)
e. Fragmentation of forest types (b)

Species diversity

a. The number of forest-dependent species (b)
b. The status (rare, threatened, endangered, or extinct) of forest-dependent species at risk of not maintaining viable breeding populations, as determined by legislation or scientific assessment (a)

Genetic diversity

a. Number of forest-dependent species that occupy a small portion of their former range (b)
b. Population levels of representative species from diverse habitats monitored across their range (b)

---

1 Indicators followed by an "a" are those for which most data are available. Indicators followed by a "b" are those which may require the gathering of new or additional data and/or a new program of systematic sampling or basic research. No priority is implied in the "a" and "b" designations.

2 IUCN categories include: (1) Strict protection. (II) Ecosystem conservation and tourism, (III) Conservation of natural features, (IV) Conservation through active management, (V) Landscape/Seascape conservation and recreation, (VI) Sustainable use of natural ecosystems.

5. The Task for Research

The criteria used should be such that they produce objective information on sustainability. An example of a criterion that works in practice is the assessment of forest health. The measurement methods used are the result of many years of research, discussion and co-operation at the international level. The same techniques are applied throughout Europe. Annual assessments provide the basis for indicating general trends in forest health. Weakening in forest health condition is a sign to political decision makers to introduce for instance emission restrictions. There is a need to develop a parallel system for monitoring forest biodiversity on a practical scale.

In order to practise silviculture at a sufficiently high level of diversity throughout the forest ecosystems, we must learn about the development processes in undisturbed forests. The clarification of these processes require specific research on the structure, succession, biodiversity and long-term monitoring of natural forest ecosystems (Leibundgut 1978; Schmidt-Vogt 1991a and 1991b; Remmert 1992). Protected forest areas or forest reserves are the main, still remaining samples of undisturbed forests, and are therefore the basic experimental forests for research on biodiversity (Schuck et al. 1994).

Due to modern silvicultural practices many important regimes in temperate and boreal forests have disappeared or have become a rare sight (Angelstam and Rosenberg 1993; Kouki 1994):

- Burnt areas with large amounts of dead wood
- Dying trees, snags and down logs
- Deciduous components in forests
- Natural development processes: human intervention in the natural dynamics and development of forests in spatial and temporal respects
- Size: Fragmentation of existing natural and semi-natural forests in total.

Trees and stand structure are in close interaction with all other parts of the forest ecosystem e.g. soil, flora and fauna. Therefore, studies on stand structure and dynamic development processes at the stand level play a key role in the description of biodiversity indicators in different forest vegetation types. Parameters of structure, giving valuable information on dynamics and development processes, are easy to measure. A large number of permanent sampling plots already exist for long periods of time in many European countries, providing useful data (Schuck et al. 1994).

When measuring sustainability, attention is paid to the difference in the development of production forests as compared to that of natural forests. A central point is that of recognising the regular functional features peculiar to individual forest zones; e.g. the clear developmental difference between boreal forests and temperate forests. In the case of Finland, for example, the way in which silvicultural measures are applied is steered by historical cultural developments of forest use (slash-and-burn cultivation, diameter-based selection felling) and, from the viewpoint of natural forests, by forest fires and other natural catastrophes (Parviainen 1993 and 1994; Parviainen and Seppänen 1994).

The following key topics for research on biodiversity can be distinguished:

Research on stand level biodiversity aims at developing silvicultural guidelines. Trees and stand structure are in close interaction with all parts of the forest ecosystems e.g. soil, flora and fauna. Therefore, studies on stand structure and dynamic processes play a key role in the description of biodiversity indicators.

Key topics where further research at stand level is needed are: defining the key indicators/parameters for biodiversity with standards (e.g. dead wood component), differences in biodiversity indicators between natural forests and managed forests (naturalness of forests), influence of disturbances on biodiversity indicators and economic comparisons between biodiversity and wood production oriented silviculture.

At landscape level the most urgent need is to develop forest inventory methods and to monitor the whole forest ecosystem. This requires an efficient use of GIS and other techniques to visualise the different forest protection and management units.

The forest ecosystem inventory should include, in addition to the traditional forest stand characteristics, assessment of the protection areas (representativeness, geographical location, site classes, age structure, size), location of forest areas with regard to other land use alternatives, key biotopes and other protective zones and watershed management areas. Further development is also needed in the definitions of terms (e.g. forest protection classes, origin of forests) and in the harmonisation of classifications.

As an example of stand level biodiversity research at the European scale, a research task called "Forest Reserves research Network in Europe" EU COST Action (E4) was launched in 1995 in order to promote biodiversity research according the new requirements. The overall objective is to create a European network of forest reserves in order to collect ongoing research and unify research methodologies and have access to a central data bank for the exchange of research results. The following tasks have been outlined:

- To achieve a common understanding on definitions, terminology and management approach for forest reserves and categories of forest protection
- To survey and analyse current information and ongoing research available on forest reserves
- To collect the published research reports on the structure and dynamics of forest reserves
- To collect, compare and harmonise research methodology for monitoring the stand structure and changes
- To promote the establishment of a common permanent sampling plot system
- To create a data bank for gathering the information

The output of the action will be a report (book) for a wide audience with definitions on forest protection and research in the natural forests and forest reserves, a catalogue with a summarising introduction to the state-of-the-art in the research of natural forest, manual and computer programmes for monitoring the structure of natural forests, a recommendation to support research of natural forests and other means of international co-operation and a data bank on organisations, publications, forest resources, research and its results with free access for the participants.

Examples of international forest assessment and monitoring initiatives where biodiversity component can be integrated are FAO/ECE Global Forest Resources Assessment Programme, the UN list of National Parks and Protected Areas (IUCN), FAO Information System on Forest Genetic Resources, European Forest Information and Communication System (EFICS) and ECE/EU IPC Forests (International Cooperative Programme on the Assessment and Monitoring of Air Pollution Effects on Forests).

References

Angelstam, P. and Rosenberg, P. 1993. Aldrig, sällan, ibland, ofta. Skog & Forskning 1/93: 34-41.

Harmonization of criteria and indicators for sustainable forest management. FAO/ITTO Expert Consultation. Report. Rome. Italy. 13-16 February 1995. FAO/ITTO. 8 p. + appendixes.

Intergovernmental Seminar on Criteria and Indicators for Sustainable Forest Management. Background Document August 19-22, 1996. Helsinki. Ministry of Agriculture and Forestry. 131 p.

Interim Report on the Follow-up of the Second Ministerial Conference 1995. Ministerial Conference on the Protection of Forests in Europe, 16-17 June 1993. Helsinki. Ministry of Agriculture and Forestry in Finland. 255 p.

Kouki, J. (ed.) 1994. Biodiversity in the Fennoscandian boreal forests: natural variation and its management. Annales Zoologici Fennici 31(1). 217 p.

Leibundgut, H. 1978. Europäische Urwälder der Bergstufe. Bern und Stuttgart. 308 p.

Ministerial Conference on the Protection of Forests in Europe, 16-17 June 1993 in Helsinki 1994. Documents. Ministry of Agriculture and Forestry in Finland. 56 p. European List of Criteria and Most Suitable Quantitative Indicators. Ministry of Agriculture and Forestry in Finland. 20 p.

Oldeman R., Parviainen, J. and Stephan, K. 1994. Sustainability. Naturopa. 75: 15-19.

Parviainen, J. 1993. Tuli metsän ekologisessa kierrossa. In: Piri, E. (ed.). Tuli metsän ekologisessa kierrossa. Metsäntutkimuslaitoksen tiedonantoja. 462: 8-14.

Parviainen, J. 1994. Finnish Silviculture. Managing for Timber Production and Conservation. Journal of Forestry. 92(9): 33-36.

Parviainen, J. and Seppänen, P. 1994. Metsien ekologinen kestävyys ja metsänkasvatusvaihtoehdot. Metsäntutkimuslaitoksen tiedonantoja 511. 110 p.

Remmert, H. 1992. Ökosysteme. Springer Verlag. Berlin, Heidelberg, New York, London, Paris, Tokyo, Hongkong, Barcelona, Budapest. 363 p.

Schmidt-Vogt, H. 1991a. Die Fichte II/3. Parey Verlag. Hamburg - Berlin. 804 p.

Schmidt-Vogt, H. 1991b. Naturnahe Waldwirtschaft. Wilhelm-Munker-Stiftung (ed.). Beiträge zur Lebensqualität, Walderhaltung und Umweltschutz, Gesundheit, Wandern und Heimatpflege. Heft 31. 56p.

Schuck, A., Parviainen, J. and Bücking, W. 1994. A review of approaches to forestry research on structure, succession and biodiversity of undisturbed and seminatural forests and woodlands in Europe. Working Paper 3. European Forest Institute. Joensuu. Finland. 62 p.

Sustaining the World's Forests. The Santiago Agreement. Criteria and Indicators for the Conservation and Sustainable Management of Temperate and Boreal Forests 1995. Journal of Forestry. 93(4): 18-21.

BIODIVERSITY: A DIVERSITY IN DEFINITION

Terminological utopia and scientific reality

M. KAENNEL
Swiss Federal Institute for Forest Snow and Landscape Research
Zürcherstr. 111; 8903 Birmensdorf
Switzerland

Abstract

A quantitative and qualitative assessment of the international scientific literature of the past ten years indicates major differences in the definitions and use of the term biodiversity. This in part reflects a steady increase in the number of occurrences of the term in the titles of scientific articles from year to year, as shown by a search of various online literature databases.

In addition, a survey of a sample of articles and books focusing on biodiversity, conservation biology and sustainability provides a variety of formal and informal definitions of the term biodiversity. The use of the term has expanded to cover a wide range of different aspects, at different scales and in different fields of activity, including natural and human sciences. The result is that while biodiversity can be referred to as a concept, its acceptation as a measurable phenomenon may have become more questionable.

Alternatives to misleading use of biodiversity as a buzzword include advising the use more context-specific, specialised terms, and suggesting that a panel of experts agree on an operational definition.

1. Introduction

Biodiversity may be viewed as a concept, i.e. a mental construct, or as a term, i.e. the linguistic label of this concept. Whereas the complexity of biodiversity as a concept is at the center of current scientific discussions, terminology focuses on another approach. A simple definition of terminology distinguishes the following aspects:

1. the set of practices and methods used for the collection, description and presentation of terms;
2. the study of the relationships between concepts and terms;
3. a vocabulary of a special subject field (Sager 1990).

In this study the terms and the definitions used in relation to the concept of biodiversity are inventoried in order to explore the multiple ways from scientific reality to terminological utopia.

P. Bachmann et al. (eds.), Assessment of Biodiversity for Improved Forest Planning, 71–81.
European Forest Institute Proceedings no. 18. © 1998 *Kluwer Academic Publishers. Printed in the Netherlands.*

2. Methodology and definitions

In a utopian terminological world, the one-to-one equation between a term and a concept would always be verified. For example, the concept forest would be replaced by as many concepts as there are types of forests in the world, and the term tree would refer to a single botanical concept, and not to e.g. a mathematical diagram or a family tree. Reality, however, is a multiplicity of terms for one given concept and a multiplicity of concepts for one given term (Gouadec 1990).

Following current methods in terminology compilation (Sager 1990), this study is based on a corpus, i.e. a sample of written documents, consisting of 125 documents. These documents were selected because they included biodiversity or biological diversity in their title or because they deal with these issues. The years of publication range from 1859 to 1996 and English is the original language.

This corpus was surveyed with two objectives: to sketch a conceptual tree of biodiversity as a subject field and to collect definitions of the concept biodiversity. No information retrieval system was used to screen the corpus, i.e. I skimmed through the documents. Since most definitions were given in the beginning of each document, they could hardly be missed.

Figure 1 is a so-called domain tree, i.e. an attempt to map the terminology (in the sense of "vocabulary") of biodiversity. It includes components of biodiversity as defined by various authors, as well as logically or otherwise related concepts. Synonyms or quasi-synonyms defined in the corpus are in italics. The resulting picture is by no means exhaustive but is complex at both semantic and conceptual levels. These two levels of a very diverse reality will be commented on successively in the next section.

3. A diverse reality

3.1 SEMANTIC DIVERSITY

Semantic diversity, i.e. the diversity of terms, is known as synonymy. Terminologists distinguish between real synonyms, i.e. terms which can be substituted to each other whatever the context, and the more common quasi-synonyms, which can differ from one another by context and sometimes by subject field.

Synonymy often results from neology, i.e. term formation. In our case, scientists simply rename a concept, and synonymy can then be asserted only by comparing the old and new definitions. In some cases, the authors explicitly acknowledge that a term already exists, but propose a new term, i.e. a neologism, for the same concept (Harper and Hawksworth 1994). Of special interest to terminologists is retronymy, which consists of renaming an existing concept more precisely in order to adjust to a conceptual shift, as will be illustrated below.

I will first illustrate synonymy with a few examples then shortly comment on neology, and finally describe the emergence of "bio-neologisms."

3.1.1 Biodiversity vs. biological diversity

The most obvious example of synonymy is the co-existence of biodiversity and biological diversity. No reference could be found in the literature to a possible distinction

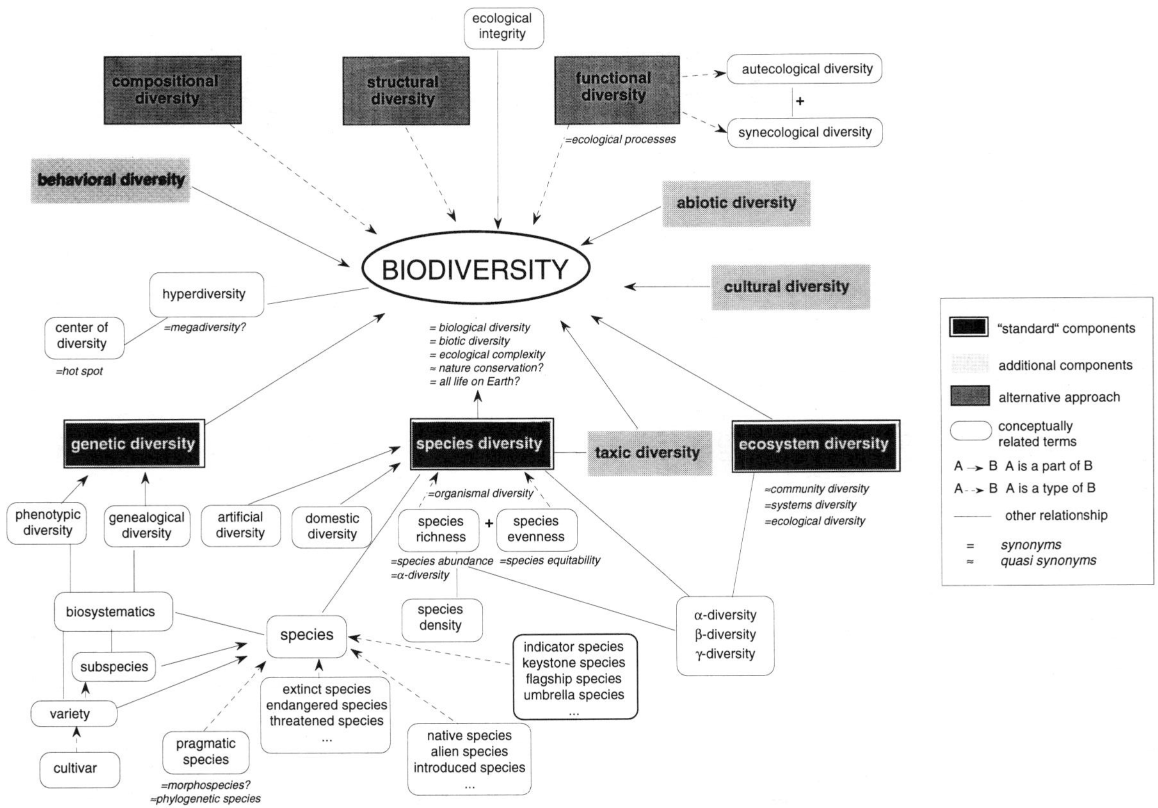

Figure 1. Provisional domain tree of biodiversity based on the survey of 125 text documents in English. Concepts used by various authors to define biodiversity are in square boxes, related concepts in rounded boxes. The type and direction of conceptual relationships are indicated by arrows. Synonyms and quasi-synonyms are in italics.

in meaning between these two terms, and several authors explicitly consider them as synonyms or quasi-synonyms, or as the contracted and expanded forms of the same term (Gaston 1996; Solbrig 1994; Heywood et al. 1995).

Other authors imply this synonymy by quoting the definition of biological diversity given in the Convention of Biological Diversity (CBD) (UNEP 1992) as a definition of biodiversity. This Convention, which is sometimes wrongly cited as "Convention on Biodiversity," never refers to the contracted form, although it had appeared in the literature four years ago (Wilson 1988). One reason for that may be that biological diversity had (and to some extent still has) a more scientific flavour than the already popular biodiversity. The Convention being aimed first at politicians and lawyers, it needed to convince them of its seriousness. In a press release of February 1996 (UNEP 1996), however, the Secretariat of the Convention referred to biodiversity: slip of the tongue or adjustment to new usage?

A similar shift in usage is also observed in the scientific literature, as confirmed by a search in Biosis, an on-line database of scientific literature (Fig. 2). The term biological diversity, which first appeared in titles in 1972 and had been discreetly present since then, benefited from the emergence of the biodiversity issue between 1988 and 1992. Since the United Nations Conference on Environment and Development (the Rio "Earth Summit") in June 1992, it seems to have reached a plateau under 50 occurrences per year, far below the 150-250 yearly occurrences of the term biodiversity. The number of references per year should be interpreted with caution as Biosis contains original titles in English as well as translated titles, and translation may introduce a bias. However, the overall trend is clear.

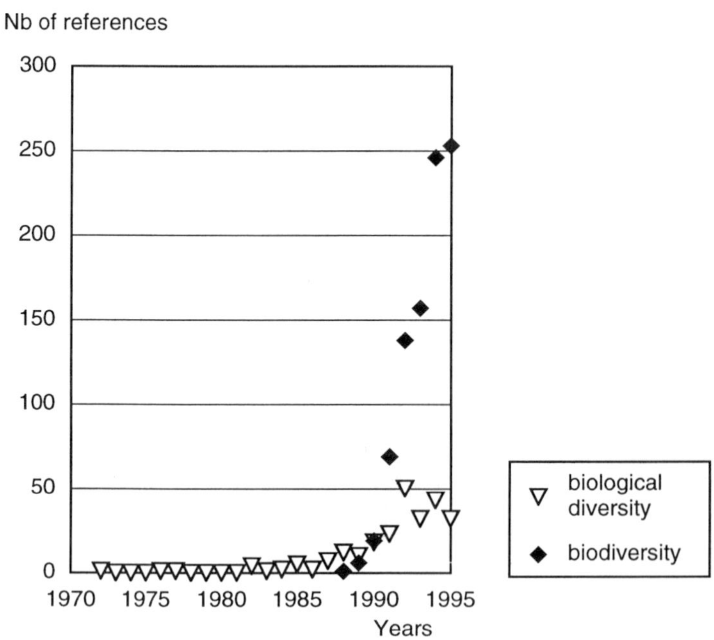

Figure 2. Number of references with the terms biological diversity or biodiversity in the title, retrieved from an online literature database for biological sciences (Biosis). Search completed in September 1996.

Synonymy is also very high around other concepts at the centre of the domain tree (see Fig. 1), i.e. concepts used by many authors, and/or "old" concepts which have been redefined and renamed several times for years. This is particularly the case for species diversity, also named ecological diversity (Eldredge 1992) or organismal diversity (Harper and Hawksworth 1994).

3.1.2 The power of term formation

The urge for imposing new terms is a well-known phenomenon in industrial terminology. For example, the fact that personal computers are usually called PCs after IBM®'s terminology gives a hint about the former vitality of this company (Gouadec 1990). Likewise, the degree of penetration of a scientific neologism, i.e. a new word or expression, coined by a group of scientists may assert or strengthen their influence. In extreme cases, negating other scientists' terminology is a means of negating or at least questioning their results or school or thought.

From a positive point of view, the impetus given to ecological studies since the emergence of biodiversity as a scientific issue gives research teams new opportunities to increase their visibility, and neology is one of the tools they may use. However, it often creates confusion, particularly when neologisms are variously defined, and overlap with one another. For example, alpha-diversity is defined either as within-area diversity of species, or as a synonym of species richness, one of the components of species diversity.

3.1.3 Buzzword, bandwagon and blunderbuss

Some neologisms can develop into a world-wide success stories and even become catchwords. The explosion of biodiversity in titles in the scientific literature both reflects and encourages the use of this term as a "buzzword" (Angermeier 1994), a "bandwagon" (Rodda 1993) or even as a "useful fad that helps many ecologists to survive doing science" because "its usage facilitates greater access to research funding" (Ghilarov 1996). This ubiquity has also been largely commented and tentatively measured (usually by counting the number of publications per year) by the scientific community itself (Dytham 1996; Gaston 1996; Haila and Kouki 1994; Harper and Hawksworth 1994). A similar concern is also observed in a few other "fuzzy" environmental concepts such as forest health (More 1996) or sustainability (Fautin 1995).

This fashion effect may be partly due to the semantic aspect of the term itself. Biodiversity belongs to the large family of neologisms formed with the bio- prefix which has appeared in the scientific, pseudo-scientific and popular writings of the past few years. The on-line Dictionary of Science and Biotechnology defines as many as 48 terms with this prefix. In the context of biodiversity itself, bio-neologisms have flourished, with or without a formal definition: biodiverse is making its way, as are bioregion and bioregional. Biodivise was coined in 1992 by journalists from The Economist reporting the disagreement of the American government on the terms of the CBD. Sociological literature abounds with biobehavioural, biocultural and bioethical issues.

In the scientific sphere, biodiplomacy refers to the pre-CBD negotiations (di Castri and Younès 1995a) and biosystematics to the systematics restricted to the biological context (Bisby et al. 1995). The concept of systematics first concerned living organisms only, but it was progressively applied to other disciplines, and the retronym biosystematics restores the original meaning.

3.2 CONCEPTUAL DIVERSITY

The text of the CBD and the earlier document of the U.S. Congress Office of Technology Assessment (OTA 1988) provide very similar definitions of biological diversity, referring to genetic diversity, species diversity and ecosystem diversity. In substance, this trilogy of standard components of biodiversity (in black boxes in Fig. 1) has become the "increasingly conventional" (Gaston 1996) definition of biodiversity. It is however far from satisfying all scientists, as shown below by its various addenda and avatars.

3.2.1 Diversity of components

In addition to the three components of biodiversity mentioned above, the following ones (in light grey boxes in Fig. 1) have been defined by one or several authors: cultural diversity (WRI et al. 1992), behavioral diversity (Stone et al. 1994), taxic diversity (Bisby et al. 1995) and functional diversity (Roberts and Gilliams 1995). Artificial diversity is either explicitly included (Bisby et al. 1995) or rejected (Angermeier 1994). Diversity among domesticated species is seldom mentioned (Dasmann 1991). In one instance, even the "abiotic matrix" (Dunster and Dunster 1996) is considered a component of biodiversity, in contradiction with the etymology of the term itself.

Such addenda were actually foreseeable, and maybe encouraged or at least made possible by the cautious wording of the CBD definition, which refers to the "variability among living organisms from all sources including, inter alia, terrestrial, marine and other aquatic ecosystems (...)." "Inter alia," latin for "among other things," opens the door to further additions to the list of biodiversity components.

3.2.2 Diversity of approaches

Several "critical deficiencies" identified by scientists in the OTA definition include "the lack of a spatial and temporal framework and the lack of any component of process" (Redford and Sanderson 1992). Various alternatives have been proposed to compensate for these lacks.

Spatial and/or temporal frameworks vary from one author to the other, or even within a definition. The spatial scale ranges from "a given place" (Dictionary of Science and Biotechnology) to "a region or the world" (WRI et al. 1992) and even to "other planets" were life to occur on them (Bisby et al. 1995). Similarly, the temporal scale ranges from "a given time" (Dictionary of Science and Biotechnology) to "living organisms rescued from fossils" and to "new elements of life" (Bisby et al. 1995).

A radically different approach is that of Crow et al. (1994). Instead of focusing on partitive relationships (genetic diversity, species diversity and ecosystem diversity as parts of biodiversity), they identify a generic relationship involving three types of biodiversity (compositional diversity, structural diversity and functional diversity, in dark grey boxes in Fig. 1), which may be applied at various integration levels. Ecological functions, structures and processes were mentioned by other authors (e.g. Bibby et al. 1992; Innes et al., in press; Sandlund et al. 1992; WRI et al. 1992), but in our sample of documents no definition prior to 1992 includes this dimension.

Still further from the standard definition is that of Solbrig (1994) stating that "biodiversity is the property of living systems of being distinct, that is, different, unlike. (...) Diversity is therefore not an entity, a resource, but a property, a characteristic of nature."

3.2.3 Diversity of disciplines

Both semantic and conceptual diversity result not only from the eagerness of biologists to bring forward neologisms or new approaches. They have also become unavoidable as biodiversity – as a term and as a concept – has raised interest in non-biological disciplines. Figure 3 illustrates the fact that the term biodiversity has relayed biological diversity in, for example, social and legal literature according to a pattern similar to what can be observed in biological literature.

For terminologists, this appropriation of a specialist's term by non-specialists (or by specialists of other disciplines) corresponds to the de-specialisation or vulgarisation stage in a normal evolutionary process. The usual stages are: (1) creation of a term, (2) specialisation, (3) equilibrium, (4) de-specialisation, and (5) emergence of new, specialised terms. The last stage is unavoidable as the co-occurence of the original meaning with de-specialised ones generates ambiguity for specialists (Gouadec 1990).

This evolution is usually very slow, but some terms skip stages (2) and (3) and become overnight fashions, for example the term interface. The emergence of biodiversity in non-biological disciplines as illustrated in Figure 3 indicates that the term is already in phase (4).

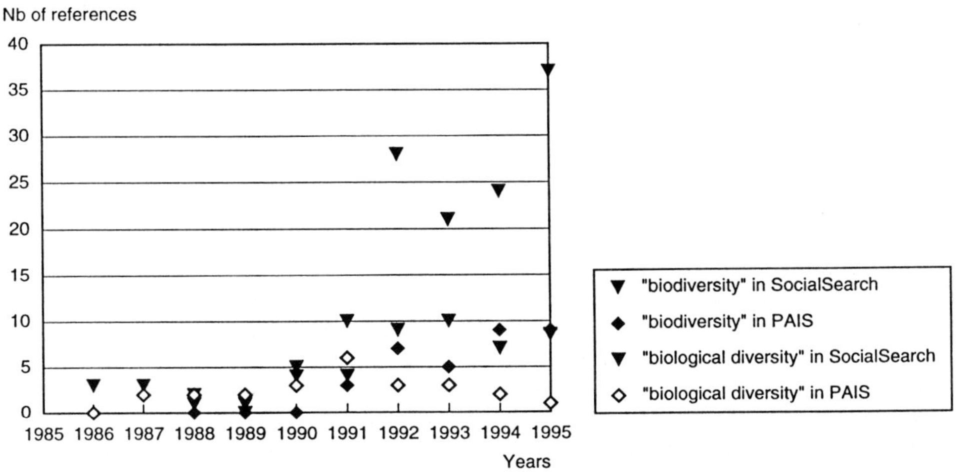

Figure 3. Number of references with the terms biological diversity of biodiversity in the title, retrieved from two non-biological online literature databases (SocialSearch, literature in social sciences; and PAIS, Public Affairs Information System, information on international law). Search completed in September 1996.

4. Between reality and utopia

How do scientists comment on this complex reality, and how can the utopia of a non-ambiguous definition be reached?

4.1 TERMINOLOGICAL CONFUSION: A TERM OF AGREEMENT

The state of confusion that may affect the reader at this point is a feeling shared by most actors on the biodiversity stage. Actually it is probably the only aspect on which most of the authors who have commented on the term or on the concept come to an agreement. This collective awareness is symptomatic of the de-specialisation phase described in the previous chapter, and of the imminence of further term formation.

Comments about this confusion range from neutrality to acceptance. Most commentators just observe that the term is "inconsistently or at times inadequately defined," "so wide as in fact to be exceedingly difficult to comprehend" (Gaston 1996), that "the public is unclear of the concept, and the scientists cannot give a simple answer" (Rodda 1993).

Others propose explanations such as the "general character of the term": biodiversity is "a descriptive, but simultaneously an abstract and descriptively complex term" (Haila and Kouki 1994), or "the fact that 'biodiversity' is a pseudocognate term in that many users assume that everyone shares the same intuitive definition" (Williams 1993, in Gaston 1996).

A few authors view this "imprecision of definition and diversity of perception" as "a strength in making biodiversity a unifying concept" (Heywood et al. 1995) or a "unifying leitmotiv" (di Castri and Younès 1995).

As for Haila and Kouki, the authors of a well-documented review of the history of the term (1994), they prognosticate that "there is no point in dreaming of general agreement on, or even rules for, biodiversity terminology."

4.2 THE QUEST FOR TERMINOLOGICAL UTOPIA

Once authors have agreed on the general bewilderment, how do they view the ideal definition? It should be "clear, concise, and widely accepted" (Angermeier 1994), "robust and closely reasoned" (Gaston 1996), "presented in a user-friendly and policy-relevant format" (Lubchenco 1995). Let us consider how these various qualities can be guaranteed.

In order to produce a widely-accepted definition, one could envisage to integrate all the components and approaches suggested by various authors (see 1.3). The resulting concept would amount to "a synonym for 'life on Earth'" as actually defined in the Explanatory leaflet about the Convention on Biological Diversity (1995). However exhaustive, such a definition would be far from satisfactory. But interestingly, in the popular acception of biodiversity the fashionable and once semantically minor prefix bio – now overshadows diversity, which was the most relevant aspect of the original concept.

Policy relevance and conciseness could apparently be achieved by resorting to an oversimplified definition. It is very likely that such a definition, however "palatable to policy agencies" (Redford and Sanderson 1992) or user-friendly for the general public, may not be considered "robust" by e.g. taxonomists.

It can reasonably be concluded that there is no single ideal definition, simply because there is no single user of language. School children, policy-makers and biometricians have different needs and different levels of abstraction, and it is illusionary to tailor a single multipurpose definition.

4.3 WHICH ALTERNATIVES?

If no consensual definition can be worded, which alternatives are left to improve communication within and among scientific disciplines and between science and its partners? Several options can be proposed, which do not necessarily contradict one another.

Firstly, considering that biodiversity has been widely accepted as a synonym of biological diversity, and if there is no expressed denial of this synonymy, the use of biodiversity as a standard term could help clarify the discussion.

Terms related to biodiversity (see Fig. 1) could similarly be standardized but this may not be applicable as neology is a natural process. For example, the concept of species diversity has inspired a "vast, divergent, and sometimes acrimonious literature" for several decades (Grassle et al. 1979), and the lack of formal definition was already commented on twenty-five years ago (Hurlbert 1971; Patil and Taillie 1979). As for species and varieties, it has long been observed that "to discuss whether they ought to be called species or varieties, before any definition of these terms has been generally accepted, is vainly to beat the air." (Darwin 1859)... But until now, "despite its universal usage as a basic unit of taxonomy, it is difficult to agree on an exact definition of what constitutes a species." (Bisby et al. 1995) Such a long-lasting lack of agreement, however, may indicate that consensus is not always a prerequisite of pragmatism.

Another suggested option is that scientists define which biodiversity they assess context-specifically, and that they use conceptual rigor within specific contexts (Haila and Kouki 1994; Gouadec 1990). This terminological breaking down is the counterpoint of the use of measurable components of biodiversity which has been recommended by several authors (Köhl and Zingg 1996). It corresponds to the "specialisation" stage of the terminological evolution described under 2.2.3.

Complementary to these two steps, Rodda's (1993) recommendation is that scientists "keep the concept from becoming tawdry through uncritical application." For example, considering how extremely wide the concept is, no doubt that numerous studies, from molecular diversity to land-use diversity, deal with it to some extent. This does not, however, justify that all these articles include biodiversity in their title. Indeed, a study on e.g. "arthropod biodiversity" is unlikely to deal with all components and types of biodiversity, and this term is here clearly used as a buzzword. Inversely, one speaks of "species diversity" or "functional diversity," not of "species biodiversity" or "functional biodiversity," since the biological context is always implicit.

Relevant use of the term would directly benefit the individual scientists: an article dealing strictly with genetic diversity but containing only biodiversity in its title and/or abstract may be missed by systematic searches in on-line databases. It would also and first of all contribute to restore the credibility of the scientific community in general. In the past few decades, other scientific issues such as forest decline (or forest health) have indeed suffered the same problem of "reification, the treatment of an abstract idea as if it were a measurable thing" (Bowman 1993), usually resulting in disappointment and anger among science's social partners.

However, the ultimate response would be a collective definition elaborated by a panel of scientists from various disciplines. This operational definition would indeed not be universal but would at least have the merit of representativeness, and could hence be referred to as a standard. The Internet offers various technical possibilities for gathering such a panel and for publishing their conclusions. The most critical aspect would probably be the selection of relevant experts.

4. Conclusion or "The strength of the term is its simplicity"

Evolutionary processes affect language just as they affect living organisms, and terminological disagreement is part of a natural process. The current diversity of definitions of the concept biodiversity mirrors a complex scientific discussion marked by a great diversity of focuses and approaches.

Nevertheless, the term biodiversity is a "brilliant piece of wordsmithing" (Bowman 1993) available to the scientific community, and "the strength of this term is its simplicity" (Cushman et al. 1995). Rightly defined, wisely used, it can become a driving unifying force to the benefit of all.

5. Acknowledgements

I am very grateful to Matthias Dobbertin for encouragements and stimulating discussions. I thank John Innes and Walter Keller for discussion and comments, and Alois Kempf and Martin Spinnler for providing the bibliometric data and commenting on them.

References

Angermeier, P.L. 1994. Does biodiversity include artificial diversity? Conservation Biol. 8: 600-602.

Bibby, C.J., Collar, N.J., Crosby, M.J., Heath, M.F., Imboden, C., Johnson, T.H., Long, A.J., Stattersfield, A.J., and Thirgood, S.J. 1992. Putting biodiversity on the map: priority areas for global conservation. International Council for Bird Preservation, Girton. 2nd printing.

Bisby, F.A., Coddington, J., Thrope, J.P., Smartt, J., Hengeveld, R., Edwards, P.J. and Duffield, S.J. (lead authors). 1995. Characterization of biodiversity. In: Heywood, V.H., Watson, R.T., and Baste, I. (eds.). Global biodiversity assessment. Cambridge University Press. Cambridge. Pp. 21–106.

Bowman, D.M.J.S. 1993. Biodiversity: much more than biological inventory. Biodiversity Lett. 1: 163.

Cushman, J.H., Dirzo, R., Janetos, A.C. et al. (lead authors). 1995. Biodiversity and ecosystem functioning: basic principles. In: Heywood, V.H., Watson, R.T. and Baste, I. (eds.). Global biodiversity assessment. Cambridge University Press. Cambridge. Pp. 275–325.

Crow, T.R., Haney, A., and Waller, D.M. 1994. Report on the scientific roundtable on biological diversity convened by the Chequamegon and Nicolet National Forests. General Technical Report NC-166. USDA Forest Service. North Central Forest Experiment Station. St. Paul, MN.

Darwin, C. 1859. The origin of species by means of natural selection or the preservation of favored races in the struggle for life. Murray, London.

Dasmann, R.F. 1991. The importance of cultural and biological diversity. In: Oldfield, M.L. and Alcorn, J.B. (eds.). Biodiversity: culture, conservation, and ecodevelopment. Westview Press. Boulder. Pp. 7–15.

di Castri, F. and Younès, T. 1995b. Introduction: Biodiversity, the emergence of a new scientific field – its perspectives and constraints. In: di Castri, F. and Younès, T. (eds.). Biodiversity, science and development: towards a new partnership. CAB International. Wallingford. Pp. 1–11.

di Castri, F. and Younès, T. 1995a. Preface. In: di Castri, F. and Younès, T. (eds.). Biodiversity, science and development: towards a new partnership. CAB International. Wallingford. Pp. xi–xiv.

Dictionary of Science and Biotechnology.
http://biotech.chem.indiana.edu/pages/ dictionary.html.

Dunster, K. and Dunster, J. 1996. Dictionary of natural resource management. UBC Press. Vancouver.

Dytham, C. 1996. O brave new words... Letter to Nature 379: 14.
Eldredge, N. 1992. Where the twain meet: causal intersections between the genealogical and ecological realms. In: Eldredge, N. (ed.), Systematics, ecology, and the biodiversity crisis. Columbia University Press. New York. Pp. 1–14.
Fautin, D.G. 1995. Sustainability – more than a buzz-word? Ann. Rev. Ecol. Syst. 26: v–ix.
Gaston, K.J. 1996. What is biodiversity? In: Gaston, K.J.(ed.). Biodiversity: a biology of numbers and difference. Blackwell Science. Oxford. Pp. 1–9.
Ghilarov, A. 1996. What does 'biodiversity' mean – Scientific problem or convenient myth? Trends Ecol. Evolution 11(7): 304–306.
Gouadec, D. 1990. Terminologie. Constitution des données. AFNOR. Paris.
Grassle, J.F., Patil, G.P., Smith, W., and Taillie, C. 1979. Preface. In: Grassle, J.F., Patil, G.P., Smith, W. and Taillie, C. (eds.). Ecological diversity in theory and practice. International Co-operative Publishing House. Fairland. MD. n.p.
Haila, V. and Kouki, J. 1994. The phenomenon of biodiversity in conservation biology. Ann. Zool. Fenn. 31: 5–18.
Harper, J.L. and Hawksworth D.L. 1994. Biodiversity: measurement and estimation. Preface. Philos. Trans. Roy. Soc. London 345, 5–12.
Heywood, V.H., Baste, I. and Gardner, K.A. (lead authors). 1995. Introduction, in Heywood, V.H., Watson, R.T. and Baste, I. (eds.). Global biodiversity assessment. Cambridge University Press. Cambridge. Pp. 1–19.
Hurlbert, S.H. 1971. The nonconcept of species diversity: a critique and alternative parameters. Ecology 52, 577–586.
Innes, J.L., Keller, W., Bösch, R., (in press). Biodiversity assessments in Swiss forest ecosystem monitoring plots. In: Measuring and monitoring forest biodiversity. Washington, D.C. Smithsonian Institute.
Köhl, M. and Zingg A. 1996. Eignung von Diversitätsindizes bei Langzeituntersuchungen zur Biodiversität in Waldbeständen. Allg. Forst.u. J.-Ztg. 167, 76–85.
Lubchenco, J. 1995. The role of science in formulating a biodiversity strategy. Bioscience Suppl.: S7–S9.
More T.A. 1996. Forestry's fuzzy concepts. An examination of ecosystem management. J. For. 94, 19–23.
(OTA) Office of Technology Assessment, U.S. Congress. 1988. Technologies to maintain biological diversity. Science Information Resource Center. Philadelphia, PA.
Patil, G.P. and Taillie, C. 1979. An overview of diversity. In: Grassle, J.F., Patil, G.P., Smith, W. and Taillie, C. (eds.). Ecological diversity in theory and practice. International Co-operative Publishing House. Fairland. MD. Pp. 3–27.
Redford, K.H., and Sanderson, S.E. 1992. The brief, barren marriage of biodiversity and sustainability. Bull. Ecol. Soc. Am. 73, 36–39.
Roberts M.R. and Gilliam F.S. 1995. Patterns and mechanisms of plant diversity in forested ecosystems: implications for forest management. Ecol. Appl. 5, 969–977.
Rodda G.H. 1993. How to lie with biodiversity. Conservation Biol. 7, 959–960.
Sager, J.C. 1990. A practical course in terminology processing. John Benjamin. Amsterdam.
Sandlund, O.T., Hindar K., and Brown A. H. D. (eds.). 1992. Conservation of biodiversity for sustainable development. Scandinavian University Press. Oslo.
Solbrig, O.T. 1994. Biodiversity: an introduction. In: Solbrig, O.T., van Emden, H.M., and van Oordt, P.G.W.H.J. (eds.), Biodiversity and global change. CAB International. International Union of Biological Sciences. Wallingford.
Stone P.A., Snell, H.I., and Snell, H.M. 1994. Behavioral diversity as biological diversity: introduced cats and lava lizard wariness. Conservation Biol. 8, 569–573.
UNEP (United Nations Environment Programme). 1992. Convention on Biological Diversity. UNEP. Nairobi.
UNEP (United Nations Environment Programme). 1992. An explanatory leaflet about the Convention on Biological Diversity. http://www.unep.ch/bio/bio-leaf.html
UNEP (United Nations Environment Programme). 1996. Biodiversity Secretariat Open for Business in Montreal (Press release). http://www.unep.ch/bio/pr2-96.html
Williams, P.H. 1993. Choosing conservation areas: using taxonomy to measure more of biodiversity. In: Moon, T.Y. (ed.) Manus. Col. ISBC KEI. Korean Entomological Institute. Seoul.
Wilson, E.O. (ed.). 1988. Biodiversity. National Academy Press, Washington, D.C.
World Resources Institute (WRI), The World Conservation Union (IUCN), United Nations Environment Programme (UNEP). 1992. Global biodiversity strategy: guidelines for action to save, study, and use earth's biotic wealth sustainably and equitably. World Resources Institute (WRI). Washington, D.C.

BIODIVERSITY AND FORESTRY IN POLAND

MALGORZATA FALENCKA-JABLONSKA
Forest Research Institute
Department of Ecology
Bitwy Warszawskiej 1920 No 3
00-973 Warsaw, Poland
(translated from German by Tamara Brügges)

Forest areas of the temperate climatic zone play an important role for the preservation of biodiversity. Almost 28% of Poland's surface is covered with forest. About 65% of this area has a highly concentrated biodiversity. According to current estimations, about 60% of the vertebrates, more than 80% of the mushroom species, most of the moss species, pteridophyte-fern species, horsetail and club-moss species live and grow in the forests of our country. It is supposed that no less than 32,000 different species live in the forests of Poland (Grzywacz 1995). This is precisely why the protection of this richness of variety is so important – especially for forestry based on the principle of sustained development.

Poland has signed the agreement on the protection of biodiversity. It has, thus, committed itself to developing a nation-wide valid strategy for the preservation of biodiversity. In Poland, every fourth hectare is a forest area, therefore, special importance is accorded to the development of a strategy for the preservation of biodiversity in the complex forest ecosystems. This strategy will contain two main directions of impact. On the one hand, it occasions the protection of the forest stands, on the other hand, it will have to claim measures to be taken for the limitation of damaging environmental influences caused by mankind. The elaboration of this strategy, as well as the sensitisation of society towards forest issues and plant regeneration are the focus of scientific research.

The significance of the problematic nature of the protection of biodiversity has clearly been recognised in Poland: In the European year of nature conservation of 1995, two Polish conferences on this issue took place. On April 6th to 7th, 1995, the Polish Forest Society and the Institute for Forestry Research organised a conference on the topic 'Protection of Biodiversity in Sustained Forestry'. On December 13th, 1995, the scientific committee 'Mankind and Environment' of the Polish Academy of Sciences (PAN) organised a conference on 'Issues of Biodiversity - Science of Biodiversity', together with the ecological committee PAN (Kozlowski 1995).

Biodiversity is a guarantor for the flexibility and the capacity for adaptation of forest ecosystems to their environment and, at the same time, favours their dynamics and development. The analyses of the external and internal correlation of the ecosystems indicate a strong mutual interrelation between biodiversity and structure and biodiversity and the function of ecosystems (Fig. 1). With regard to the external factors, the climatic conditions, migrations, current and historical developments of utilisation forms have the strongest influence on biodiversity. Indirectly, biodiversity is also influenced by the food

P. Bachmann et al. (eds.), Assessment of Biodiversity for Improved Forest Planning, 83–89.
European Forest Institute Proceedings no. 18. © 1998 *Kluwer Academic Publishers. Printed in the Netherlands.*

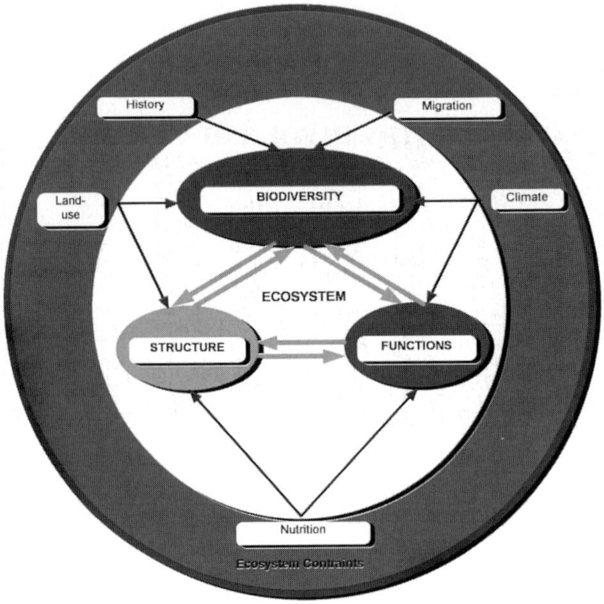

Figure 1. Schematic presentations of the interactions within an ecosystem of biodiversity, community structure, and function, and effects of external ecosystem contraints on these parameters. Source: E.-D. Schulze, H. A. Mooney (eds.) – 1993.

cycles within the ecosystems. It should be taken into consideration, however, that no direct connection exists between biodiversity and system stability (Falencka-Jablonska 1995).

Due to its geographical situation, Poland has a rather non-typical flora and fauna compared to the rest of Europe. There are no natural borders to the East or to the West, and the influences of the maritime and the continental climate have a mutual interaction. The respective mosaic of different habitats and ecosystems leads to a high biodiversity. According to Andrzejewski and Weigle (1991), Poland is characterised by one of the highest biodiversities within Europe, in particular, due to the typical forest and swamp bioceonose of the lowlands of Central Europe.

The borders of the distribution areas of numerous plants and animal species run through our country, such as 96 indigenous terrestrial vertebrates, of which 30% are mammals, 16% are birds, 22% are reptiles and 28% are amphibians. Of the higher plant species, 40% have their borders of distribution area within Poland.

With regard to the European forest areas, Poland ranks fifth behind Sweden, Finland, Spain and France. The proportion of forest of the whole land surface, however, lies below the European average. In our country, the softwoods take a significantly larger area with a proportion of 78% compared to 22% of hardwoods. The analysis of the surface development shows a significant decrease in the forest proportion after World War II. Currently, an increase in forest areas has been ascertained. A prognosis for the year 2010 estimates a forest proportion of 30% of the land surface (Fig. 2).

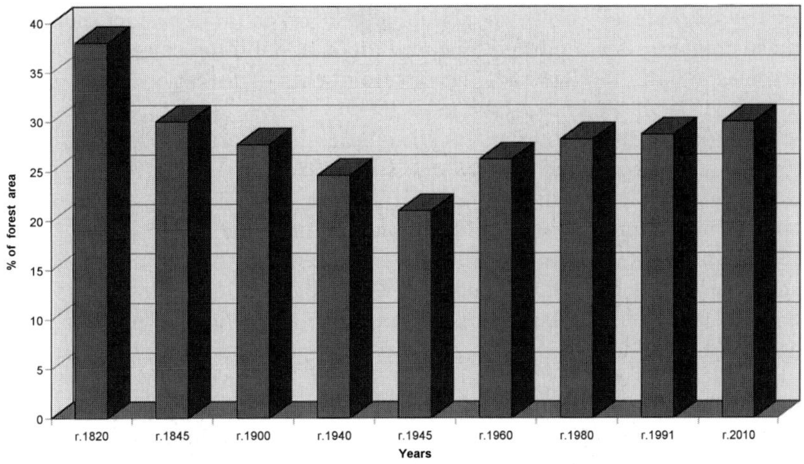

Figure 2. Dynamics of changes in forest area of Poland

The investigation of the dependence of biodiversity on the changing of ecosystems is, among others, supported by the following criteria:

- floristically: investigation of species structure and interrelations between the single elements of the tree and shrub layers. Determination of interrelated influence of site and plant layer of the respective area; determination of the phases of natural succession and regeneration.
- faunistically: investigation of the interrelation between the species within the single trophic levels as well as between them; investigation of the population structure; investigation of the composition of species as an index for the development status of the respective ecosystem.
- mycologically: analysis of the diversity of fungi in the soil, the ecto- and endomycorrhizae as well as the diversity of microflora of the aerial parts of the tree.

These criteria have already been applied in long-term investigations. They are used to investigate the changes in forest biodiversity of different management forms. The focus lay mainly on pine woods, which dominate in our country. These pine forests grow on the poorest sites, which are unsuitable for agriculture, and have a proportion of more than 75% of the forest area. The investigation included areas with different silvicultural tending methods: selective thinning on richer sites, clear fellings with subsequent newly established stands, various ways of stand conversion on highly degraded stands (Falencka-Jablonska et al. 1993 and 1994) and various protection areas.

The completed investigations led to the knowledge that all trophic levels have to be included together in the investigation on forest ecosystems. Relevant documentation often only focuses on the changes of single plants, on the population structures of single fauna species (e.g. beetles, birds), or on the occurrence of specific mycorrhizae. However, only a judgement of the whole system makes it possible to observe events and processes as they occur with the changes of forest ecosystems. The complex correlation between the single levels, however, should also be taken into consideration. At the same time, these processes are modified by environmental pollution caused by man, rendering this factor especially decisive for the developments of forest ecosystems. (Falencka-Jablonska et al. 1994; Falcncka-Jablonska 1995).

For the judgement of biodiversity, all criteria characterising the three main links of the food chain are being taken into consideration. This is the reason for rendering possible recognising and counteracting deterioration processes in the forests. A sustained development in the forestry of Poland can thus be ensured.

The question of site quality for the tree species, the differentiating of forest ecosystems according to their development within the natural succession as well as to the degree of their synanthropisising is particularly important for the floristic analysis. Furthermore, the inventory of rare, endangered and protected plant species is also of importance. Repeated inventory at an interval of 3-5 years allow statements on the dynamics and development tendencies in forest ecosystems. It remains difficult, however, to judge the influence of tending on the natural development.

The proportions of alteration of the character species and the accompanying species of the forest societies can be consulted as a measure of site changes (Fig. 3).

The dynamics of the fauna, however, can be characterised by means of the species' composition of the soil macro-fauna (Fig. 4), the coleoptera as well as the tree-dwelling birds (Sokolowski et al. 1985). The biodiversity also changes under the pressure of industrial emissions (Falencka-Jablonska 1994). One criterion for the judgement of the latter can be drawn from the population structure and density as well as from the species' composition of the secondary pests in the forest stands. Furthermore, the correlation between the economic development, the respective increase of energy consumption, the development of spatial structure and the heterogeneity and diversity of ecosystems should also be taken into consideration (Nienartowicz 1996).

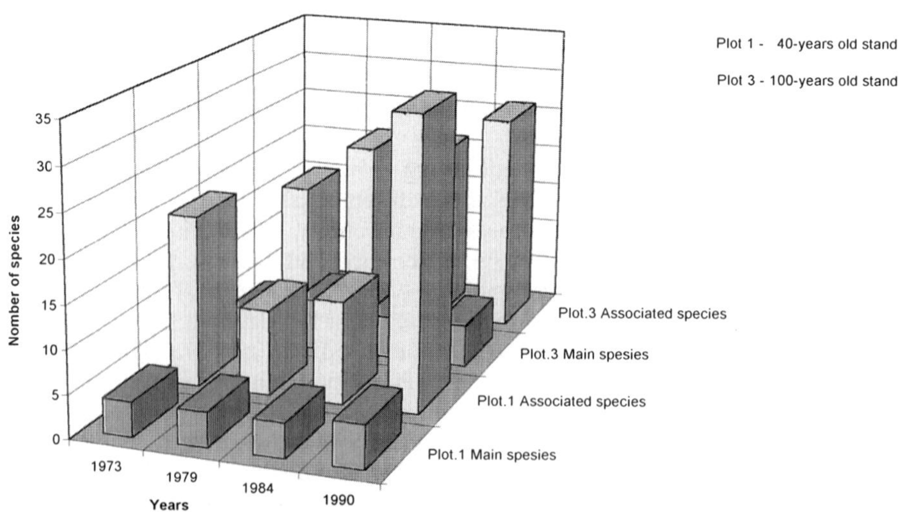

Figure 3. Changes in participation of main and associated species in herb layer of *Leucobryo-Pinetum* in zone of medium air pollution

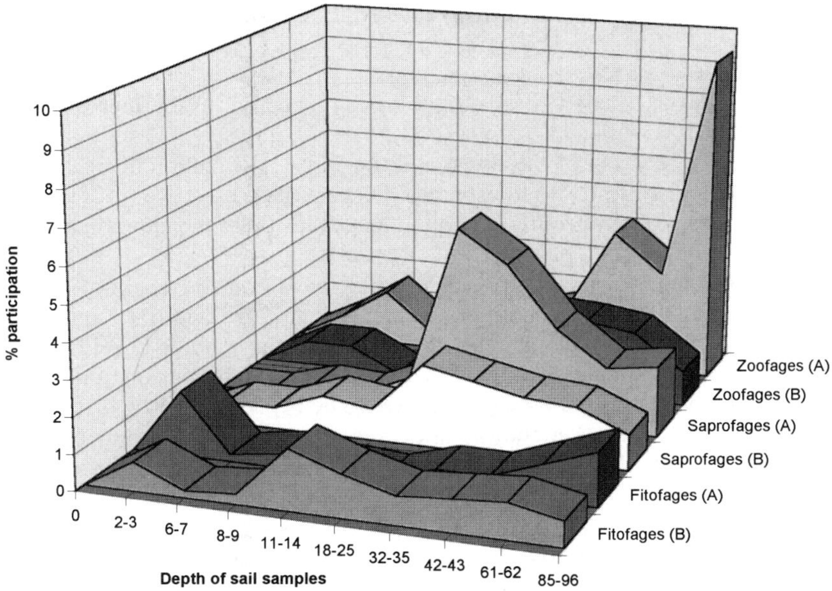

Figure 4. Differentiation in participation of group species in soil. (Szujecki et al. 1993 – modified Falencka-Jablonska et al. 1993)

A comparison of the surface proportions of the single forest types and their development during the last 50 years show the following result: the proportion of meso- and eutrophic forests has doubled, that of open woodlands, however, was halved (Table 1). The prognosis shows the result of 3 different scenarios:

A Optimistic variant, the economic crisis is gradually overcome.
B Realistic variant, based on a constant development, together with a stagnating economic crisis.
C Pessimistic variant, based on variant B together with a world wide climatic change.

TABLE 1. Proportions of the forest ecosystems of the whole surface of Poland 1946-1990 and the prognosis for the year 2030. (According to 'Prognosis of Environmental Changes' 1995).

	% of surface				
groups of ecosystems	1946	1990	A2030	B2030	C2030
swamp forests	0.6	1.0	1.1	1.5	0.0
open woodlands	0.2	0.1	0.2	0.0	0.0
eutrophic hardwoods	2.2	4.8	8.0	4.5	5.0
mesotrophic hardwoods	2.7	6.6	8.7	6.0	10.0
softwoods and mixed forests	15.1	15.3	12.0	15.0	10.0
total of forest ecosystems	20.8	27.8	30.0	27.0	25.0

According to the respective scenario, different developments of the forest ecosystems are to be expected:

1. A proportional expansion of all forest types respective to the increase of forest areas.
2. A reinforced utilisation of the forest and the forest soil, with a decrease of forest areas, in particular of private forest areas (Olaczek 1995).
3. Additionally, a global warming leading to an expansion of the unproductive areas.

The above-mentioned scenarios are additionally based on the assumption that the existing environmental pollution could be decreased and, therewith, its significance in the analysis of a further development of the ecosystems reduced. With emissions remaining constant as measured in the 1980s, a destabilisation and a further disappearance of 48-69% of the forest areas have to be expected (Graczyk 1991).

The aim of short-termed economic successes in forestry has contributed to the decrease in biodiversity. Therefore, the demand for sustained forest utilisation occurs in the strategy for the protection of biodiversity 1996-2000. The following elementary demands shall be the basis of this strategy:

The promotion of indigenous tree species, raising site-specific forest stands, harmonising the reforestation policy with that of nature and landscape conservation.

As measures taken in order to reach the goals of biodiversity protection and the promotion of ecological forestry, seven special forest complexes (LKP) were excluded. These areas consist of a total of 333,712 ha forest, owned by the state, (decree no. 30 of the general directory of the forest enterprises owned by the state of 19 December 1994). These forest complexes are dedicated to the promotion of environmentally friendly process engineering as well as to the aims of the 'Polish policy of the entire protection of forest stands'. The main goals of the economy within these forest complexes are:

- Sustained conservation and restoration of natural forest structures taking into consideration natural succession.
- Application of efficient work procedures together with the consideration of ecological principles.
- Compliance with the principles of sustained forestry and active environmental protection.
- Protection of biodiversity of the natural coexistence of plants, animals and micro-organisms and the conservation of genetic diversity.

References

Andrzejewski, R. and Weigle, A. (eds.) 1991. Polskie studium róznorodnosci biologicznej. (Polish Study of Biological Diversity). Warszawa. Narodowa Fundacja Ochrony Srodowiska. Pp. 1-186.

Falencka-Jablonska, M. 1995. Zmiany róznorodnosci biologicznej a typ zagospodarowania lasu. In: Ochrona róznorodnosci biologicznej w zrównowazonej gospodarce lesnej. (Changes of Biological diversity realed to forest managment type. In: Protection of Biological Diversity in the Balanced Forest Managment). Materialy z Sympozjum. Warszawa. April 6-7 1995. Pp. 119-140.

Falencka-Jablonska, M., Sokolowski, A.W., Kowalski, S., Kowalski, T., Stepniewska, H., Domanska, Z. 1993. Stale obserwacje zmian róznorodnosci biologicznej lasów dla wybranych sposobÛw zagospodarowania. (Permanent Observation on Biodiversity Changes for Chosen Types of Forest Managment). Technical Report. 2U. 5/93.

Falencka-Jablonska, M., Matras, J., Sulkowska, M., Medrzykowski, M. and Zielony, R. 1994. Stale obserwacje zmian róznorodnosci biologicznej lasów dla wybranych sposobów zagospodarowania. (Permanent Observation on Biodiversity Changes for Chosen Types of Forest Managment). Technical Report. NCR-653.

Graczyk, A. 1991. Prognoza zagrozenia lasów polskich zanieczyszczeniem atmosfery w roku 2000. (The Forecast of the Threat of Atmospheric Pollution on Polish Forest in the Year 2000). Sylwan. 11: 5-10.

Grzywacz, A. 1995. In: Ochrona róznorodnosci biologicznej w zrównowazonej gospodarce lesnej. (In: Protection of Biological Diversity in the Balanced Forest Managment). Materialy z Sympozjum. Warszawa. April 6-7 1995.

Kozlowski, S. (ed.) 1995. Prognoza ostrzegawcza zmian srodowiskowych w warunkach zycia czlowieka w Polsce na pocatku XXI wieku. (The Warning Forecast of Environmental Canges Related to Human Lite in Poland at the Beginning of XXI Century). Warszawa. Ecol. Inst. PAN.

Nienartowicz, A. 1995. Energetyka duzych systemów ekologicznych. (Energetics of large ecological systems). Univ. M. Kopernica. Torun. Pp. 1-145.

Olaczek, R. 1995. Prognoza zmian ekosystemów i fitocenoz Polski.. In: Prognoza ostrzegawcza zmian srodowiskowych w warunkach zycia czlowieka w Polsce na pocatku XXI wieku. (The Forecast of changes of ecosystems and phytocenosis of Poland. In: The Warning Forecast of Environmental Canges Related to Human Lite in Poland at the Beginning of XXI Century). Warszawa. Ecol. Inst. PAN. Pp. 161-178.

Sokolowski, A.W., Wolk, K., Gutowski, J. 1981. Badania wplywu rodzaju rebni na zmiany roslinnosci i warunków odnowienia w lasach Polski pólnocno-wschodniej. (Studies on the Form ot Cutting on Changes in Vegetation and Regeneration Conditions in north-east Poland). Technical Report

THE APPROACH OF THE BRITISH FORESTRY COMMISSION TO THE CONSERVATION OF FOREST BIODIVERSITY

S. J. HODGE[1], G. PATTERSON[2] & R. MCINTOSH[3]
[1] *Forestry Commission Research Agency, Northern Research Station, Roslin, Midlothian, Scotland, EH25 9SY*
[2] *Forestry Authority Forestry Practice Division.*
[3] *Forest Enterprise Business Enterprise Division.*

Abstract

This paper describes Britain's approach to the conservation of forest biodiversity and aims to show how the Forestry Commission is evolving systems of sustainable forest management and biodiversity evaluation suited to Britain's substantial plantation forest resource. A Biodiversity Research Programme is an important part of this strategy, and is formulated to provide the guidance and decision support tools required by forest managers to follow an ecologically based forest design and management planning process.

1. Priorities for the Conservation of Biodiversity in Britain's Forests

Britain has an intensively managed and profoundly man-modified rural landscape. At the end of the 19th century forest cover in Britain was only 5%; increasing to 10 % in the 20th century (Forestry Commission 1995). British forests are classified according to continuity of forest cover on the site (ancient or secondary), whether the forests have ever been replanted (semi-natural or planted), and whether composed of predominantly broadleaf or conifer species.

1.1 PROTECTING THE LEGACY OF FOREST BIODIVERSITY IN ANCIENT AND SEMI-NATURAL FORESTS

Of the current 2,381K ha of forest, 13% is ancient semi-natural (mainly broadleaved forests predominantly of birch and oak in the north and west, and oak, beech, ash, and lime in the south and east, as well as 16,000 ha of pine forest in Scotland) (Spencer and Kirby 1992; Roberts et al. 1992). Most ancient semi-natural forests have been managed intensively in the past, using predominantly coppice and coppice-with-standards systems in lowland forests, and wood-pasture and high-forest systems in upland forests. They are an irreplaceable part of Britain's ecological heritage, and in many cases past management practices have helped to maintain important plant and animal communities in these often small and isolated refugia by maintaining stable and/or structurally diverse conditions. 9% of forest cover is of ancient forest sites which have been directly

P. Bachmann et al. (eds.), Assessment of Biodiversity for Improved Forest Planning, 91–101.
European Forest Institute Proceedings no. 18. © 1998 *Kluwer Academic Publishers. Printed in the Netherlands.*

replanted, often with non-native species, in order to increase productivity. Continuity of forest cover on these sites means the biological legacy may still be great, particularly if some features of the semi-natural forest cover remain, but this legacy is vulnerable to the increasing dominance of the planted tree crops.

This legacy of biodiversity is being protected by: ensuring a high level of statutory protection and management for the conservation and enhancement of biodiversity; where feasible restoring replanted ancient forest sites; creating new native forests linked to ancient and semi-natural forests; and positive management of the impact of deer and recreational users.

1.2 ENHANCEMENT OF BIODIVERSITY IN COMMERCIALLY IMPORTANT PLANTATION FORESTS

58% of Britain's forest is secondary coniferous forest, mainly of introduced species such as Sitka spruce (*Picea sitchensis*), Douglas fir (*Psuedotsuga menzieseii*), and Corsican pine (*Pinus nigra* var *maritima*); and 10% is secondary broadleaved forest. These forests were established on open grazing land and have little legacy of forest biodiversity. Incorporated semi-natural open habitats such as upland and lowland heath can be of conservation importance, as are incorporated remnants of semi-natural forest vegetation. However, as stands mature and a diverse forest age structure is developed, these new conifer dominated forests are developing valuable species assemblages of their own and benefiting species requiring large forested home ranges or forest interior conditions (Hambler and Speight 1995). There are three broad strategies for enhancing biodiversity in these forests:

- protecting incorporated semi-natural open habitats, and linking these into permanent open space networks, along with open habitats on restock sites and beyond the forest;
- protecting remnants of semi-natural woodland (e.g. in steep gullies where protection was afforded from sheep and deer), and expanding them by promoting natural colonisation of native woodland or by planting native stock;
- identifying conifer stands where natural processes such as disturbance and succession can be left to create new forest habitats, and adjusting forest design to favour important species or communities that use the conifer stands (e.g. red squirrel (*Sciurus vulgaris*), capercaillie (*Tetrao urogallus*), arboreal lichens in some coastal dune forests, and bryophyte communities in west Scotland).

Conflict can occur between conservation aims (e.g. increasing the area of large-seeded broadleaves provides favoured habitat for the introduced grey squirrel (*Sciurus carolinensis*) which is displacing the native red squirrel) and must be resolved by clear prioritisation of conservation objectives at the regional, forest and stand scale.

1.3 MAXIMISING THE ECOLOGICAL BENEFIT OF NEW FORESTS

The contribution of new forests to the conservation of biodiversity is promoted by using new forests to ameliorate the effects of forest fragmentation; creating 'new native woodlands' composed of native species and subject to minimum intervention (Rodwell and Patterson 1994); discouraging forest creation on valuable open semi-natural habitats; and promoting forest creation on improved agricultural land.

2. Forest Biodiversity: Policy and Responsibilities

Rio (UNCED 1993) and Helsinki (Second Ministerial Conference on the Protection of Forests in Europe 1993) commitments are reflected in 'Sustainable Forestry: the UK Programme' (UK Government 1994a) and 'Biodiversity the UK Action Plan' (UK Government 1994b). The UK Sustainable Forestry Programme lays out the UK Government's approach to sustainable forestry and provides a brief for research into the 'measurement and cost-effective enhancement of biodiversity', and into the impacts of various forestry practices on biodiversity, including 'methods of management modelled on natural ecosystems'.

The UK Biodiversity Action Plan aims to conserve and where practical enhance the overall populations and natural ranges of native species and the quality and range of wildlife habitats and ecosystems. The Action Plan approach focuses on threatened and vulnerable habitats and species (UK Biodiversity Steering Group 1995a), for which costed Habitat and Species Action Plans are being prepared. Upland oakwoods and native pinewoods are priority habitats, for which action is being led by the Forestry Commission. Also included are semi-natural habitats such as lowland heath, upland heathland, and acid grasslands, which may be threatened by forestry expansion, and in some circumstances justify restoration from forest cover to the previous open habitat.

The primary responsibility for biodiversity conservation in Britain lies with the three conservation agencies: English Nature, Scottish Natural Heritage and the Countryside Council for Wales which administer the domestic systems of statutory protection for sites and species of high conservation importance. The Forestry Commission is the Government forestry department with responsibility for promoting the conservation of forest biodiversity through its main arms: the Forestry Authority, Forest Enterprise and Forest Research.

3. Achieving Forest Biodiversity Conservation Priorities: Controls and Incentives

3.1 REGULATION

Forest regulation in Britain derives from the Forestry Act 1967 and subsequent amendments. This allows the Forestry Commission to control tree felling and prevent it if it would seriously damage biodiversity or other values of forests. Conditions are used to specify whether and how the area should be replanted. In practice most tree felling and regeneration is carried out in accordance with, and in the context of, forest management plans approved by the Forestry Authority. For the state forests this is done by means of Forest Design Plans (see below). For privately owned forests similar plans are required under the Woodland Grant Scheme. Both types of plan are subject to local consultation with statutory and voluntary environmental interests. Where new planting is likely to have significant environmental impact, for example on the biodiversity value of semi-natural open habitats, there is a requirement for an Environmental Assessment.

The draft UK Forestry Standard, which is currently subject to consultation, will draw together, and where possible quantify, current guidance and set minimum design and management standards for the main forest types in Britain.

3.2 INCENTIVES

Private forest owners are given financial incentives within the Woodland Grant Scheme to conserve and enhance biodiversity. These include per hectare grant payments for planting and regenerating areas which meet conservation and other relevant guidelines. The grants allow for a proportion of open areas, normally up to 20%, to be left unplanted and managed as open ground within all forests. Management grants can be targeted to provide biodiversity benefits in special sites and recently a special biodiversity grant has been made available to help to deliver the habitat and species action plans prepared under the UK Biodiversity Action Plan.

3.3 ADVICE

Nature Conservation, Forest and Water, and Landscape Design Guidelines have been produced, as well as a set of eight guides to the management of the main British semi-natural forest types. New biodiversity guidelines are planned for the near future. In addition, the Forestry Authority organises training on all aspects of environmental policy often in co-operation with partner organisations (for more information see http://www.forestry.gov.uk/).

4. Achieving Forest Biodiversity Conservation Priorities: Managing State Forests

In its early years the Forestry Commission was charged with the task of developing a strategic timber reserve. Changing perceptions of the value of semi-natural habitats and the effect of plantations on visual amenity and biodiversity caused a significant re-appraisal in the late 1970's and early 1980's resulting in the development of a forest design and planning process by the Forestry Commission to realise multi-purpose objectives in its forests

4.1 KEY BIOTOPES

Of highest priority is the protection of valuable semi-natural forest and open habitats. The most important of these are contained within the 371 Sites of Special Scientific Interest, covering 70K ha, managed by the Forest Enterprise, including EU Habitats Directive habitats such as Native Pinewoods, Ancient Semi-Natural Broadleaved Forest, Lowland Heath, Raised Bogs and Limestone Pavements.

4.2 NEW CONIFER HABITATS

First rotation, even-aged, thicket stage conifer plantations were thought to have a low value for biodiversity because they were new ecosystems, and lacked structural and spatial diversity. However, it is now becoming clear that ecosystems incorporating introduced conifer species can be of high conservation value. Stands retained to grow out of the pole-stage towards maturity can support distinctive and valuable species assemblages of lower plants, invertebrates and vertebrates (UK Biodiversity Steering Group 1995b; Butterfield et al. 1995; author's data), and certain sites have been designated Sites of Special Scientific Interest to protect these values. Extensive areas of coniferous forest have become important for a number of endangered species such as the red squirrel, Scottish crossbill (*Loxia scotica*) and capercaillie. For economic reasons management by patch clear-felling predominates in most extensive conifer forests, and rotation lengths are limited by wind-throw hazard. Temporary open habitats and young stands are therefore a significant component of the forest ecosystem and have been found to be important for a range of mobile open land and forest edge species. Their ecological value is enhanced by practices such as the retention of standing deadwood, encouragement of broadleaves and increasing the width and structural diversity of permanent edges.

4.3 LANDSCAPE SCALE FOREST PLANNING

In the 1980s the Forestry Commission increasingly recognised the importance of landscape scale forest planning, although at this time visual amenity was the primary focus. A forest design process was developed (Forestry Authority 1994) which has now been applied to 440K ha of forest and aims to: diversify of the age structure of plantations; ensure the size and shape of felling coupes is in scale and sympathy with the land form; increase tree species diversity in the second rotation, particularly with the use of native broadleaves; increase the area of permanent open space; as well as to provide a framework within which the protection and management of special conservation sites takes place.

The forest design process starts with a visual, ecological and cultural site appraisal. From this basis the felling plan is formulated to identify the location, shape and size of all coupes intended for felling in the next 20 years. The restocking plan shows the future distribution of tree species along with the extent and location of major areas of open space; usually accounting for 15-20% of the forest area. The main productive tree species will not normally exceed 65% of the forest area. Restocking plans are visually appraised from key vantage points using computer aided design. Plans are subject to a comprehensive system of consultation with local land-use and environmental interests.

This visually driven approach to landscape scale forest planning often leads to ecologically beneficial designs, but work is underway in the Forestry Commission to formally draw ecological considerations into the process at the landscape as well as the stand scale; initially using an adaptation of the approach used in the NW USA (Diaz and Apostol 1993). This broadening of emphases is leading to the increasing adoption of ecological concepts such as permanent habitat networks whereby biodiversity 'hot spots' are linked by permanent open and wooded linear habitats.

5. Achieving Forest Biodiversity Conservation Priorities: Monitoring

Biological recording in Britain is undertaken by a plethora of organisations, loosely co-ordinated by the conservation agencies and the Biological Records Centre. National censuses of forests and trees have been undertaken by the Forestry Commission every 10 years since 1924 (Locke 1987). A national inventory of forest and trees is currently under way, based on random sampling rather than permanent sample plots. As well as information on tree species composition, management status and productivity, the current inventory includes assessment of forest structure, deadwood and natural regeneration to provide information on the ecological value of British forests (Forestry Commission unpublished). The current inventory will be complete in 2001 and will then proceed on a rolling basis, incorporating measures of biodiversity being developed in the Biodiversity Research Programme (see below).

6. Achieving Forest Biodiversity Conservation Priorities: Research

The Forestry Commission has responded to forest policy developments by instigating two new research programmes; the Native Woodlands Research Programme and the Biodiversity Research Programme.

6.1 THE NATIVE WOODLANDS RESEARCH PROGRAMME

The overall objective of this programme is to understand the structure and dynamics of native forests and provide advice on their management, restoration and extension. This programme is currently small and will develop in support of the Forestry Commission's role as co-ordinators of the native forest Habitat Action Plans; with the current emphasis on support of the Habitat Action Plans for Upland Oakwoods and Native Pinewoods (UK Biodiversity Steering Group 1995b).

6.2 THE BIODIVERSITY RESEARCH PROGRAMME

The Biodiversity Research Programme was established by the Forestry Commission in 1994 to focus primarily on the 1570K ha of planted secondary forests in Britain. It aims to: develop monitoring protocols and collect base-line information on species, structural and habitat diversity in secondary forests; identify biodiversity criteria and indicators for secondary forests at the ecological unit (a) and landscape (b) scale; identify and recommend practical standards by which to appraise biodiversity in secondary forests; and identify and recommend silvicultural systems and management practices that maintain and enhance biodiversity in secondary forests.

Research must be clearly focused on priority information needs, and to ensure that outputs are of practical use to forest managers. The research programme has therefore been planned around the decision making process forest managers are being encouraged to follow (Figure 1). This decision making process will also be the framework for formulation and delivery of research outputs. The complexity of biodiversity issues precludes outputs in the form of simple and widely applicable recommendations; forest managers need access to the information and expertise that will allow them to make ecologically sound management decisions in their own particular situation. For this reason, one important output will be a computer based decision support system designed to lead the forest manager through this process.

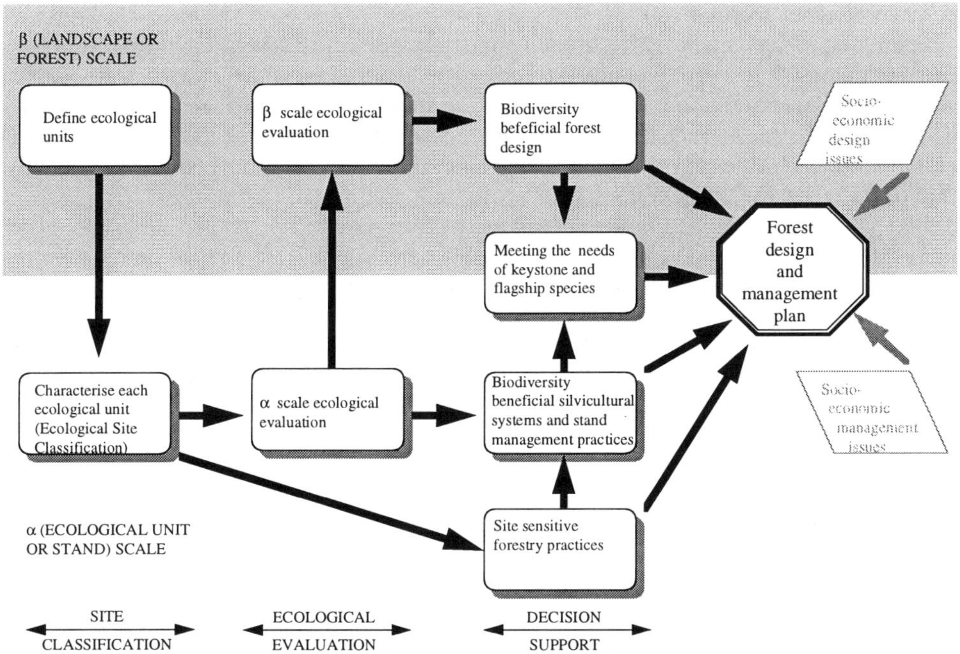

Figure 1: Decision process framework for the Biodiversity Research Programme

6.2.1 Site classification

Forest managers need the means to delineate land into basic ecological units, and to establish the inherent characteristics and capability of each unit. The Ecological Site Classification project is fulfilling this need by providing a site delineation and classification system for forestry in Britain, based on five ecologically meaningful factors: accumulated temperature, moisture deficit, windiness, soil nutrient regime and soil moisture regime (Pyatt 1995). Soil nutrient regime is the most difficult factor to measure and a system of plant indicator species is being developed to facilitate this, along the lines of Ellenberg (1988). These factors are currently being used to predict crop tree species suitability and most appropriate native forest type, either for individual ecological units, or at the landscape and regional scale using GIS. A software version is currently being developed for use by forest managers, initially to inform choice of crop species and promotion of more natural forest stands, but this will be extended by development of additional modules for minimising the ecological impact of forest operations (e.g. cultivation and drainage, use of fertiliser) and the management of semi-natural vegetation in open habitats and riparian corridors.

6.2.2 Ecological Evaluation

Climate and soil characteristics alone cannot provide a full basis for guidance on the conservation of forest biodiversity. An ecological evaluation, at the a and b scale, is also required to determine the current conservation importance (biological legacy) of each ecological unit, and to reconsider the value of ecological units once they are aggregated at the landscape scale (e.g. increasing the value of favourably juxtaposed or particularly rare units). The tools for undertaking ecological evaluation (Biodiversity Evaluation Tools) are being developed in the Biodiversity Assessment Project (Ferris-Kaan et al. in press). Biodiversity Evaluation Tools for each forest type will be a set of indicators plus methodologies that offer a reliable picture of biodiversity for that forest type. Without a practical and reliable assessment methodology a biodiversity indicator will be of little value. Furthermore, no single indicator can provide all the information required on biodiversity status and individual indicators can offer a misleading picture, so packages of indicators are needed to account for the different important facets of each ecosystem, and to provide resilience against confounding factors (Noss 1990).

These two projects will provide the means for objective and consistent classification and evaluation of the site as a basis for decision making over forest planning (primarily b scale) and management (primarily a scale).

6.2.3 Decision support

Forest managers need to understand the impact on biodiversity of forest design options, silvicultural systems and management operations. In ecological terms, this requires an understanding of natural and management processes, and how they affect the composition and structure of forest ecosystems. Whilst the importance of primary nutrient, carbon and water related processes is recognised, three secondary natural processes are of key and practical importance in British forestry (Figure 2). The focus of our research is to better understand the significance and nature of these natural processes, and their management process equivalents. This is necessary to provide guidance on how foresters can work closer to natural ecological processes, and how to promote characteristics of forest structure and composition that confer ecological and conservation benefits, as well as meeting other management objectives.

Much work has already been done on the value and management of edges, patches of broadleaves, open space and riparian areas in plantation forests (e.g. Ferris-Kaan 1995; Petty et al. 1995; Patterson 1993). This work continues and is complemented by research into the conservation management of forest stands, particularly in situations where the potential for enhancement of biodiversity is high (e.g. areas where there has been partial continuity of woodland cover, stands currently with diverse canopy structure due to partial crop failure or wind damage) and where the commercial potential of the stand is low (e.g. stands at high altitude and on deep peat).

Research on forest herbivores, particularly deer, has been underway for some time, with the focus being to provide a basis for sustainable management of deer populations (e.g. Ratcliffe and Mayle 1992) ostensibly to limit damage to forest crops (Gill 1992). Since that time, the importance of deer and other herbivores as wider modifiers of forest ecosystems has been recognised, particularly as deer densities have continued to increase and a detrimental impact on woody regeneration, vegetation structure and field layer composition is being seen in many forest ecosystems. The challenge of current research is to develop recommendations for target deer densities in different forest types to achieve different management objectives. A simulation modelling approach is being

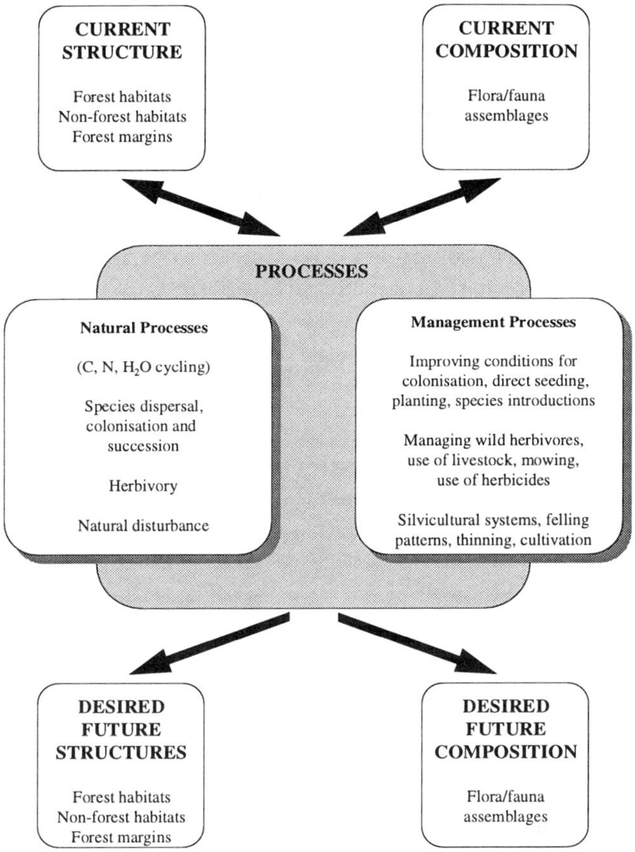

Figure 2. Natural processes, and their management equivalents, upon which current research is focusing

favoured due to the complexity and costs involved with comprehensive field trials, and this is being pursued within an EU collaborative group.

Like many other research institutes, the Forestry Commission Research Agency has realised the importance of ecological processes at the b scale, as well as a scale. The role of natural disturbance regimes in British conditions is being considered and current work is seeking to characterise disturbance agents in British conditions, to understand how they impact on British forest types, and to consider likely impacts on biodiversity of management regimes that mimic natural disturbance (because plantation forests are predominantly composed of introduced conifers, are profoundly man-influenced, and are fundamentally different to the predominantly broadleaved native forest cover, there is no *a priori* reason to mimic natural disturbance patterns in plantation forests unless benefits to biodiversity can be gained relative to other possible management scenarios). In parallel with this research, the value of landscape metrics (Diaz 1996) as a means of characterising b scale patterns and processes is being considered, although ultimately such measures can only be validated with reference to specific species or communities. This is where autecological research becomes important.

The final stage of the forest design and management decision process is to test the emerging design and management plan for its capacity to support keystone and flagship species. By developing population models for certain vertebrate species it is intended to define key habitat requirements as a set of parameters which can, through use of GIS, be used to test the capacity of a forested landscape to support each of these species, and the communities they represent. Current work is focusing on the modelling of field voles (*Mycrotus agrestis*) and tawny owl (*Strix aluco*) populations in relation to habitat change, and on modelling the habitat requirements of the red squirrels. Other species or groups, such as song-birds, corvids, woodland grouse and woodpeckers will be added to cover other important plantation ecosystem guilds and habitats.

7. Closing Comments

In Britain priority is given to the protection and enhancement of biodiversity in the 310K ha of ancient semi-natural forests. However, the majority of Britain's forests are conifer plantations. Whilst an important economic resource, the potential importance of these forests for biodiversity is being increasingly recognised. Forest planning, management and research effort is focusing on how to increase the 'non-market value' of this resource for biodiversity, landscape and recreation without compromising wood supply requirements. Historical factors led to the development of an internationally recognised process of forest landscape design driven by aesthetic considerations. Research effort is now increasing to formulate design principles and management regimes that offer the best potential for enhancing the biodiversity of plantation forests, both in protection and enhancement of open habitats within the forest, protection and enhancement of native woodlands within the forest, facilitating the development of 'old growth' conifer habitats, and by the ecologically optimal combination of all these elements at the landscape scale. It is only by clear priority setting and provision of site sensitive decision making tools through research that conflicts between competing demands will be resolved, the maximum benefits to biodiversity will be accrued per unit sacrifice in wood production, and the means will be secured to determine and monitor the impact on biodiversity of British forestry practice.

References

Butterfield, J., Luff, M.L., Baines, M. and Eyre, M.D. 1995. Carabid beetle communities as indicators of conservation potential in upland forests. In: Petty, S.J., Garson, P.J. and McIntosh, R. (eds.). Kielder – the ecology of a man-made spruce forest. Forest Ecology and Management. 79(1-2): 63-78.

Diaz, N.M. 1996. Landscape metrics: a new tool for forest ecologists. Journal of Forestry 94(12): 12-16.

Diaz, N. and Apostol, D. 1993. Forest landscape analysis and design: a process for developing and implementing land management objectives for landscape patterns. USDA Forest Service. Pacific Northwest Region.

Ellenberg, H. 1988. Vegetation ecology of Central Europe. 4th ed. Cambridge University Press. Cambridge. 731p.

Ferris-Kaan, R. 1995. Management of linear habitats for wildlife in British forests. In: Saunders, D.A., Craig, J.L. and Mattiske, E.M. (eds.). Nature Conservation 4: the Role of Networks. Surrey Beatty & Sons. Chipping Norton. NSW. Australia.

Ferris-Kaan, R., Peace, A.J. and Humphrey, J.W. in press. Assessing biodiversity in managed forests. Proceedings of the Monte Verità conference on assessment of biodiversity for improved forest planning. October 7-11 1996.

Forestry Authority 1994. Forest Landscape Design. HMSO. London. 28p.
Forestry Commission 1995. Facts and figures 1994-1995. HMSO. London. 2p.
Forestry Commission unpublished. Forest inventory training handbook.
Gill, R.M.A. 1992. A review of damage by mammals in north temperate forests: 1. Deer. Forestry. 65: 145-169.
Hambler, C. and Speight M.R. 1995. Biodiversity conservation in Britain: science replacing tradition. British Wildlife. 6(3): 137-147.
Locke, G.M. 1987. Census of forest and trees. Forestry Commission Bulletin 63. HMSO. London. 123p.
Noss, R.F. 1990. Indicators for monitoring biodiversity: a hierarchical approach. Conservation Biology. 4(4): 355-364.
Patterson, G.S. 1993. The value of birch in upland forests for wildlife conservation. Forestry Commission Bulletin 109. HMSO. London. 34p.
Petty, S.J., Garson, P.J. and McIntosh, R. 1995. Kielder – the ecology of a man-made spruce forest. Forest Ecology and Management. 79(1-2). 160p.
Pyatt, D.G. 1995. An ecological site classification for forestry in Great Britain. Research Information Note 260. Forestry Commission. Farnham. 5p.
Ratcliffe, P. R. and Mayle, B. A. 1992. Roe deer biology and management. Forestry Commission Bulletin 105. HMSO, London. 28p.
Roberts, A.J., Russell, C., Walker, G.F. and Kirby, K.J. 1992. Regional variation in the origin, extent and composition of Scottish forests. Bot. J. Scotl. 46(2): 167-189.
Rodwell, J.S. and Patterson, G.S. 1994. Creating new native forests. Forestry Commission Bulletin 112. HMSO. London. 74p.
Spencer, J.W. and Kirby, K.J. 1992. An inventory of ancient forest for England and Wales. Biol. Cons. 62: 77-93.
UK Biodiversity Steering Group 1995a. Biodiversity: the UK Steering Group Report. Volume 1: meeting the Rio challenge. HMSO. London. 103p.
UK Biodiversity Steering Group 1995b. Biodiversity: the UK Steering Group Report. Volume 2: action plans. HMSO. London. 324p.
UK Government 1994a. Sustainable forestry: the UK programme. HMSO. London. 32p.
UK Government 1994b. Biodiversity: the UK action plan. HMSO. London. 192p.

BIODIVERSITY INFORMATION NEEDS IN FINLAND: A QUESTIONNAIRE

TIINA TONTERI
Finnish Forest Research Institute, Helsinki Research Centre
P.O. Box 18; 01301 Vantaa
Finland

Abstract

A questionnaire was made to clarify the needs of future biodiversity information users in Finland. It was sent to 90 organisations, of which 48 submitted an answer. Results from four principal questions posed in the questionnaire are presented. According to the answers, the most important objectives of biodiversity measurement in the National Forest Inventories should be assessing the state and monitoring the changes of biodiversity. The respondents supported different approaches, which were direct measurement of biodiversity and assessment of biodiversity using indicators. The answers of different respondent groups were, however, highly overlapping. In respondents' opinions, the most important subjects to measure were subjects related with key habitats or key biotopes, abundance and quality of dead and decaying wood and endangered species. The estimated benefits from biodiversity information varied with the organisation groups.

1. Introduction

The most challenging task in planning an adequate forest biodiversity monitoring system is to choose the right targets for monitoring, because both biological considerations and needs of information users should be taken into account. Since international agreements and the new forestry law of Finland (Maa- ja metsätalousministeriö 1995) aim at preserving biodiversity, new needs have arisen in two kinds of organisations, the ones that influence biodiversity, like those practising forestry, and the others that direct the observance of the agreements and laws. New forestry practices assumed to maintain biodiversity have rapidly been introduced into practical forestry in Finland (Metsäkeskus Tapio 1994a), but there is yet little scientific evidence of their success, even though some research has been started (Kaila 1995). Also the environmental discussion has led to a situation where both consumers and industrial producers need some information of the sustainability of the production and use of raw materials in the forest industries.

The National Forest Inventory (NFI) will be a part of the national biodiversity monitoring system in Finland. To aid its planning, it was important to find out the needs of different groups which will use biodiversity information in the future. This was performed by means of two questionnaires: one was directed to different organisations, and another to their staff. The main results of the questionnaire addressed to organisations will be presented here.

P. Bachmann et al. (eds.), Assessment of Biodiversity for Improved Forest Planning, 103–110.
European Forest Institute Proceedings no. 18. © 1998 *Kluwer Academic Publishers. Printed in the Netherlands.*

2. Material and methods

2.1 THE QUESTIONNAIRE

The questionnaire was meant to cover all major national organisations which are likely to need biodiversity information in the future. It was sent to 90 organisations in late December 1995 - early March 1996. The organisations that did not reply received the questionnaire for tree times during the study period. The organisations were asked to identify themselves in one of the following groups: forest industry companies; governmental forestry organisations; other forestry organisations; ministries; university departments with teaching or research in ecology or forest ecology; governmental research organisations; and independent nature conservation organisations.

The questionnaire, following the guidelines given by Oppenheim (1992), contained 18 questions about the respondent organisation, opinions about biodiversity research, biodiversity monitoring, publication of monitoring results, taking biodiversity into account in forestry planning and the National Forest Inventory. The results of answers to four principal questions will be presented here. The four questions were:

1. What kind of objectives should the measurement of biodiversity have in the National Forest Inventories?
2. If the finances are insufficient, which biodiversity subjects are the most important to measure in the National Forest Inventories?
3. When focusing on needs of your organisation, in what extent is the measurement of following subjects necessary in the National Forest Inventories? (50 biodiversity-related subjects were listed, see Fig. 1.).
4. Which benefits do you think information on state and changes of biodiversity will bring to your organisation?

The respondents were asked to answer in their own words to questions 1, 2 and 4. For question 3, the respondents were supposed to choose a value on a numeric scale for each subject. The scale ranged from 1 to 5, where 1 = not at all necessary and 5 = strongly necessary.

2.2 DATA ANALYSIS

The different subjects mentioned in answers to questions 1, 2 and 4 were classified. For each question, the ten classes most frequently mentioned by the respondents were tabulated.

To illustrate the differences in information needs of different organisations, a principal components analysis (PCA) (see e.g. ter Braak 1987) on a correlation matrix of answers to all 50 subjects in question 3 was performed using CANOCO v. 3.10. (ter Braak 1990). Two respondents with no answers to the question were omitted from the analysis. If there were only few opinions missing in respondent's answers, they were replaced by the neutral or "do not know" value 3. Results were presented as a correlation biplot adjusted to subjects' variance (ter Braak 1990).

3. Results

48 (53 %) of the organisations submitted an answer. In addition, 6 organisations contacted and announced that the topic is not relevant for them, or that they had no opinion.

In respondents' opinions, the most important objectives (question 1) of biodiversity measurements in the National Forest Inventories should be producing information about the state of biodiversity and changes in biodiversity (Table 1). Some of the subjects mentioned in question 3 seemed to be so important that they were mentioned in answers to this question as well. It is also interesting that five of the respondents stated that the monitoring should be performed using biodiversity indicators.

A principal components analysis showed that there were two principal directions of variation in the answers to the question 3 (Fig. 1). The general importance of measurements changed from high to low scores of the first axis, low values meaning high importance. A gradient ranging from the already currently measured tree and site variables (lower left corner of Fig. 1) to more time demanding subjects, like taxonomic species groups (higher left corner of Fig. 1), was slightly diagonal between axes 1 and 2. Conceptual diversity components and environmental variables were located between these two extremes. Thus there were opinions which suggested concentrating the inventory on site and stand characteristics, and others that considered taxonomic species groups to be more important targets for monitoring. Subject groups like stand

TABLE 1. The ten most common subjects mentioned in answers to the question "What kind of objectives should the measurement of biodiversity have in the National Forest Inventories?" The answer frequencies are listed by organisation groups. Explanations for abbreviations: For. ind. comp. = forest industry companies; Gov. for. org. = governmental forestry organizations; Oth. for. org. = other forestry organisations; Min. = ministries; Univ. dept. = university departments; Gov. res. org. = governmental research organisation; Ind. nat. cons. = independent nature conservation organisations.

	For. ind. comp	Gov. for. org.	Oth. for. org.	Min.	Univ. dept.	Gov. res. org.	Ind. nat. cons.	Total
The state of biodiversity	3	2	8	1	2	0	0	16
Biodiversity changes	1	1	7	1	2	0	1	13
Subjects related with key habitats/biotopes	1	0	4	0	1	1	0	7
Information for developing environmentally friendly forestry	3	1	1	1	0	0	0	6
Information about influences of forestry practices to biodiversity	0	1	3	0	1	0	0	5
Use of biodiversity indicators	1	0	2	0	1	1	0	5
Taxonomic species groups	1	0	2	0	1	0	0	4
Reliability of results	1	0	1	0	1	1	0	4
Abundance and quality of dead and decaying wood	0	0	2	0	0	1	0	3
Area of nature conservation areas	1	0	2	0	0	0	0	3
Number of organisations which answered the question	9	5	19	2	6	1	1	43

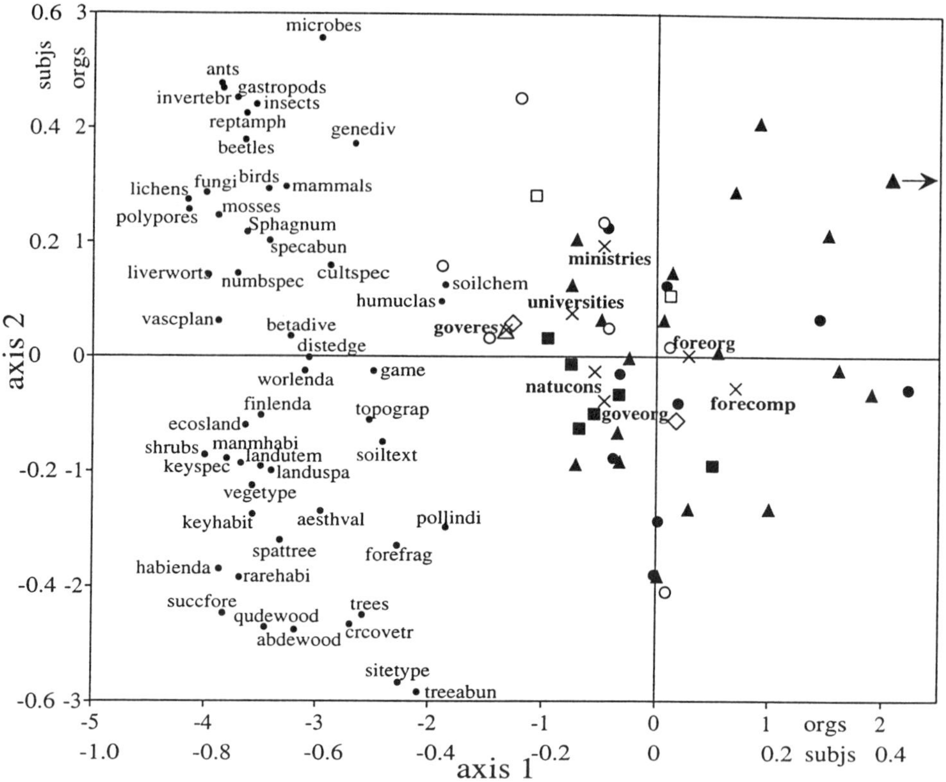

Figure 1. A correlation biplot of a principal components analysis (PCA) of respondents' opinions on necessity to measure 50 different subjects relative to biodiversity in National Forest Inventories. Axes 1 and 2 have two scales, one for organisations (orgs) and the other for subjects (subjs). The centroids of the organisation classes are marked with crosses. Explanations for abbreviations and markers of individual organisations: forecomp and large black circles = forest industry companies; goveorg and black squares = governmental forestry organisations; foreorg and black triangles = other forestry organisations; ministries and open squares = ministries; universities and open circles = university departments; goveres and open triangle = governmental research organisation; natucons and diamonds = independent nature conservation organisations. Explanations for abbreviations of subjects: abdewood = abundancy of dead wood; aesthval = aesthetic landscape values; ants = ants; beetles = beetles; betadive = beta diversity; birds = birds; crcovetr = crown coverage of trees; cultspec = species favoured by man; distedge = distance to the nearest edge of the forest compartment; ecosland = variation of ecosystems at landscape level; finlenda = species endangered in Finland; forefrag = forest fragmentation; fungi = fungi; game = game species; gastropods = gastropods; genediv = genetic diversity; habienda = potential habitats for endangered species; humuclas = humus class; insects = other insects; invertebr = other invertebrates; keyhabit = key habitats or biotopes; keyspec = key species; landuspa = spatial land use history; landutem = temporal land use history; lichens = lichens; liverworts = liverworts; mammals = mammals; manmhabi = traditional cultural biotopes; microbes = microbes; mosses = mosses; numbspec = number of species; pollindi = air pollution indicator species; polypores = polypores; qudewood = quality of dead wood; rarehabi = rare habitats or biotopes; reptamph = reptiles and amphibians; shrubs = shrubs; sitetype = site type; soilchem = soil chemistry; soiltext = soil texture; spattree = spatial organisation of trees; specabun = species abundancies; Sphagnum = *Sphagnum* mosses; succfore = successional stage of forest stand; topograp = topography; treeabun = relative abundancies of tree species; trees = trees; vascplan = vascular plants; vegetype = vegetation type; worlenda = species endangered in the world scale.

TABLE 2. The ten most common subjects mentioned in answers to the question: "If the finances are insufficient, which biodiversity subjects are the most important to measure in the National Forest Inventories?". The answer frequencies are listed by organisation groups. Explanations for abbreviations are the same as for Table 1.

	For. ind. comp	Gov. for. org.	Oth. for. org.	Min.	Univ. dept.	Gov. res. org.	Ind. nat. cons.	Total
Subjects related with key habitats / biotopes	5	2	6	1	0	1	1	16
Abundance and quality of dead and decaying wood	4	2	4	0	2	1	0	13
Endangered species	4	1	3	1	3	0	0	12
Taxonomic species groups	1	1	5	0	1	0	0	8
Biodiversity indicators	0	0	4	1	1	1	0	7
Rare biotopes or ecosystems	1	3	1	1	0	0	0	6
Air pollution indicator species	1	1	1	0	0	0	1	4
Key species	2	0	2	0	0	0	0	4
Number of species	1	1	1	0	1	0	0	4
Potential habitats for rare or endangered species	2	0	2	0	0	0	0	4
Number of organisations which answered the question	10	5	17	2	7	1	2	44

TABLE 3. The eleven most common benefits mentioned in answers to the question: "Which benefits do you think information on state and changes of biodiversity will bring to your organisation?" The answer frequencies are listed by organisation groups. Explanations for abbreviations are the same as for Table 1.

	For. ind. comp	Gov. for. org.	Oth. for. org.	Min.	Univ. dept.	Gov. res. org.	Ind. nat. cons.	Total
Development of environmentally friendly forestry	2	1	6	0	0	0	0	9
Development of research	0	0	2	0	6	1	0	9
Development of own information services	2	1	3	0	0	0	0	6
Background information, information for comparison	0	1	3	0	2	0	1	7
Improvement of image	4	1	0	0	0	0	0	5
Information to use in marketing products	4	0	1	0	0	0	0	5
Development of usual activities	1	0	3	1	0	0	0	4
Improvement of education	1	0	1	0	2	0	0	4
Independent, objective information	1	0	3	0	0	0	0	4
Criteria for ecocertification	2	1	0	0	0	0	0	3
Information about the state of biodiversity	1	0	2	0	0	0	0	3
Number of organisations which answered the question	9	5	19	2	7	1	1	44

characteristics, habitat subjects and taxonomic species groups were mostly strongly correlated within each group (Fig. 1). This means that, for instance, if an organisation was interested in obtaining information about abundance of dead wood, it was likely to be interested in other characteristics describing a tree stand as well.

The centroids of different organisation groups revealed that the governmental research organisation, ministries and universities were the most eager supporters of inventorying the taxonomic groups (Figure 1). Governmental forestry organisations, independent nature conservation organisations and the governmental research organisation were the most interested ones to obtain biodiversity information based on tree and stand characteristics. However, the overlap of the organisation groups was large, and no distinct subject characteristics for any of the organisation groups could be observed.

In answers to question 2, the most important subjects, especially for the respondents within forestry, were key habitats or biotopes, followed by abundance and quality of dead and decaying wood and endangered species (Table 2.).

The benefits obtained from biodiversity information differed widely among the organisation classes (Table 3). In forestry, an important benefit was the possibility to use the information for developing environmentally friendly forestry. For forest industry companies, the most expected benefits were related with marketing, while research organisations planned to use the information for development of research.

4. Discussion

4.1 THE NEEDS OF THE ORGANISATIONS

In the questionnaire answers there were two main approaches to measuring biodiversity. The organisations ranged from ones which strictly supported extensive monitoring of different taxonomic or ecological species groups and communities to others which stated that a sufficient way to measure biodiversity is indirect assessment using properties of sites, stands and trees. Some of the organisations with the latter attitude commented in paper margins that there will probably not be enough resources to carry out an extensive monitoring program with direct measurements of species and their abundance. However, some of these respondents listed subjects related to endangered species as some of the most important subjects for monitoring in the NFI. The answers may be strongly influenced by the opinions of the person who wrote them. But it is also likely that the results illustrate the real needs of biodiversity information: the research in the universities is often concentrated into specialised aspects of forest ecology, while organisations in practical forestry often want answers to general questions, like "Have we succeeded in taking biodiversity into account in our forests?".

It is hardly surprising that, key habitats or biotopes and dead and decaying wood were considered the most important targets for monitoring. Key habitats are protected by the new forestry law (Maa- ja metsätalousministeriö 1995), and the current forest management recommendations (Metsäkeskus Tapio 1994a and 1994b) encourage forest owners to leave dead and living decaying wood into forests during thinning and clear-cutting. There are also strong social and economic pressures to develop environmentally friendly forestry in Finland.

4.2 OPINIONS AND THEIR APPLICATIONS

When setting maintenance of forest biodiversity as the goal, biodiversity monitoring should provide tools for assessing its possibilities to preserve. From an ecological point of view, a good biodiversity monitoring system should take the following topics into account (modified from Jeffers 1996):

a) All relevant biological organisation levels.
b) Population and community processes central to biodiversity.
c) Spatial processes central to biodiversity.
d) Succession and other temporal changes.
e) Abiotic factors that influence biodiversity.
e) Influence of man to biodiversity.
f) Relevant combinations of the above mentioned subjects.
g) Unbiased sampling and sound methods of data analysis.

Taking these things into consideration leads to a vast need of resources. An inevitable problem in all biological monitoring is optimising the costs and accuracy of data simultaneously (see e.g. Margules and Austin 1991). In assessments of landscape level diversity, it will probably be possible to successfully apply cost-effective remote sensing methods in the future. But in species, population and community level assessments there generally are two monitoring approaches. Biodiversity can either be observed directly, using sample sites or sample areas (Jeffers 1996), or with indirect methods, using indicators (Angelstam 1996). The indicators can be species known to have a good indicator value, or environmental factors which directly influence biodiversity.

The advantages of the direct monitoring approach are:

a) The data are accurate and reliable, if sampling is done properly.
b) It is possible to use the data even for ecological research.
c) It is possible to use the data for ecological modelling, which can further be used for improving the monitoring system.

The disadvantages of the direct monitoring approach are:

a) The approach is laborious, and often demands a heavy background organisation.
b) High costs.
c) It may be difficult to obtain a sufficient amount of specialists to determine the species.
d) Species determination, sample counting etc. is slow, so getting results will take time.

The advantages of the indicator approach are:

a) It is easy to obtain information from large areas.
b) There is no need to employ as much staff as in the direct approach.
c) A smaller number of specialists is sufficient.
d) Lower costs.
e) Results are obtainable fast.

The disadvantages of the indicator approach are:

a) To use indicators for prediction, calibration or simulation models are needed. This often means establishing a large research project before monitoring is started.
b) The calibration or simulation models should be adequate. In practice this means that, 1. the models have a reasonable accuracy over large geographic areas; 2. the indicators have a good indicator value; 3. the indicators must be easily observed or measured; 4. the models take all factors that influence the predicted subject into account.
c) Does not provide data for ecological research.
d) Cannot be self-improving. Separate research is needed for developing the monitoring system.

When weighing these approaches, the direct approach provides more information, more possibilities for development and research, and a more reliable base for monitoring itself, but only if it is performed skilfully. Choosing a monitoring approach is largely a matter of money. In the 9th Finnish National Forest Inventory, measurement of dead and decaying wood, assessment of key habitats or biotopes and occurrence of new forest management practices were added to the pre-existing inventory, and the development of a multi-source inventory system utilising new remote sensing methods in biodiversity monitoring has begun (Tomppo 1996).

References

Angelstam, P.K. 1996. Forest landscape management for maintenance of biodiversity - a Swedish perspective. In: Bachmann, P., Kuusela, K. and Uuttera, J. (eds.). Assessment of biodiversity for improved forest management. EFI Proceedings 6. Pp. 69-86.

Jeffers, J.N.R. 1996. Measurement and characterisation of biodiversity in forest ecosystems – new methods and models. In: Bachmann, P., Kuusela, K. and Uuttera, J. (eds.). Assessment of biodiversity for improved forest management. EFI Proceedings 6. Pp. 59-67.

Maa- ja metsätalousministeriö. 1995. Metsälakitoimikunnan mietintö. Helsinki.

Metsäkeskus Tapio. 1994a. Luonnonläheinen metsänhoito. Metsänhoitosuositukset. Metsäkeskus Tapion julkaisuja. No. 6.

Metsäkeskus Tapio. 1994b. Good forest for the good of all. Helsinki.

Margules, C.R. and Austin, M.P. (eds.) 1991. Nature conservation: cost effective biological surveys and data analysis. CSIRO. Australia.

Kaila, S. 1995. Monimuotoisuus talousmetsien uudistamisessa -yhteistutkimus. Folia Forestalia. 4: 335-342.

Oppenheim, A.N. 1992. Questionnaire design, interviewing and attitude measurement. Pinter Publishers. London.

Ter Braak, C.J.F. 1987. Ordination. In: Jongman, R.H., ter Braak, C.J.F. and van Tongeren, O.F.R. (eds.). Data analysis in community and landscape ecology. Pudoc. Wageningen.

Ter Braak, C.J.F. 1990. Update notes: CANOCO version 3.10. Agricultural Mathematics Group. Wageningen.

Tomppo, E. 1996. Biodiversity monitoring in Finnish forest inventories. In: Bachmann, P., Kuusela, K. and Uuttera, J. (eds.). Assessment of biodiversity for improved forest management. EFI Proceedings 6. Pp. 87-94.

SHADOW PRICING TRADE-OFFS BETWEEN TIMBER SUPPLY AND ENVIRONMENTAL QUALITY[1]

DIETMAR W. ROSE AND SYED A. HUSAIN
College of Natural Resources, University of Minnesota
1530 N. Cleveland Avenue
St. Paul, Minnesota, USA

Abstract

This paper summarizes results of a statewide study that analyzed strategies that can enhance environmental quality while maintaining forest productivity. Two strategies were examined: intensified management on selected forest sites, and expansion of timber production to marginal agricultural lands. Both these practices can reduce the acres required to meet specific timber demands. Preliminary results show that environmental restrictions imposed on timber harvesting without increased timber productivity will lead to timber supply shortages and sharply rising timber prices for forest industries in Minnesota. On the other hand, wood production on marginal agricultural sites can eliminate timber supply shortfalls, reduce the upward pressure on timber prices and reduce harvest of forest land. Critical forest sites can thus be protected while aforestation of agricultural sites can provide additional environmental benefits.

1. Introduction

In Minnesota as in other parts of the US, the public is asking for greater restrictions on timber harvesting because of the potential negative impacts on biodiversity, wildlife, water and recreation values. The recent Minnesota Generic Environmental Impact Statement (GEIS) requested by the Minnesota Environmental Quality Board was designed to assess the environmental impact of all forest activities and to examine the trade-offs between different levels of timber harvesting (Jaako Pöyry 1992). In the GEIS, site-specific mitigation strategies designed to enhance wildlife, recreation, biodiversity and water values were modeled (Rose and McDill 1996). The GEIS resulted in a number of major forestry policy changes as well as the implementation of the first state law regulating forest management activities. Additional policy changes that call for immediate set-aside of extended rotation forests of 25-50% of state owned forests are being considered, however, without fully understanding their impact on forest productivity and social, economic, and environmental values.

Until impacts of timber harvesting are fully understood, opportunities for reducing the environmental pressure of timber harvesting while maintaining forest productivity should be explored. The harvest restrictions recommended by a panel of experts for the GEIS led to infeasibilities in meeting specified timber production targets and caused

1 Research supported by Minnesota Center for Transportation Studies and the Consortium for Plant Biotechnology research

P. Bachmann et al. (eds.), Assessment of Biodiversity for Improved Forest Planning, 111–119.
European Forest Institute Proceedings no. 18. © 1998 *Kluwer Academic Publishers. Printed in the Netherlands.*

substantial increases in the cost of producing timber. The current study was designed to explore how increasing the timber production base to marginal agricultural lands could help sustain timber production and help protect the environment by reducing the pressure on forest land from timber harvesting because fewer acres will be required to meet industrial demands.

2. Increasing Forest Productivity

The potential for increasing wood supplies from existing forest lands in Minnesota through management intensification appears to be substantial. A study by Kapple and Hoganson (1995) examined timber supply changes through more intensive forest management on forest lands located close to markets. This study concluded that management intensification in selected sites could reduce timber supply shortages and pressures on timber prices.

Additionally, large acreage of marginal agricultural lands, currently idled under the federally funded Conservation Reserve Program (CRP), have the potential to produce 5-10 times as much wood on one acre than under the current extensive management practices in Minnesota. Timber production on CRP land not only can reduce the need for harvesting from existing forest lands, but also contribute environmental benefits associated with growing a perennial crop on agricultural lands. There are almost 1.8 million acres of CRP lands set aside in Minnesota (Taff 1993). Many of these acres could support fast growing hybrid tree species such as hybrid poplar while protecting soil resources, providing wildlife habitat and help reduce the pressure on critical ecosystems such as riparian zones, highly erodible soils and certain locations where public demand for non-timber services are strong. Inclusion of CRP lands in the forest resource base can also create significant savings in public expenditures currently incurred in order to make contract payments. Growing short-rotation hybrid poplar on these lands could provide an additional source of revenue for farmers.

3. Study Description

3.1. STUDY OBJECTIVES

The current study addresses three major questions: 1) what contributions could the expansion of the timber production base to CRP lands make to timber supplies, 2) what economic and environmental benefits would result from this practice, and 3) what economic incentives exist for farmers to produce timber.

3.2 STUDY AREA

The study included most of the northern forested landscape in Minnesota where most of the forest industries are concentrated. The southern portion of the state with mostly mixed hardwoods and small sawmills was not included. The basic inventory data used for the GEIS were the 1990 Minnesota Forest Inventory and Analysis (FIA) data provided by the USDA Forest Service and the Minnesota Department of Natural Resources. Inventory plots were used as analysis areas since this was the most disaggregated form of the data. These FIA sample plots represent, on average, 1,300 acres of a given ownership, covertype, size and density. The largest model run involved 11,184 forest analysis areas and over 3,000 CRP land parcels. Table 1 summarizes major forest statistics.

TABLE 1 .Commercial Forest Acres by Covertype and Owner in Northern Minnesota in 1990

Covertype/ Owner	National Forests	Misc. Public	State	County	Other Private	Forest Industry	All Owners
Pines	220,100	38,800	158,000	129,200	248,800	86,400	881,300
Balsam Fir	185,900	20,000	167,000	181,900	182,500	71,900	809,200
N. White Cedar	92,200	71,300	267,500	97,700	70,700	49,000	648,400
Tamarack	29,800	46,500	334,000	127,800	170,700	9,100	717,900
Spruce	282,000	65,900	599,400	218,100	200,700	75,500	1,441,600
N. Hardwoods	345,600	148,200	435,200	615,300	1,786,500	138,900	3,469,700
Aspen	642,100	222,500	887,200	1,034,100	2,086,200	293,400	5,165,500
Balsam Poplar	23,400	29,800	127,700	88,500	206,700	25,700	501,800
All Covertypes	1,821,100	643,000	2,976,000	2,492,600	4,952,800	749,900	13,635,400

3.3. CRP DATABASE

The CRP data set was available from the Minnesota Department of Agriculture. A CRP land database was added to the forestry database already used in the existing planning models. These agricultural sites were described to differentiate soil quality, erosion potential, opportunity costs based on agricultural crop production potential, and geographic location. The new "stands" were identified with a UTM coordinate to be able to connect it with the network utilized by the planning package. The current road network was extended to incorporate the road network for all counties in Minnesota to determine the timber transportation costs associated with each road.

In accordance with the goals of CRP, acreage reported as highly erodible in the CRP database was exempted from the analysis. Steep slopes or other severe soil limitations associated with these lands render them unsuitable for any type of farming activity including biomass production. Furthermore, researchers and environmentalists agree that wetlands should not be further exploited for the purpose of agricultural production. Wetlands provide several environmental, wildlife and economic benefits and their destruction can have significant long term impact on the ecosystem itself. Therefore, acreage consisting of confirmed non-converted wetlands was also exempted from further analysis. This left about half a million acres of the total 1.8 million acres of CRP lands for potential timber production. Key characteristics of each CRP parcel relevant for the analysis were soil productivity class and annual market land rents which represent opportunity costs which are considered when forestry investment options are being analyzed by the scheduling model.

3.4 TIMBER DEMAND

Six pulpwood markets and 10 timber product classes were recognized. Quantities demanded at major market centers were set for each planning period to reflect the requirements of the mills at each location. They coincide with the major forest industry plants and represent the majority of the total timber demand. Table 2 shows timber demands by market and product for 10 planning periods and incorporates projections of industry expansion.

TABLE 2. Annual forest products demand (thousands of cords) for 6 major Minnesota timber markets

Product \ Market	Bemidji	Brainerd	Cook	Duluth	G. Rapids	I. Falls	Total
Aspen	580	319	203	590	519	458	2669
Pine Pulpwood	57.78	-	-	162.78	-	42.78	263.34
Spruce	-	100	-	379.5	163.5	-	643
N. Hardwoods	89	198	59	355	69	49	819
Pine Bolts and Sawlogs	102.61	27.61	27.61	27.61	27.61	27.61	240.66

3.5 DEVELOPING MANAGEMENT ALTERNATIVES

RxWrite, a collection of computer programs was used to develop from 10 to 80 management alternatives for each of the more than 11,000 forestry plots. RxWrite (McDill and Rose in preparation) simulates clear cutting and three types of thinning: random, from above, and from below. Where stands are thinned or selectively cut, the model grows the remaining trees in the tree list. A wide range of options concerning thinning intensity, timing and frequency can be specified by the user. RxWrite utilizes stand-level inventory data with individual tree records. The stand and tree evaluation and modeling system (STEMS), developed by the USDA Forest Service (Belcher et al. 1982), was used to model tree growth in RxWrite. Standard regeneration tree lists are applied by RxWrite following clear cutting. The regeneration lists vary by cover type, site index, and type of regeneration, either natural or planted. The species composition and number of trees in these regeneration stands were derived from averages of Minnesota sample plots. In addition, the transition of stands after clear cutting through natural regeneration was modeled using an empirical matrix of cover type transition probabilities for major ecological regions in the state.

Each of the initial stand alternatives and alternatives of all possible regeneration linkages were simulated to generate the timing, costs and product flow information associated with each alternative over the planning horizon, essentially all the information needed to carry out stand-level cash flow analyses. Management alternatives for each CRP parcel were developed separately for each soil productivity class recognized in the CRP database. Only one 10-year rotation of hybrid poplar with an identical regeneration alternative was considered for each parcel in addition to agricultural production.

3.6 ESTIMATING BIOMASS YIELDS ON CRP LANDS

Research on intensively managed plantations has provided us with excellent information on the production potential of short-rotation hybrid poplar clones. In Minnesota, biomass yield rates for hybrid poplar are expected to range between 2 to 5 dry tons/acre, given different soil and climatic conditions. These estimates were mostly derived from a network of research plantations which has been established in a five state region of the north central U.S during the mid-eighties (Hansen et. al 1994). In the CRP database, land productivity is indicated by land capability classes and sub-classes. In general, it is expected that soils with a higher productivity rating for the agricultural crops will also have a higher rating for hybrid poplar. However, this relationship will not hold for soils which have a lower rating for agricultural crops because of poor drainage conditions, since hybrid poplar tends to grow well under such conditions (Berguson 1994).

For the purposes of this study, yield rates for all CRP land parcels were determined by consultation with experts from the US Forest Service hybrid poplar project at Rhinelander, Wisconsin. Yield rates were modeled as a function of land capability classes and sub-classes, rainfall, and soil types.

3.7 SCHEDULING MODEL

A scheduling algorithm developed by Hoganson and Rose (1984) was used to develop optimal solutions for a number of management scenarios. The computer program DTRAN (Kapple and Hoganson 1991) was used to implement the algorithm. First, the product flows to be constrained over time are identified and the desired output levels specified. Then initial estimates of the marginal costs (dual variable values) of producing the desired output levels were specified for each product for each period. Using the estimated marginal costs of production as prices, the optimal management strategy (prescription) for each analysis area is determined separately using the standard economic approach of maximizing the net present value (NPV) of the stand. The product (and input) flows for all analysis areas are then aggregated to determine the total forest-wide production levels corresponding to the current set of dual value estimates. These aggregate production levels are then compared with the desired output levels to determine whether the schedule is acceptable. If it is acceptable, the solution found is reported and the process stopped. Otherwise, the results of the last iteration are used to re-estimate the marginal costs of production for each constrained product in each period, and the process returns to the problem of identifying optimal prescriptions for each stand using the new dual variable estimates.

A key advantage of this approach is the ability to handle many analysis areas or stands. The method also provides integer solutions for stands. This is a direct consequence of the nature of the solution algorithm – a single optimal prescription is identified for each analysis area. This characteristic of the solutions provided by the Hoganson/Rose algorithm can be extremely useful for implementing solution schedules on the ground. Another advantage of the algorithm is its intuitive interpretation. The dual variable values estimate the marginal cost of meeting the production targets specified by the flow constraints. One can then assess whether each marginal cost is acceptable for the targeted output level. If the costs are not acceptable, the targets should be lowered or new, lower-cost ways to achieve the targets must be found.

3.8 MODEL SCENARIOS

Four scenarios were analyzed. In the first scenario, which served as a benchmark for comparison, it was assumed that all commercial forest land was potentially available for harvest without specific concern for other commodities or non-commodities (see the total column in Table 3). For the second scenario, management restrictions were imposed on certain timber stands to reduce environmental impacts of timber harvesting. These restrictions were modeled after the GEIS.

In the GEIS, wildlife, recreation, water, soils, and biodiversity experts were provided with detailed descriptions of the forest management plans from an initial set of runs without management restrictions. These descriptions included the timing and location of all scheduled forest management activities and the effects of these activities and normal stand development on the forest inventory at each planning interval. The experts analyzed the stand-level management schedule using discipline-specific models and expert judgment to develop recommendations on how to mitigate possible negative environmental impacts associated with each cutting level.

Based on these inputs, all commercial forest plots were assigned to one of 5 treatment classes: 1) plots for which all standard (unconstrained) silvicultural options were permitted; 2) plots within a certain distance of water, where only random thinnings at specified intervals were permitted (no clear cutting); 3) extended rotation plots, where minimum cutting ages were increased; 4) old growth plots where no harvesting was allowed, and 5) plots not available for any harvesting activity for economic, environmental, and social reasons. Stands assigned to a specific treatment class could only be linked to regeneration options by treatment class, e.g., stands with extended rotations had to be linked to regeneration alternatives with extended rotations. Table 3 summarizes acres by treatment class and owner. The table shows that in effect 9.61 percent of the commercial land base or 1,055,500 acres were not available for any harvesting activities while on 14.55 percent or 1,597,900 acres harvesting restrictions were imposed.

Since schedules are developed for a 100-year planning horizon, aspects of uncertainty in the predictions are naturally of concern. The growth model itself has been extensively tested and been shown to have small errors even for 20-30 year projections. For the major covertype aspen that is at the lower end of its rotation. Furthermore, the major purpose of the study is a comparison of scenarios in terms of shadow prices, not the development of a 100-year management schedule.

Optimal management schedules that reflected these restrictions on management provided the basis for assessing the trade-offs between managing with or without management restrictions. Both the unrestricted and restricted forest management scenarios were repeated after adding the CRP parcels described above to the timber base, adding two additional scenarios.

TABLE 3. Commercial Forest Acres by Owner and Treatment Class

Owner/Treatment	Normal No harvest restrictions	Buffers No clearcuts thinning only	Extended Rotations	Old Growth No harvesting	Reserved No harvesting	All Acres
Nat. Forests	1315800	112900	364900	27500	0	1821100
Misc. Public	402600	10200	0	0	230200	643000
State	2127200	99800	597300	0	151700	2976000
County	2224300	141800	0	0	126500	2492600
Other Private	4188400	263600	0	0	500800	4952800
Forest Industry	723700	7400	0	0	18800	749900
All Owners	10982000	635700	962200	27500	1028000	3635400

Optimal management schedules that reflected these restrictions on management provided the basis for assessing the trade-offs between managing with or without management restrictions. Both the unrestricted and restricted forest management scenarios were repeated after adding the CRP parcels described above to the timber base, adding two additional scenarios.

4. Results

The estimated marginal costs (shadow prices) of producing various levels of forest products over time is the most direct output of the scheduling model. These marginal production costs reflect the required production levels, the initial inventory, and changes over time in the inventory as a result of management activities. Differences in production costs between alternate model runs measure the impact of changing forest product requirements and management constraints as defined in the treatment classes. Additionally, the model provides an optimal, stand-level operational plan.

We will show major results only for the aspen covertype which is the most critical for timber supply in the State (Figures 1-4). Shadow price increases reflect shortages in specific timber products brought on by age class imbalances in the inventory. In general,

Fig. 1 Aspen Shadow Prices - 6 Markets
Unrestricted Forest base Only

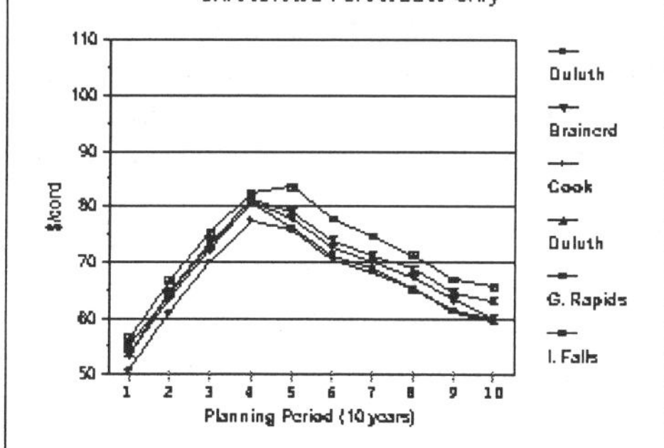

Fig 2. Aspen Shadow Prices - 6 Markets
Restricted Forest Base Only

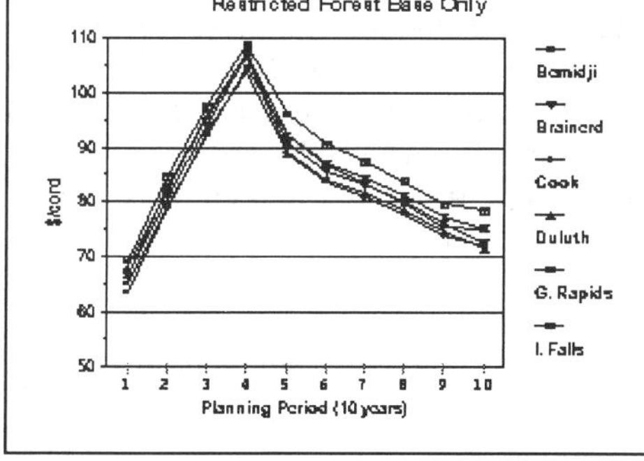

Fig. 3 Aspen Shadow Prices - 6 Markets
Unrestricted Forest Base + CRP

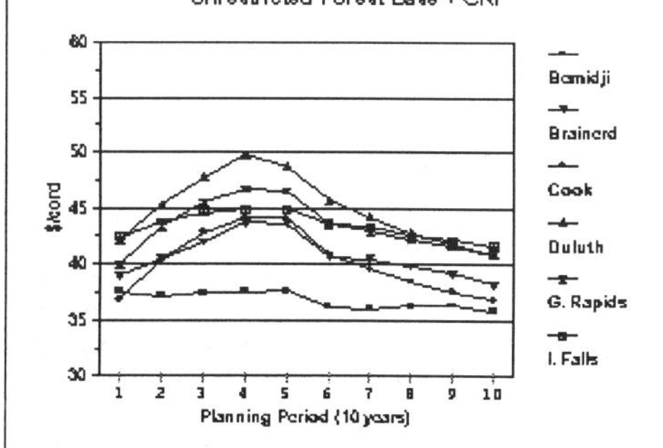

Fig. 4 Aspen Shadow Prices - 6 Markets
Restricted Forest Base + CRP

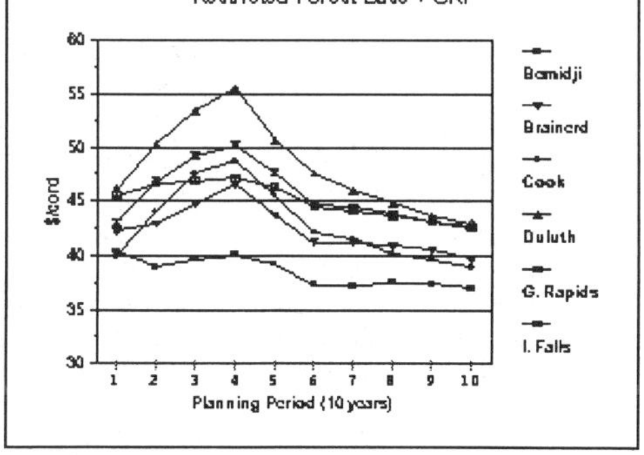

shadow prices in all four scenarios increase for the first 4-5 decades before they begin to decline. These figures also illustrate the advantages between individual markets, the impact of a reduced commercial forest land base due to management restrictions and the effect of expanding the productive base to CRP lands in reducing these shadow prices.

Restrictions on forest management to reduce environmental damages associated with timber harvesting have a substantial impact on the cost of timber production. Since in both scenarios without CRP acreage timber demands could not be met due to physical supply shortages, the associated shadow prices cannot be directly compared with shadow prices of the other two scenarios that include CRP acres. For the latter two, timber demands for each product, market and planning period were met within 5 percent. It is clear, however, that physical shortages of timber, especially for aspen will exert further upward pressure on timber prices and that the expansion of the productive land base to CRP is one option that deserves serious attention.

The use of CRP lands for production of short-rotation hybrid poplar has far-reaching impacts for timber prices and timber harvest acreage. Shadow prices or marginal production costs show significant reductions. Harvest acreage due to the higher productivity of hybrid poplar on agricultural land can be reduced substantially and thus help protect sensitive forest areas. Not every CRP parcel was allocated to timber production because of the opportunity for agricultural production. The combination of the opportunity cost for a CRP parcel, the transportation costs to timber demand centers, and the parcels soil productivity influence the economics of timber production and determine the relative advantage of one agricultural site versus another one.

Additionally, locations with the greatest potential for biomass production were identified. Harvest acreage reductions and the potential environmental mitigation effects were also estimated as shown in table 4. This table displays harvested acreage by planning period on forest and CRP lands for the different scenarios. While harvest acres from forest land are quite similar for all scenarios, the CRP acres harvested increase with the mitigation run because the higher shadow prices for timber in various locations make wood production profitable on more CRP acres.

TABLE 4. Comparison of Total Acres Harvested by Management Scenario

Scenario 1 Comm. Forest no restrictions		Scenario 2 Comm. Forest with restrictions	Scenario 3 Comm. forest no restrictions plus CRP			Scenario 4 Comm. Forest with restrictions plus CRP		
Planning Period	Forest Acres	Forest Acres	Forest Acres	CRP Acres	Total Acres	Forest Acres	CRP Acres	Total Acres
1	2,139,200	2,131,700	2,058,600	0	2,058,600	2,063,200	0	2,063,200
2	1,896,900	1,912,700	1,630,700	172,355	1,803,055	1,670,200	204,654	1,874,854
3	1,811,600	1,763,700	1,582,600	172,355	1,754,955	1,595,800	204,654	1,800,454
4	1,829,700	1,811,600	1,590,000	172,355	1,762,355	1,625,700	204,654	1,830,354
5	2,032,800	2,058,800	1,701,100	172,355	1,873,455	1,752,800	196,313	1,949,113
6	1,994,300	1,985,400	1,547,000	149,725	1,696,725	1,522,100	167,681	1,689,781
7	1,843,900	1,826,900	1,431,000	146,348	1,577,348	1,462,100	164,059	1,626,159
8	1,822,800	1,806,300	1,435,200	139,158	1,574,358	1,468,700	156,327	1,625,027
9	1,694,400	1,825,100	1,432,000	134,919	1,566,919	1,459,500	150,034	1,609,534
10	1,756,200	1,689,500	1,364,400	122,433	1,486,833	1,389,700	142,375	1,532,075

5. Summary and Conclusions

The modelling of timber supply with additional stands and short-rotation production alternatives has added a new dimension to previous planning efforts and has provided important insights for policy decisions as related to subsidies for marginal farm lands and impacts on traditional forest industries. The linking of agricultural and forestry production decisions in a model has enhanced our ability to make better resource decisions.

At this time, relatively little is known about best management guidelines that can deal with the complex requirements of wildlife, water, recreation, biodiversity, and timber production issues. This project identified specific forest lands that could be set aside from harvesting or be managed with restricted harvest options (extended rotation options, uneven-aged management etc.). The study identified existing and future supply problems and made recommendations for their mitigation. Agricultural lands for the most efficient production of biomass were identified.

References

Belcher, D.W., Holdaway, M.R. and Brand, G.J. 1982. A description of STEMS: The Stand and Tree Evaluation and Modeling System. USDA Forest Service Gen. Tech. Report NC-79. 18 p.

Berguson, W. 1994. Evaluation of land suitability and production economics of hybrid poplar. Natural Resources Research Institute. University of Minnesota. Duluth

Hansen, E.A., Ostry, M.E., Johnson, W.D., Tolsted, D.N., Netzer, D.A., .Berguson, E. and Hall, R.B. 1994. Field performance of Populus in short-rotation intensive culture plantations in the north-central U.S. Res. Pap. NC-320. St. Paul MN. U.S. Department of Agriculture. Forest Service. North Central Forest Experiment Station. 13 p.

Hoganson, H.M. and Kapple, D.C. 1991. A multi-market timber supply model. University of Minnesota Department of Forest Resources Staff Series Paper No. 83.

Hoganson, H.M. and Rose, D.W. 1984. A simulation approach for optimal timber management scheduling. For. Sci. 30(1): 220-238.

Jaakko Pöyry Consulting Inc. 1992. Maintaining productivity and the forest resource base: A technical paper for a generic environmental impact statement on timber harvesting and forest management in Minnesota. Jaakko Pöyry Consulting Inc. Tarrytown. NY. 305 p. + 15 appendices.

Kapple, D.C. and Hoganson, H.M. 1995. Close-to-market intensification of forest management. Final report to Minnesota Center for Transportation Studies. 139 p.

McDill, M. and Rose, D.W. in preparation. RxWrite users guide. Staff Paper Series. St. Paul. MN. University of Minnesota. Department of Forest Resources.

Rose, D.W. and McDill, M. 1996. Incorporating non-commodity considerations in strategic planning of natural resources. Invited paper in "Caring for the Forest: Research in a Changing World". Congress Report. Volume II of IUFRO World Forestry Congress in Tampere. Finland. Pp. 248-255.

Taff, Steven J. 1993. The CRP in the upper midwest: what should we do next? Minnesota Extension Service. University of Minnesota. St. Paul. MN.

Quantification of Forest Diversity

OPTIMIZING THE MANAGEMENT OF UNEVEN-AGED SPRUCE-FIR STANDS WHILE PRESERVING STRUCTURAL DIVERSITY

JEFFREY H. GOVE
USDA Forest Service
Northeastern Forest Experiment Station
P. O. Box 640
Durham, NH USA 03824

Introduction

The National Forest Management Act of 1976 (NFMA) final rule for the National Forest System of the United States contains the following mandate for forest managers:

> Management prescriptions, where appropriate and to the extent practicable, shall preserve and enhance the diversity of plant and animal communities, including endemic and desirable naturalized plant and animal species, so that it is at least as great as that which would be expected in a natural forest and the diversity of tree species similar to that existing in the planning area (NFMA, §219.27(g) p. 43051–43052).

The phrases "...shall preserve and enhance..." and "... so that it is at least as great..." connote the idea of diversity maintenance and present a motivation for this study. Specifically, with NFMA in mind, the question to be addressed is how to determine an optimal sustainable diameter distribution for an uneven-aged stand while simultaneously maintaining diversity.

To answer this question, the type of diversity that is to be maintained must be clearly defined. When discussing diversity or the omnibus concept of biodiversity, it is common to think in terms of species diversity. Yet the diversity of plant and animal species is not the only type of diversity that is important to the community; indeed, species diversity may be very dependent upon other components of community diversity, viz. genetic and structural diversity. A well-known illustration of this is found in the classic studies on bird-species diversity and foliage-height diversity (MacArthur and MacArthur 1961, MacArthur et al. 1962, MacArthur 1964). These authors found that bird-species diversity was very dependent on the diversity of the vertical vegetative profile and that adding information on plant-species diversity after foliage-height diversity was accounted for contributed no new information on bird-species diversity.

In this study, structural diversity within a forest stand is of primary interest; structural diversity is taken to be composed of two components: diameter-class and foliage-height diversity. Diameter-class diversity encompasses the horizontal aspect of structural diversity and can be thought of as being linked to spatial arrangement. Similarly, the vertical component of structural diversity can be thought of as being largely made up of

P. Bachmann et al. (eds.), Assessment of Biodiversity for Improved Forest Planning, 123–134.
European Forest Institute Proceedings no. 18. © 1998 *Kluwer Academic Publishers. Printed in the Netherlands.*

foliage-height diversity. As mentioned, foliage-height diversity has been found to be an efficient predictor for bird-species diversity. Thus, it can be thought of as a surrogate, or indicator, for bird-species diversity, which is more difficult to measure. Therefore, the dual result follows that maintaining foliage-height diversity should maintain bird-species diversity as well.

Gove et al. (1995) showed how constraints can be added to a commonly used optimization model of uneven-aged stand structure first proposed by Adams and Ek (1974) to maintain diameter-class diversity. The resulting mathematical program determined the optimal diameter distribution that maximizes the desired constrained objective function and ensures that diversity is »at least as great« in the optimal stand as it was in some target stand. That model is extended here to encompass the vertical component; tradeoffs between diversity maintenance and economics are discussed.

In keeping with the spirit of NFMA as quoted above, the »target« condition or stand is defined as any stand in the planning area whose current structural diversity may be used as a goal for management practices on other nearby stands. Therefore, it seems logical that if the stand before cutting (antecedent condition) is representative of the natural condition to be preserved, it may be used as the target; if not, an undisturbed neighboring stand deemed to represent the natural condition may be used. As in Gove et al. (1995), the antecedent stand is used in this study as the target condition.

1. Stand-level Optimization Model

The generic Adams and Ek (1974) paradigm yields a diameter distribution for an uneven-aged stand which maximizes some physical stocking, growth, or economic quantity subject to certain constraints of similar type. The economic formulation used here is from Adams (1976), which yields investment-efficient stocking guides at optimality. The formulation is given in the following

$$\begin{aligned} &\underset{\{\gamma,\beta,\tau\}}{Max} && LEV \text{ or } BMS \\ &\text{St:} && N_j(t) - N_j(t-1) \geq 0,\ j = 1,\ldots,s_d + 1 \\ & && BA = PSL\ \{\text{optional}\} \\ & && \gamma,\beta > 0, \tau \geq 0 \end{aligned} \tag{1}$$

$$T_j(\underline{\pi}^{\#}) - T_j(\underline{\varphi}^{\#}) \geq 0,\ \text{j} = 1,\ldots,\text{s}_\text{d}\text{-}1 \tag{2}$$

$$T_j(\underline{\tilde{\pi}}^{\#}) - T_j(\underline{\tilde{\varphi}}^{\#}) \geq 0,\ \text{j} = 1,\ldots,\text{s}_\text{h}\text{-}1 \tag{3}$$

where

s_d = The number of diameter classes; $s_d = 9$ in this study. The diameter-class width used was approximately 5 cm, with a minimum diameter of 1.27 cm.

s_h = The number of vertical height strata; $s_h = 3$ in this study as described in the section that follows.

$N_j(t-1)$ = The number of trees in diameter class j in the optimal stand at the beginning of the 5-year growth period; given the optimal values for γ, β, and τ, $N_j(t-1)$ is found from the following truncated two-parameter Weibull model

$$\frac{\tau\left(\exp\left(-\left(DBHL_j / \beta\right)^{\gamma}\right)-\exp\left(-\left(DBHU_j / \beta\right)^{\gamma}\right)\right)}{\left(1-\exp\left(-\left(DBHMAX / \beta\right)^{\gamma}\right)\right)}$$

$N_j(t)$ = The number of trees in diameter class j in the optimal stand at the end of the 5-year growth period given the growth dynamics predicted by the growth model.

$DBHL_j$ = The adjusted lower limit for the jth diameter class.

$DBHU_j$ = The adjusted upper limit for the jth diameter class.

DBHMAX = The adjusted maximum allowable diameter class in the optimal distribution ($DBHMAX \approx 47.0$ cm in this study). Note that the diameters $DBHL_j$, $DBHU_j$, and DBHMAX are adjusted so that $DBHL_1$ =0 and $DBHU_1$ =6.35 by shifting the diameter distribution to the left 1.27 cm. Thus, ($DBHL_2$, $DBHU_2$) = (6.35,11.43)....

BA = The total basal area per hectare in the optimal stand (taken at the midpoint of the 5-cm diameter classes).

τ = The total number of trees per hectare.

γ , β = The Weibull probability density function shape and scale parameters, respectively.

$LEV = VG / \left((1+arr)^{cc}-1\right) - VGS$ = The land expectation value for the optimal stand in U.S. dollars.

$VGS = \sum_{j=1}^{s} V_j N_j(t-1)$ = The value of initial growing stock.

$VG = \sum_{j=1}^{s+1} V_j N_j(t) - VGS$ = The 5-year value growth.

cc = The cutting cycle (time between harvests) length in years (cc = 5 in this study).

V_j = The individual-tree value in dollars for the jth diameter class.

arr = The alternative rate of return (5% in this study).

PSL = Some target physical stocking level of basal area per hectare.

BMS = Total above ground biomass fresh weight in kg per hectare.

The notation presented describes all of the model components except those in constraint sets (2) and (3), which are described in detail elsewhere. Constraints (1) are known as the sustainability constraints (Adams and Ek, 1974) and simply ensure that the optimal diameter distribution can be sustained in the steady-state condition since the number of trees in any diameter class at the end of the 5-year growing cycle must be at least as great as the number at the beginning. If these constraints fail to be met, given enough cutting cycles, the distribution eventually would degenerate under the steady-state assumption.

The optimization model without constraint sets (2) and (3) included will be termed Model I.

Including equation set (2) in Model I ensures that diameter-class diversity will be at least as great in the optimal stand as in the antecedent stand. Some explanation of right tail-sum diversity profiles is required to apprehend this, further details and examples are found in Gove et al. (1995).

Assume that the antecedent community has relative abundance vector $\underline{\varphi}$ based on its 5-cm diameter distribution. The ranked relative abundance vector $\underline{\varphi}^{\#}$ is constructed by ranking the components of $\underline{\varphi}$ in descending order to obtain $\underline{\varphi}^{\#} = (\varphi_1^{\#},\ldots,\varphi_s^{\#})$ where $\varphi_1^{\#} \geq \ldots \geq \varphi_s^{\#}$. Similarly, let the relative abundance vector for the optimal stand be given by $\underline{\pi}$; the individual components of $\underline{\pi}$ are given by $\pi_j = N_j(t-1)/\tau$, where $\tau \equiv \sum_{j=1}^{s_d} N_j(t-1)$. It follows that the ranked relative abundance vector for the optimal stand is $\underline{\pi}^{\#}$.

The right tail-sum (T_j) or intrinsic diversity profiles of the relative abundance vector have been defined by Patil and Taillie (1979a) with further theory found in Patil and Taillie (1982). These profiles are defined for the antecedent and optimal stand-diameter distributions as

$$T_j(\underline{\varphi}^{\#}) = \sum_{i=j+1}^{s_d} \varphi_i^{\#}, \quad j = 1,\ldots,s_d-1$$

and

$$T_j(\underline{\pi}^{\#}) = \sum_{i=j+1}^{s_d} \pi_i^{\#}, \quad j = 1,\ldots,s_d-1$$

respectively, where $T_0(\underline{\varphi}^{\#}) = T_0(\underline{\pi}^{\#}) = 1$ and $T_{s_d}(\underline{\varphi}^{\#}) = T_{s_d}(\underline{\pi}^{\#}) = 0$. Having computed the right tail-sums for this pair of "communities," any intrinsic diversity ordering of the communities, if they exist, can be determined using the following defintion.

In general, given communities C_1 and C_2 with ranked relative abundance vectors $\underline{\varphi}^{\#}$ and $\underline{\pi}^{\#}$, respectively, a comparative plot of the (j, T_j) pairs yields an intrinsic diversity profile for each community (Patil and Taillie1979b; Gove et al., 1994, 1995). If the profile for $T_j(C_1)$ lies everywhere above the profile for $T_j(C_2)$, then community C_1 is intrinsically more diverse than community C_2 (written $C_1 \overset{imd}{>} C_2$). This graphical concept can be expressed mathematically: if all of the right tail-sum differences $d_j = T_j(C_1) - T_j(C_2), j = 1,\ldots,s-1$ between communities C_1 and C_2 are nonnegative, then it also can be inferred that $C_1 \overset{imd}{>} C_2$. If the profiles for the two communities cross at any point, or equivalently, if there is a change in sign in the right tail-sum differences, then the two communities are not comparable in an intrinsic sense.

Employing this definition for the antecedent and optimal stand-diameter distributions (communities), we immediately find that

$$d_j = T_j(\underline{\pi}^{\#}) - T_j(\underline{\varphi}^{\#}), j = 1,\ldots,s_d - 1 \tag{4}$$

yields the right tail-sum differences between the two stands. Therefore, if it is desired that the optimal stand-diameter distribution be required to be at least as diverse (diversity maintenance) as the antecedent stand-diameter distribution, then (4) is rewritten as an inequality such that the d_j's are all greater than or equal to zero. This leads immediately to constraint set (2), which assures that right tail-sum profile for the optimal stand is everywhere equal to or above that for the antecedent stand. Thus, the optimal stand is required through constraint set (2) to be at least as diverse as the antecedent stand, and may, in fact, be intrinsically more diverse than the antecedent stand. Including constraint set (2) in the Model I formulation will be termed Model II.

The same argument can be developed for constraint set (3). However, in this case, interest lies in the vertical component of structural diversity. MacArthur and MacArthur (1961) measured foliage density for stands studied and experimentally determined the height strata divisions for calculating foliage-height diversity that best accounted for the variation in bird-species diversity. They found that groupings of 0–0.61 m, 0.61–7.62 m, 7.62 m+, or 0–0.91 m, 0.91–9.14 m, 9.14 m+ worked equally well. Inasmuch as these authors' data spanned stands over a great geographical extent and numerous forest types, it seems reasonable to use their height divisions for calculating foliage-height diversity as a surrogate for bird-species diversity in this study. The divisions adopted here are the latter set (0–0.91 m, 0.91–9.14 m, 9.14 m+). Predicted fresh foliar biomass in kg for all trees taller than 0.15 m was partitioned into each of the these three vertical strata and used to calculate relative abundances. Let $\underline{\tilde{\varphi}}$ and $\underline{\tilde{\varphi}}^{\#}$ be the relative and ranked relative abundance vectors for the foliage height distribution in the antecedent stand. Similarly, let $\underline{\tilde{\pi}}$ and $\underline{\tilde{\pi}}^{\#}$ be the corresponding vectors for the optimal stand. Then with $s_h = 3$ vertical height strata, constraint set (3) assures that the foliage-height distribution for the optimal stand will be at least as diverse as in the antecedent stand. Including constraint set (3) in the Model II formulation will be termed Model III.

It is not obvious from the previous discussion why the intrinsic diversity profiles were used rather than simply using one or several of the more common diversity indices as constraints for maintaining diversity. The answer has its origins in Hurlbert (1971), who pointed out that the popular diversity indices can rank communities inconsistently. Examples of this inconsistent ordering also are given in Patil and Taillie (1982) and Gove et al. (1994). This has led to the development of numerous diversity profile approaches in an effort to quantify consistent (intrinsic) rankings among communities if they exist. For example, the Δ_β profiles of Patil and Taillie (1979a, 1982) incorporate infinitely many diversity indices as point estimates along the profiles. Specific point estimates include the popular species count (Δ_{-1}), Shannon (Δ_0), and Simpson's (Δ_1) indices. If the Δ_β profiles cross between $\beta = 0$ and $\beta = 1$, for example, then Shannon's and Simpson's indices inconsistently order the two communities being compared. The link between right tail-sum and Δ_β profiles is that if there is an intrinsic ordering between any two communities, then the Δ_β profiles will not cross. If there is no intrinsic ordering then the Δ_β profile will cross somewhere. However, since $-1 \le \beta < \infty$, the profiles may cross far out in the tail (β large) rather than in the "popular" range $-1 \le \beta \le 1$. Therefore, using even one or all of the more popular diversity indices (species count, Shannon, Simpson) as constraints to maintain diversity will not ensure intrinsic diversity ordering; as the Δ_β profiles illustrate, infinitely many constraints would be required to accomplish this same objective.

2. Stand-level Growth Model

The optimization model requires an underlying growth model to calculate the future stand 5 years hence. The data to develop the model come from remeasurements of permanent plots on the Penobscot Experimental Forest (PEF) located in the townships of Bradley and Eddington, Maine USA. Two compartments, or stands (9 and 16) on the PEF have been managed by intensive single-tree selection system since 1957. The cutting cycle for these compartments is 5 years with measurements taken pre- and postharvest and a residual basal area goal of 23 $m^2 ha^{-1}$ and a 5-cm q value (Davis and Johnson, 1987 p. 58) of 2.1. The species composition is 85–90 percent softwoods and 10–15 percent hardwoods. Softwood species on these compartments include eastern hemlock (Tsuga canadensis (L.) Carr.), red spruce (Picea rubens Sarg.), white spruce (P. glauca (Moench) Voss), balsam fir (Abies balsamea (L.) Mill.), northern white cedar (Thuja occidentalis L.), and eastern white pine (Pinus strobus L.). The most common hardwood species are red maple (Acer rubrum L.) and paper birch (Betula papyrifera Marsh.), with small numbers of other northern hardwood species scattered throughout. A detailed description of these and other compartments on the PEF is found in Brissette (1996).

A whole-stand, distance-independent growth model for the 1977–81 cutting cycle in compartment 16 (6.6 ha) was developed from data taken on 20 remeasured concentric growth plots. All trees 11.43 cm dbh and larger were measured on 0.08-ha circular plots; all trees with $1.27\text{cm} \le \text{dbh} < 11.43\text{cm}$ were measured on 0.02-ha circular plots. The growth dynamics for the stand are assumed to follow the familiar accounting

$$N_j(t) = N_j(t-1) + I(t) - U_j(t) - M_j(t) \quad \text{, for } j = 1$$

$$N_j(t) = N_j(t-1) + U_{j-1}(t) - U_j(t) - M_j(t) \quad \text{,for } j = 2,\ldots,s_d \text{ and}$$

$$N_{s_d+1}(t) = U_{s_d}(t) \quad \text{,for the last diameter class,}$$

where

$I(t) =$ Ingrowth: the number of trees per hectare growing into the 5-cm diameter class during the growth period.

$U_j(t) =$ Upgrowth: the number of trees per hectare growing from diameter class j to $j+1$ during the growth period.

$M_j(t) =$ Mortality: the number of trees dying in diameter class j during the growth period.

The following equations were fit to the growth data to predict the required ingrowth, upgrowth, and mortality quantities

$$\hat{I}(t) = -3243.5230 + 14805.5160\widehat{MSD}(t-1)^{-1} + 14184.8246\left(\log_e\left(\widehat{BA}(t-1)\right)\right)^{-1}$$

$$\hat{U}_j(t) = \hat{D}_j^{0.477678} \times \hat{N}_j(t-1)^{0.725770} \times \widehat{MSD}(t-1)^{-1},\ j = 1,\ldots,s_d$$

$$\hat{M}_j(t) = \hat{N}_j(t-1)^{1.90044} \times \exp\left(-4.67241\hat{D}_j\right),\ j = 1,\ldots,s_d$$

where D_j is the midpoint of the jth diameter class and MSD(t-1) is the quadratic mean stand diameter at the beginning of the 5-year growth period.[1]

For all trees smaller than 1.27 cm dbh, only count data were available by $n_h = 4$ height classes (0.15–0.30 m, 0.30–0.76 m, 0.76–1.37 m and 1.37 m–1.27 cm dbh) taken on three 4.05-m^2 circular plots laid out equidistant on the boundary of the 0.02-ha plots. Poisson regression using generalized linear models (McCullagh and Nelder, 1989) was used to fit the following model to predict the number of stems ha^{-1}/1000 in the kth height class ($N_k'(t-1)$; i.e., $N_k(t-1) = N_k'(t-1) \times 1000$) at the beginning of the growth period

$$\hat{N}_k'(t-1) = 2.2650 - 0.3412\Im_2 - 1.0164\Im_3 - 2.0714\Im_4 + 0.00015\hat{N}_{k+1}(t-1) - 0.0117\widehat{BA}(t-1).$$

In this model, the $\Im_k, k = 2,\ldots,4$ are 0,1 indicator variables denoting the kth height class and $N_{k+1}(t-1)$ is the number of trees ha^{-1} in the next highest height or diameter class. For example, if $k = 4$ then the $k+1$ is the 5-cm diameter class. This same equation is used to predict $N_k(t)$ using $BA(t)$ and $N_{k+1}(t)$ after the 5-year growth equations have been applied.

The alternative objective function of total above ground biomass in kg fresh weight now can be calculated based on Young et al.'s (1980) equations for red spruce

$$BMS = 1000^{-1}\sum_{k=1}^{n_h} N_k(t-1) \times B_k(t-1) + 2.2046^{-1}\sum_{j=1}^{s_d} N_j(t-1) \times B_j(t-1)$$

where $B_k(t-1)$ and $B_j(t-1)$ are the predicted above-ground biomass for the $n_h = 4$ height classes in grams and $s_d = 9$ diameter classes in pounds, respectively.

These equations satisfy the requirements for Model II. However, to use Model III, additional information is required on the vertical stand components. Equations to predict total tree height (H), height to the base of the live crown (h), and crown radius (r) were fit to data taken from several stands on the PEF

$$\hat{H} = \exp\left(4.4621 - 4.6025\hat{D}_j^{-1}\right)$$
$$\hat{h} = 0.3653\hat{H}$$
$$\hat{r} = 4.49174 + 0.67958\hat{D}_j - 0.12569\hat{H} + 0.09423\left(\hat{H} - \hat{h}\right)$$

These equations were fitted for data measured on trees larger than 11.43 cm dbh. As no data were available on trees smaller than 11.43 cm, these equations were used to extrapolate H, h, and r to the 5-cm class except that a lower limit of 1.82 m was placed on total tree height; also, for all trees in the 5-cm class, r = 0.25(H - h) was used. This latter relationship was developed from a regression through the origin on crown widths for larger trees. In lieu of data on crown length and width for the four regeneration height classes, a crown length of $0.9H_k$ was used, where H_k is the midpoint height for the kth height class. The same relationship was used for crown radius in all four height classes as in the 5-cm dbh class.

A simple cone was used to model crown shape for all trees. This seemed reasonable as most of the species were spruce or fir and have approximately conical crown shapes. Foliar biomass (including leaves and branches) in kg fresh weight was calculated for all trees from equations developed by Young and Carpenter (1967) for red spruce. The foliar biomass was assumed to be distributed proportional to the volume of the crown. Biomass

1 These formulas were fitted based on the English system of units; therefore, all hatted quantities are with respect to this system—the predicted values can be converted to metric in standard fashion.

was allocated to each of the $s_h = 3$ height strata by determining the proportion of crown volume falling within each of these strata and apportioning biomass, respectively. This latter calculation for the vertical distribution of crown biomass clarifies why it is required to model regeneration as well as trees of sapling and larger sizes classes. With the allocation of crown biomass into the vertical height distribution, the components for Model III are complete.

3. Results

Each of the three optimization models discussed in the previous section were solved with a particular goal in mind. That goal was to come as close as possible to the residual basal area recommendations of the current guide for uneven-aged spruce-fir (Frank and Bjorkbon, 1973). The guide calls for a 5-year cutting cycle with a residual basal area of 27.6 m^2 ha^{-1}, a 2.5-cm q of approximately 1.3, and 2.47 trees ha^{-1} in the largest (48.3 cm) diameter class. The stand supports a total of 1,852.5 trees ha^{-1} comprising approximately 243.8 t ha^{-1} of above-ground biomass and a LEV of -\$993.32 ha^{-1}. This is a multiproduct, whole-stand guide. Because of the constraints imposed on the different models, it was not possible to find feasible solutions at the desired residual basal area in all cases. For Models II and III, the antecedent stand enters through the $T_j(\varphi^{\#})$ and $T_j(\tilde{\varphi}^{\#})$ profiles but it does not enter into the Model I formulation. The antecedent stand used is simply the diameter distribution for Compartment 16 on the PEF at the beginning of the cutting cycle for which the growth equations were developed (Table 1). All optimal models presented in this section have been constrainted to follow a reverse J-shaped or q distribution by setting the Weibull shape parameter equal to one (actually one was the lower bound and it never moved from this in any of the solutions). GRG2 (Lasdon and Waren 1986) was used to find solutions to all of the models.

The Model I optimal sustainable diameter distributions for Compartment 16 based on the 5-year period for which the growth equations were developed are shown in Table 1. The most striking result is that for the LEV solution, the optimal distribution has a much higher q value than the antecedent stand. This is a result of the interplay between economics (the objective function) and biology (the growth equations through the sustainability constraints). The objective function can be written as (Martin 1982): LEV = VG - HC, where HC are the holding costs associated with the residual (optimal) diameter distribution. The holding costs assign a penalty to the objective:If they are larger than the value of the 5-year growth, then the land expectation value will be negative. Therefore, a solution is sought that effectively minimizes the holding costs while still satisfying the sustainability constraints. Minimizing the holding costs basically translates to minimizing the number of higher value, larger diameter trees in the stand. This is exactly what has been done in the optimal economic solution. The solution puts as many trees into the smaller diameter classes as possible while still meeting the basal-area and sustainability constraints. In this solution, there are effectively no trees in the optimal stand larger than 40 cm.

In the case of the physical stocking objective, the trend is essentially the opposite. To maximize biomass, it is logical to include as many larger trees in the stand as possible, again, while still satisfying the sustainability and basal-area constraints. In this distribution there is a pronounced decrease in the q factor and an associated increase in residual trees in the larger diameter classes. Indeed, there are approximately 41 trees in the 35-cm or greater class compared with only 5 in the maximal LEV stand.

TABLE 1. Comparison of antecedent stand and Model I optimal stands.

Diameter class (cm)	j	Antecedent[a] $N_j(t-1)$	Max LEV $N_j(t-1)$	Max LEV d_j^b	Max BMS $N_j(t-1)$	Max BMS d_j^b	Value per tree[c]
5[d]	1	1403.5	4022.3	-0.069	1162.0	0.147	0.0
10	2	268.1	895.5	-0.106	465.0	0.064	0.17
15	3	140.2	294.6	-0.093	253.2	0.016	0.60
20	4	79.1	96.9	-0.073	137.8	-0.009	1.32
25	5	59.3	31.9	-0.050	75.0	-0.015	2.43
30	6	57.4	10.5	-0.024	40.8	-0.060	5.63
35	7	40.8	3.4	-0.005	22.2	0.004	8.62
40	8	9.9	1.1	0.0	12.1	0.003	11.86
45	9	0.6	0.4	0.0	6.6	0.0	15.57
Total trees (ha^{-1})		2058.9	5356.6		2174.7		
BA (m^2 ha^{-1})		22.9	26.8		26.8		
LEV (\$ ha^{-1})		-389.6	283.2		-365.0		
BMS (t ha^{-1})		190.5	187.4		223.9		
q (5-cm)		2.1[e]	3.0		1.8		
Weibull shape (γ)			1.0		1.0		
Weibull scale (β)			1.8		3.3		

[a] Compartment 16, PEF, 1977 postharvest.
[b] $d_j = T_j(\pi^{\#}) = T_j(\varphi^{\#})$.
[c] Tree values in U.S. dollars are from 1993 sale information on the PEF; value for the 50 cm dbh class is \$20.94.
[d] The 5 cm dbh class includes trees 1.27 cm to 7.62 cm.
[e] Fitted with least squares (Leak 1963).

The biomass has increased in this solution by nearly 40 metric tons per hectare. However, the cost of carrying this extra biomass in large trees (through the holding costs) is borne out in the negative LEV. This interplay between economics and biology is typical and will become even more pronounced when the diversity constraints have been included in the model.

No diversity constraints were included in the Model I solutions, but comparisons of intrinsic diversity still can be made. The right tail-sum profile differences (4) are included in Table 1 for both solutions. All of these differences are negative for the optimal LEV stand, implying that the antecedent stand-diameter distribution is $\overset{imd}{>}$ than that for the optimal stand. However, the d_j profile changes sign for comparison with the optimal BMS stand, implying that there is no intrinsic ordering with respect to horizontal diversity between these two stands. The vertical profiles are shown in Table 2; in both cases, the antecedent stand is $\overset{imd}{>}$ than the optimal stands.

The results for Model II were nearly identical for both objective functions, with the greatest basal area attainable in the optimal stand at 26.2 m^2 ha^{-1}. The cost of adding diameter-class diversity constraints to the model are shown in Table 3. In general, for a given number of "species," the community of maximal diversity is the completely even community where all species are equally represented. Therefore, when diversity maintenance is a goal in the model, constraints (2) will force the optimal stand to have a relative abundance vector that is as even, or more so, than the antecedent stand while still satisfying the other constraints. Comparison of Model II results with those for either of the Model I solutions shows that adding diameter-class diversity constraints has led to a substantial evening of the diameter distribution. As a result, there are even more trees in

TABLE 2. Relative abundances and d_j profiles for the vertical height distribution.

Height class (m)	j	Ante-cedent[a]	LEV	BMS	Model II LEV[b]	LEV	BMS
				Relative Abundance			
0.0–0.91	–	0.0754	0.0546	0.0463	0.0482	0.0762	0.0754
0.91–9.14	–	0.6228	0.8498	0.6830	0.6499	0.6227	0.5310
9.14+	–	0.3018	0.0956	0.2707	0.3019	0.3011	0.3936
				d_j Profile			
–	1	–	-0.227	-0.060	-0.027	0.0	0.092
–	2	–	-0.021	-0.029	-0.027	0.0	0.0
–	3	–	0.0	0.0	0.0	0.0	0.0

[a] Compartment 16, PEF, 1977 postharvest.
[b] Maximal BMS results were identical at this basal area ha^{-1}

the larger diameter classes in this stand than in the previous solutions (direct comparison is hindered because of the differing residual basal areas, but relative comparisons remain valid). In addition, the d_j profile for Model II shows all positive values, indicating that the constraints have reversed the intrinsic ordering for diameter-class diversity found in the Model I solution; now the optimal stands are $\overset{imd}{>}$ than the antecedent stand. The d_j profiles for the vertical height strata (Table 2) still order the antecedent stand $\overset{imd}{>}$ than the optimal stands.

Adding the vertical diversity constraints in the Model III results (Table 3) shows that the evening effect becomes even more pronounced. In addition, this vertical diversity maintenance inflicts higher cost on the residual basal area with maximum feasible value at approximately 21 m^2ha^{-1} for both objectives. As with Model I, the optimal BMS stand is the most even and, again, this is reflected in the lower LEV. The d_j profiles for the foliage-height diversity have now become positive for both solutions, implying that the optimal stands are $\overset{imd}{>}$ than the antecedent in the vertical component as well.

4. Discussion and Concluding Remarks

Three different optimization models have been presented that illustrate the conflict between diversity maintenance and long-run economic criteria when the latter are used in the objective function. Three additional models are presented in which the physical stocking objective has replaced the economic one, and economics has been rendered ineffectual with regard to the model solution. In all six optimal stands, only the Model I economic objective gave a positive return. This model is unencumbered by the equalizing effect imposed by both the diversity constraints and the alternative objective. In addition, even the maximal BMS stand in Model I was better economically than either the antecedent stand or the spruce-fir guide, though it still had a negative return.

Imposing diversity constraints of either type causes an evening of the optimal diameter distribution and an associated increase in the holding costs. The »hidden« costs associated with the diversity maintenance constraints are that only lower residual basal area stocking levels can be supported while still meeting the sustainability constraints. Thus, diversity is at odds with economics and sustainability is at odds with both. In most cases, this is further compounded by trying to maximize above-ground biomass as well.

TABLE 3. Optimal stand results for Models II and III.

		Model II		Model III			
		Max LEV[a]		Max LEV		Max BMS	
Diameter class (cm)	j	$N_j(t-1)$	d_j	$N_j(t-1)$	d_j	$N_j(t-1)$	d_j
5[b]	1	878.3	0.188	731.2	0.200	250.1	0.344
10	2	380.8	0.105	324.1	0.117	141.5	0.283
15	3	222.3	0.048	192.8	0.058	104.2	0.211
20	4	129.7	0.014	114.7	0.021	76.8	0.145
25	5	75.7	0.0	68.2	0.005	56.6	0.098
30	6	44.2	0.003	40.6	0.006	41.6	0.070
35	7	25.8	0.008	24.1	0.010	30.7	0.048
40	8	15.1	0.005	14.4	0.005	22.6	0.022
45	9	8.8	0.0	8.5	0.0	16.6	0.0
Total trees (ha^{-1})		1780.7		1518.6		740.8	
BA (m^2 ha^{-1})		26.2		21.1		20.7	
LEV ($ ha^{-1})		-493.5		-391.5		-773.2	
BMS (t ha^{-1})224.1		184.1		194.0			
q (5-cm)	1.7	1.7		1.4			
Weibull shape (γ)		1.0		1.0		1.0	
Weibull scale (β)		3.7		3.8		6.5	

[a]Maximal BMS results were almost identical at this basal area ha^{-1}.
[b]The 5 cm dbh class includes trees 1.27 cm to 7.62 cm.

The use of intrinsic diversity profiles for diversity maintenance has been extended from previous work to the vertical dimension. If the work of MacArthur and MacArthur (1961) is correct, the addition of constraints allowing the forest manager to maintain foliage-height diversity, also will ensure the maintenance of bird-species diversity. There are two caveats, however. First, MacArthur and MacArthur (1961) considered all vegetation in their vertical profiles where only trees are considered here. It is possible that ground-cover layers also play an important part in habitat selection by birds. However, modeling these vegetative types in a practical way is problematic and probably would add little to the biomass of the lower stratum relative to the majority, which is consistently found in the two upper strata (Table 2). Second, certain warbler species in spruce-fir forests are known to be niche partitioners (MacArthur, 1958). This activity may confound the foliage-height and bird-species diversity correlation somewhat in spruce-fir forests (MacArthur and MacArthur, 1961), but generally will not prevent these results from being applied to other forest types. In any case, the foliage-height diversity of a stand is more easily estimated or measured than bird-species diversity, and its use as a surrogate is secure.

It should be remembered that the results of this study are for a specific managed spruce-fir stand. The growth model, prices, and other predictive components, as well as the antecedent condition are stand specific. Therefore, although the modeling approach itself is quite general given the required data, the results should not be extrapolated loosely even to other spruce-fir stands. Finally, these models also fit into the stochastic optimization framework developed by Gove and Fairweather (1992) and could be used to develop a distribution of possible optimal stands.

References

Adams, D.M. (1976) A note on the interdependence of stand structure and best stocking in a selection forest, Forest Science 22, 180–184.

Adams, D.M. and Ek, A.R. (1974) Optimizing the management of uneven-aged forest stands, Canadian Journal of Forest Research 4, 274–286.

Brissette, J.C. (1996) Effects of intensity and frequency of harvesting on abundance, stocking and composition of natural regeneration in the Acadian forest of Eastern North America, Silva Fennica 30, 301–314.

Davis, L.S. and Johnson, K.N. (1987) Forest management. McGraw-Hill, New York.

Frank, R.M. and Bjorkbom, J.C. (1973) A silvicultural guide for spruce-fir in the Norheast. USDA Forest Service General Technical Report NE-6, Radnor, PA

Gove, J.H., and Fairweather, S.E. (1992) Optimizing the management of uneven-aged forest stands: A stochastic approach, Forest Science 38, 623–640.

Gove, J.H., Patil, G.P., Swindel, B.F., and Taillie, C. (1994) Ecological diversity and forest management, in G.P Patil and C.R. Rao (eds.) Handbook of Statistics, Volume 12: Environmental Statistics, Elsevier Science, B.V., North-Holland, New York, pp.409–462.

Gove, J.H., Patil, G.P., and Taillie, C. (1995) A mathematical programming model for maintaining structural diversity in uneven-aged forest stands with implications to other formulations, Ecological Modelling 79, 11–19.

Hurlbert, S.H. (1971) The nonconcept of species diversity: a critique and alternative parameters, Ecology 52, 577–586.

Lasdon, L.S., and Waren, A.D. (1986) GRG2 user's guide. Department of General Business, School of Business Administration, Univ. Texas at Austin.

Leak, W.B. (1963) Calculation of q by the least squares method, Journal of Forestry 61, 227–228.

MacArthur, R.H. (1958) Population ecology of some warblers of northeastern coniferous forests, Ecology 39, 599–619.

MacArthur, R.H. (1964) Environmental factors affecting bird species diversity, American Naturalist 98, 387–397.

MacArthur, R.H. and MacArthur J.W. (1961) On bird species diversity, Ecology 42, 594–598.

MacArthur, R.H., MacArthur, J.W., and Preer, J. (1962) On bird species diversity. II. Prediction of bird census from habitat measurements. American Nataturalist 96, 167–174.

Martin, G.L. (1982) Investment-efficient stocking guides for all-aged northern hardwood forests. University of Wisconsin, Madison, Agricultural & Life Sciences Research Report R3129.

McCullagh, P. and Nelder, J.A. (1989). Generalized linear models. Chapman and Hall, NY.

National Forest Management Act of 1976.} (1982) National forest system land and resource management planning, Federal Register 47, 43026–43052.

Patil, G.P. and Taillie C. (1979a) An overview of diversity, in J.F. Grassle, G.P. Patil, W. Smith, and C. Taillie (eds), Ecological diversity in theory and practice, International Cooperative Publishing House, Fairland, MD, 3–27.

Patil, G.P., and Taillie, C. (1979b) A study of diversity profiles and orderings for a bird community in the vicinity of Colstrip, Montana, in G.P. Patil, and M.L. Rosenzweig (eds). Contemporary quantitative ecology and related ecometrics, International Cooperative Publishing House, Fairland, MD, 23–48.

Patil, G.P., and Taillie, C. (1982) Diversity as a concept and ist measurement, Journal or the American Statistical Association 77, 548–567.

Young, H.E. and Carpenter, P.M. (1967) Weight, nutrient element and productivity studies of seedlings and saplings of eight tree species in natural ecosystems. Maine Agricultural Experiment Station, Orono, Maine, Technical Bulletin 28.

Young, H.E., Ribe, J.H., and Wainwright,K. (1980) Weight tables for tree and shrub species in Maine. Maine Agricultural Experiment Station, Orono, Maine, Miscellaneous Report 230.

MULTISCALE ANALYSIS OF THE SPATIAL DISTRIBUTION OF BREEDING BIRD SPECIES RICHNESS USING THE ECHELON APPROACH

GLEN JOHNSON[1], WAYNE MYERS[2],
GANAPATI PATIL[1], AND DAVID WALRATH[2]
[1]Center for Statistical Ecology and Environmental Statistics, Department of Statistics
[2]School of Forest Resources and Environmental Resources Research Institute
Penn State University, University Park, PA 16802 USA

Abstract

As an important component of biodiversity monitoring, species richness is a response variable of considerable interest to conservation biologists. Monitoring over a large spatial extent (such as statewide) requires an objective method for characterizing the spatial distribution of species richness. Objectivity is needed for delineation of areas that are relatively species rich or poor, and to also provide a way to compare other spatial distributions, as with temporal monitoring for change detection. The echelon approach introduced by Myers, Patil and Taillie (1995) provides a way to objectively characterize spatial structure of response variables that are at least ordinal. We apply this method to the characterization of breeding bird species richness across the state of Pennsylvania, using US-EPA EMAP hexagons which constitute a very coarse grain (635 km^2). After identifying relative species rich areas at the coarse measurement scale, the hexagons covering such areas are analyzed in much finer detail, using the same echelon protocol but with a measurement scale (grain) equal to approximately 24 km^2 rectangles corresponding to one sixth of a 7.5 minute USGS topographic quadrangle.

1. Introduction

In order to minimize expansion of the endangered species list, preventative medicine is required. This includes protection of critical habitat areas, as identified from a landscape to regional perspective. In the United States, the national GAP Analysis program responds to this need by aiming at the determination of "gaps" in protection of critical habitat areas (Scott, et al., 1993). As part of the Pennsylvania Gap program, species richness mapping has been performed statewide using data that was compiled at the scale of 635 km2 hexagons that constitute a subset of the USEPA-EMAP national tessellation (Myers, Brooks, Storm and Bishop, 1995).

Following the protocol introduced by Myers, Patil and Taillie (1995), and expanded in greater detail by Myers, Patil and Joly (1996), the statewide species richness for various vertebrate groups was spatially organized into hierarchically related echelons using the EMAP hexagons as the primary mapping units. Of these different groups, bird

P. Bachmann et al. (eds.), Assessment of Biodiversity for Improved Forest Planning, 135–150.
European Forest Institute Proceedings no. 18. © 1998 *Kluwer Academic Publishers. Printed in the Netherlands.*

species were compiled from the Pennsylvania Breeding Bird Atlas (cf. Brauning, 1992) database which is based on mapping units that equal one sixth of a 7.5 minute USGS quadrangle (approximately 24 km^2).

In this paper, we use the coarse-grained (EMAP hexagon) statewide echelon map of breeding birds to identify areas of the state that exhibit regionally high species richness from a statewide perspective. We then apply echelon mapping to a selection of these areas using the much finer-grained original data from the Breeding Bird Atlas (Brauning and Gill, 1983-1989).

Results are presented with some assessment of the spatial structure of species richness using the finer resolution mapping units within those areas of regionally high species richness. A primary research objective is to decipher what echelon patterns at the finer scale may lead to different conclusions between the two measurement scales about locally species-rich areas within a regionally rich area.

2. The echelon concept

Consider a geographic area that is either systematically tessellated into equal size and shape sectors or is partitioned into irregular size sectors, such that a variable of interest has a realized value within each sector. Often, a primary interest is to identify »hot spots«, which are areas of concern due to either high or low response values.

For certain applications, a benchmark for hot spot determination may be provided by a numerical criterion such as a chemical action level or cleanup standard, or perhaps by a percentile of the data (Johnson, Nussbaum, Patil and Ross, 1995). In the absence of a numerical criterion, hot spots may be defined as areas having high or low responses "relative to their surroundings". Indeed, a hazardous waste site manager may want to identify areas of relatively high contaminant concentrations, regardless of a numerical criterion.

In other applications, such as identification of areas across a given landscape that reveal high or low biodiversity, there simply may not exist objective criteria for defining an area of concern. As an example, consider the thematic presentation of species richness in Figure 1. While human visual perception can see a pattern of relative highs and lows, the delineation of boundaries for local to regional areas of high (or low) species richness becomes a subjective exercise. The visual cues for subjectively distinguishing relative highs are influenced by the arbitrarily chosen number of greyscale categories in this particular map.

Amongst the potential applications of the echelon approach (Myers, Patil and Joly, 1996), this concept can serve to objectively delineate areas of relative high or low values of a regionalized response variable by constructing hierarchically related spatial objects from the original data. First, local peaks and plateaus of the response variable are identified amongst all sectors (regular cells or irregular polygons) of a spatial partitioning. Then, echelon objects of the first order are identified by moving outward and downward from the local peaks and plateaus until sectors are reached that yield a change in the response variable towards increasing values. Sectors that constitute such local minima in the virtual topography serve to delineate the boundaries of first order echelons. All of the sectors within each first order echelon then become a member of that echelon. This process is similarly repeated to identify higher order echelon objects which

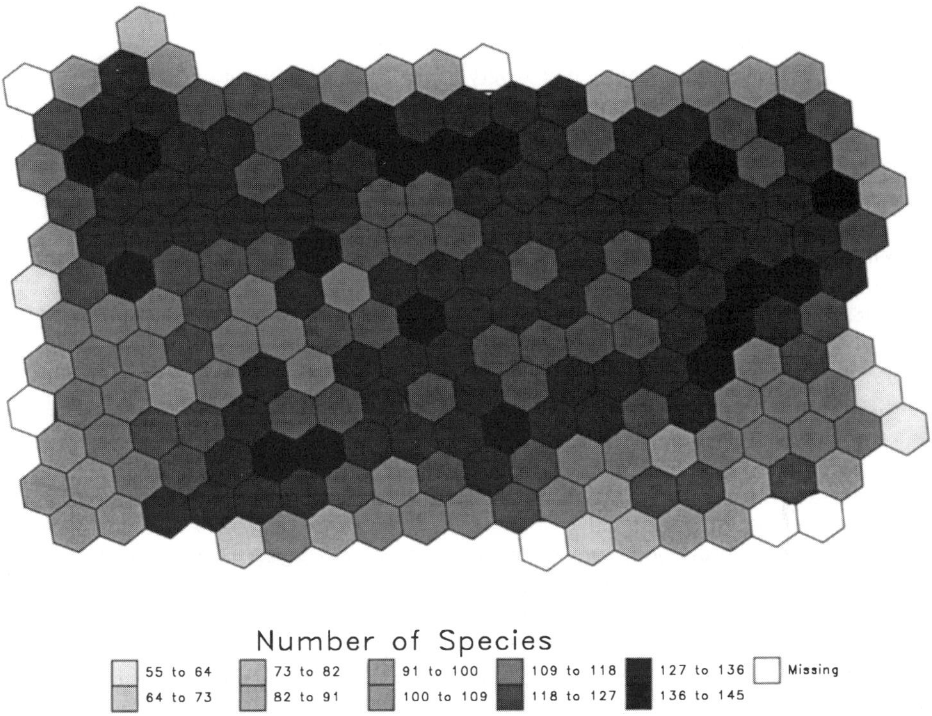

Figure 1. Bird richness in the hexagons

are founders (parents) of lower order echelon objects (children). A top-view perspective can then be mapped where each sector is assigned to a unique echelon object which in turn has an order ranging from one (1) to the highest order echelon object. This spatial structure can then be represented in a similar manner as streams are organized in a drainage basin, where first order streams have no tributaries, second order streams are formed by the confluence of first order streams, etc.. By this protocol, different observers will arrive at the same echelon structure.

The hierarchical nature of echelons further provides a means of delineating high and low areas at various spatial scales. In other words, once a large regionally high area is delineated, then sub-regional to local highs can be delineated within this region. This is especially important for large scale monitoring since environmental management is done at various scales (such as Statewide → County → Township; or Chesapeake Bay Watershed → Susquehanna → Conestoga → etc.). The purpose is to *objectively* delineate relative high/low areas at multiple scales, whereas simply viewing a thematic map or a response surface can result in multiple interpretations of relative high/low, depending on who the interpreter is.

3. Data

In support of the Pennsylvania Gap Analysis program, the Nature Conservancy (TNC) was contracted by the Environmental Protection Agency (EPA) to compile species lists within Environmental Monitoring and Assessment Program (EMAP) hexagons. Each hexagon is 635 km^2 in area. Information was provided for the major vertebrate groups, along with some invertebrate groups and trees. The most accurate group is breeding birds, since they were compiled from the Pennsylvania Breeding Bird Atlas (cf. Brauning, 1992) which is a statewide census of almost 5000 blocks, where each block is equivalent to one sixth of a 7.5 minute U.S.G.S. quadrangle map.

Species were codified according to the degree of subjective probability that a given species occurred, based on available evidence. Focusing on breeding birds for our analysis, we treated a species as present in a block or hexagon if it was listed as confirmed or probable (at least an 85% chance of breeding based on evidence).

4. Statewide Mapping with EMAP Hexagons

The statewide echelon map based on the coarse grained EMAP hexagons is presented in Figure 2. Regions of high species richness, relative to their surroundings in a statewide perspective, were delineated by complexes of first and second order echelons. General correspondence can be seen with the thematic map in Figure 1; however, the boundaries of regionally high areas in Figure 1 are not clearly defined, whereas the echelon objects of Figure 2 serve to objectively define such boundaries.

For the purpose of exploratory data analysis, we focused on the two largest complexes, one in the northwest and one in the northeast. Since we know the underlying land use is quite diverse for both of these complexes, for comparison we also chose a smaller complex that appears in the northcentral part of Pennsylvania which has more homogeneous land use, consisting largely of forest interior with some agricultural land.

5. Regional Mapping with Atlas Blocks

The regional areas that were further analyzed are presented in isolation from the remaining state, along with the finer resolution echelon maps within these areas, in Figures 3 to 8.

In all three regions, the finer scale mapping revealed a complexity of surface response that could not necessarily be expected from what was observed at the coarser scale. Therefore, while the coarse scale of EMAP hexagons may provide a way to identify regionally high areas from a statewide perspective, finer scale analysis is necessary for determining possible locations and networks of high species richness within a region.

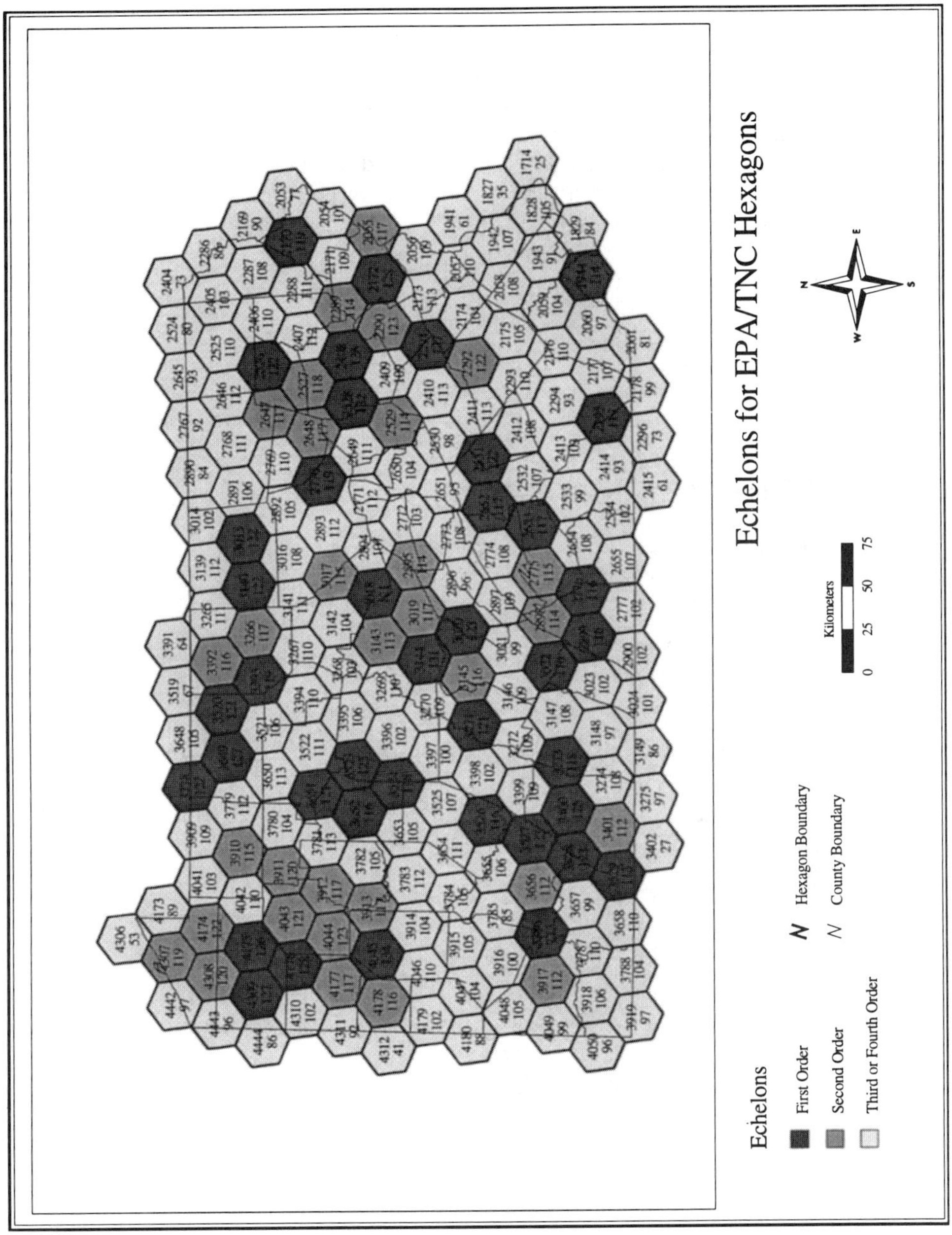

Figure 2. Statewide echelon map based on EMAP hexagons. The 4 digit number in each hexagon is the EPA-EMAP identifier, while the number below is species richness.

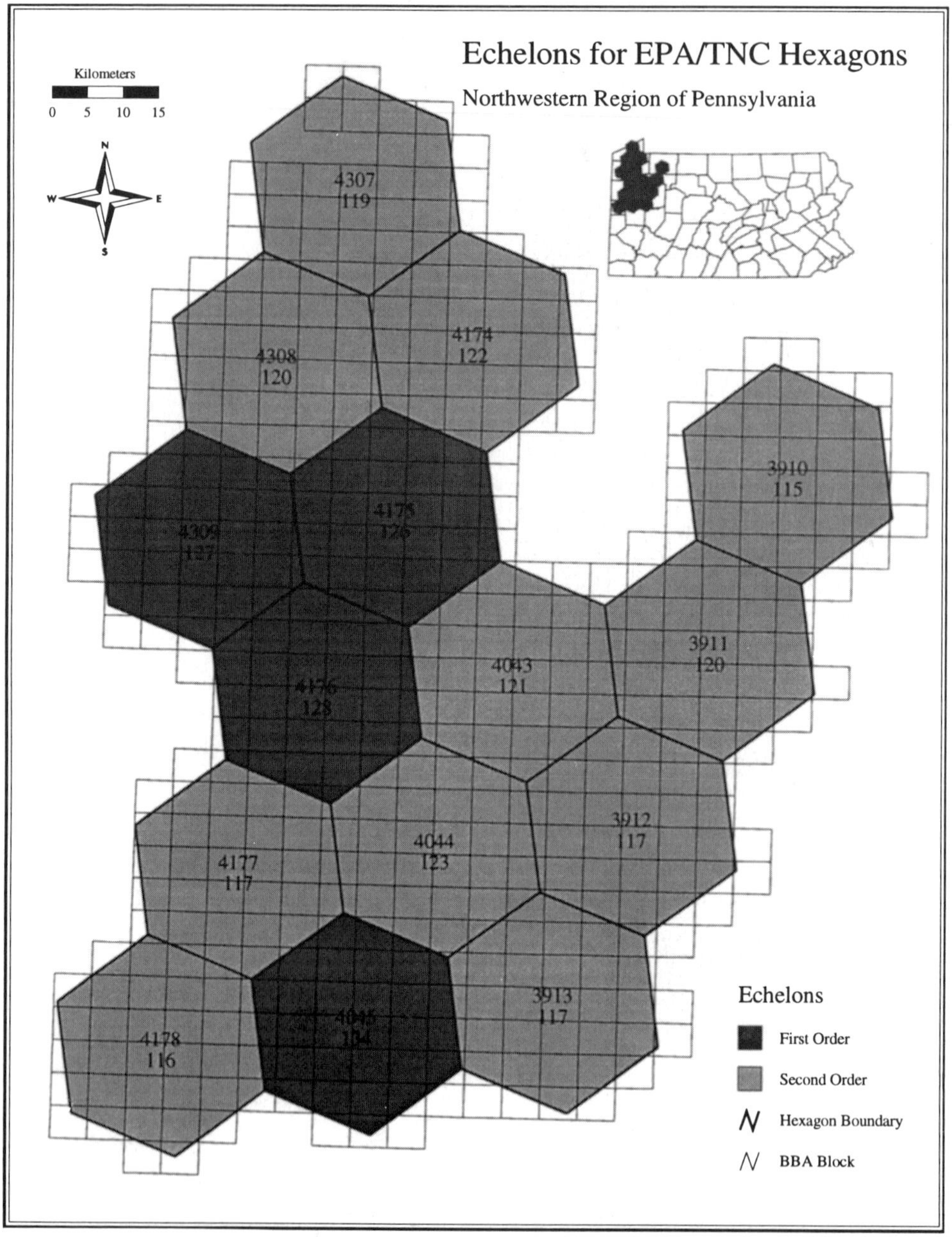

Figure 3. Northwest region EMAP hexagons. The 4-digit number in each hexagon is the EPA-EMAP identifier, while the number below is species richness.

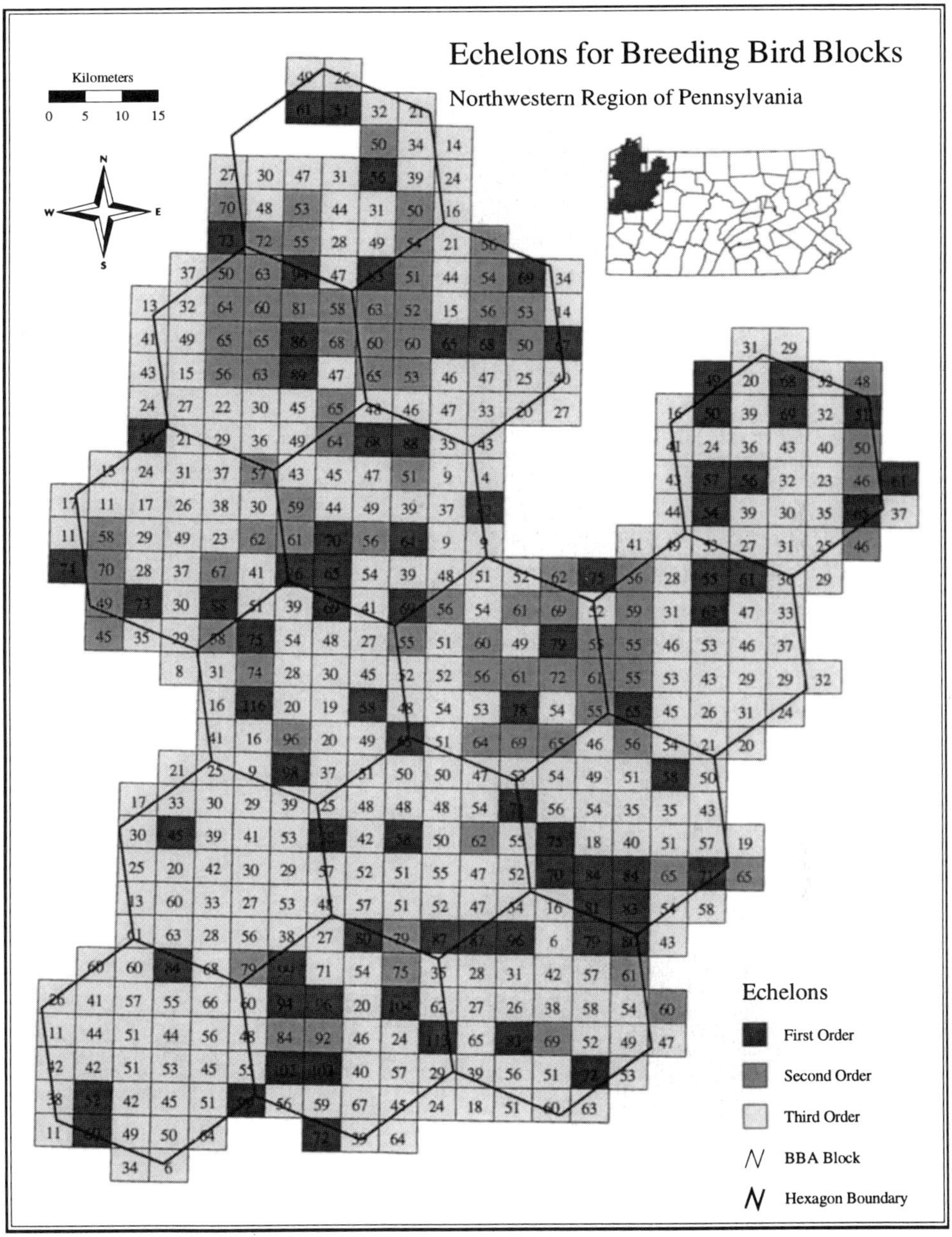

Figure 4. Northwest region Breeding Bird Atlas blocks. The number in each block species richness.

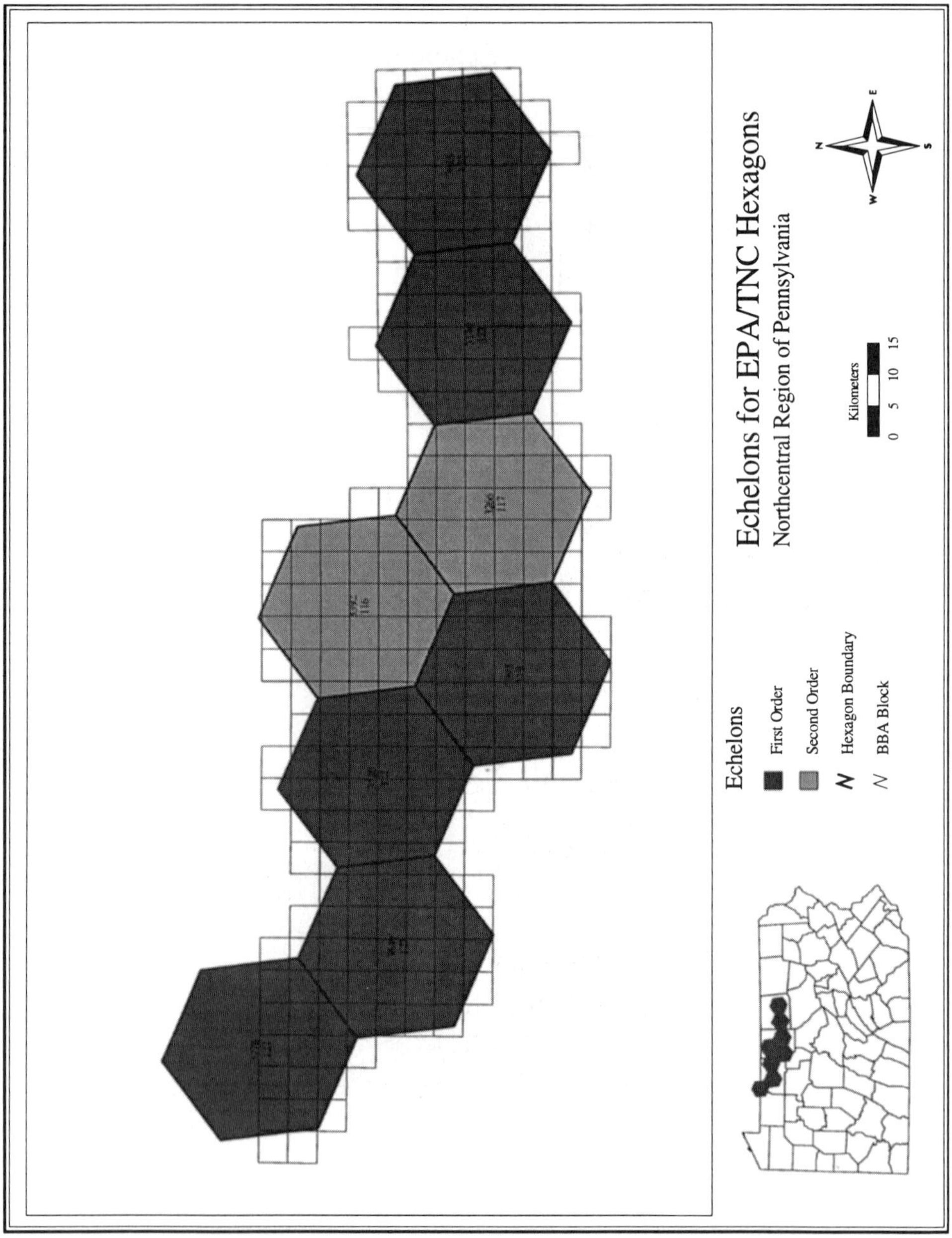

Figure 5. Northcentral region EMAP hexagons. The 4-digit number in each hexagon is the EPA-EMAP identifier, while the number below is species richness.

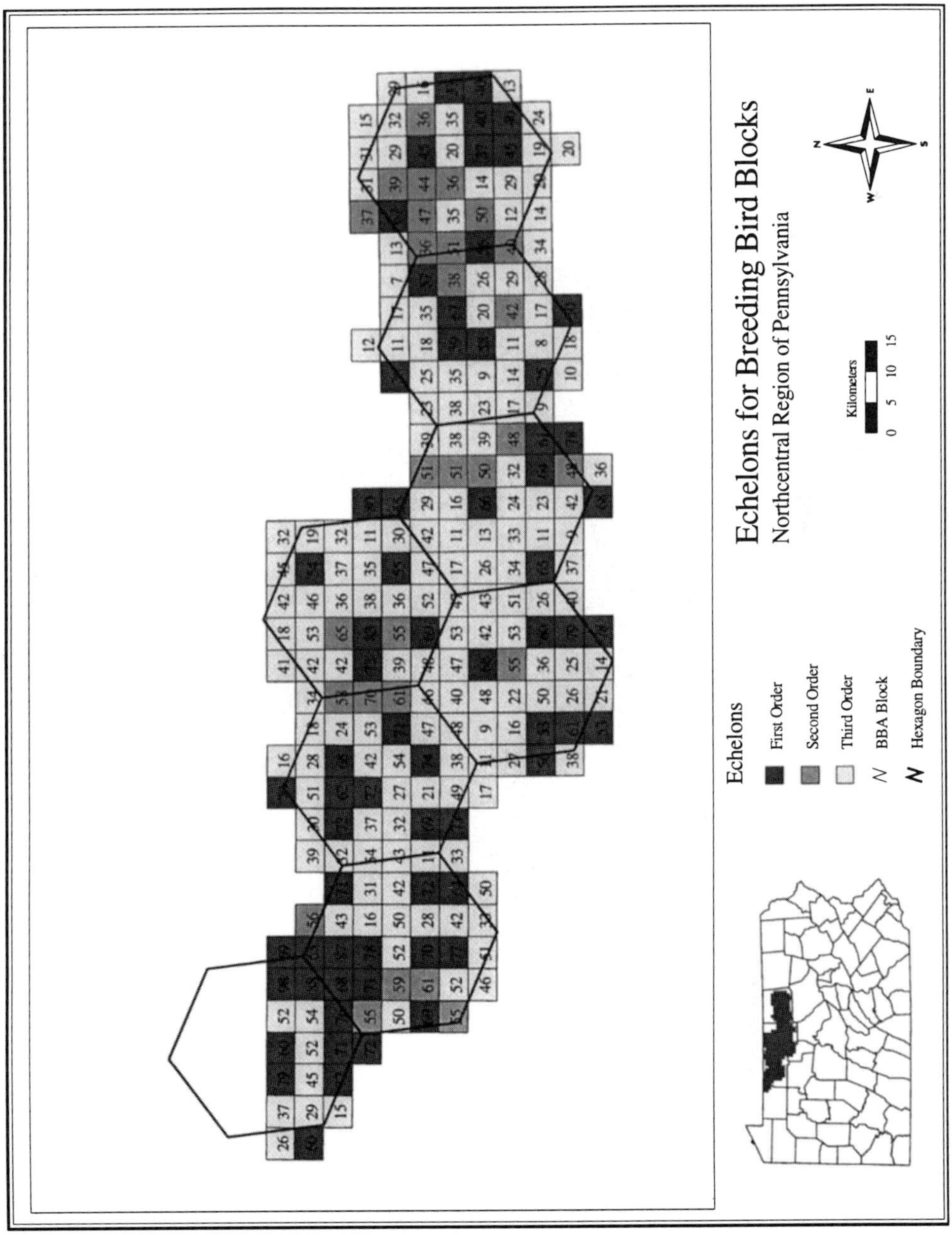

Figure 6. Northcentral region Breeding Bird Atlas blocks. The number in each block species richness.

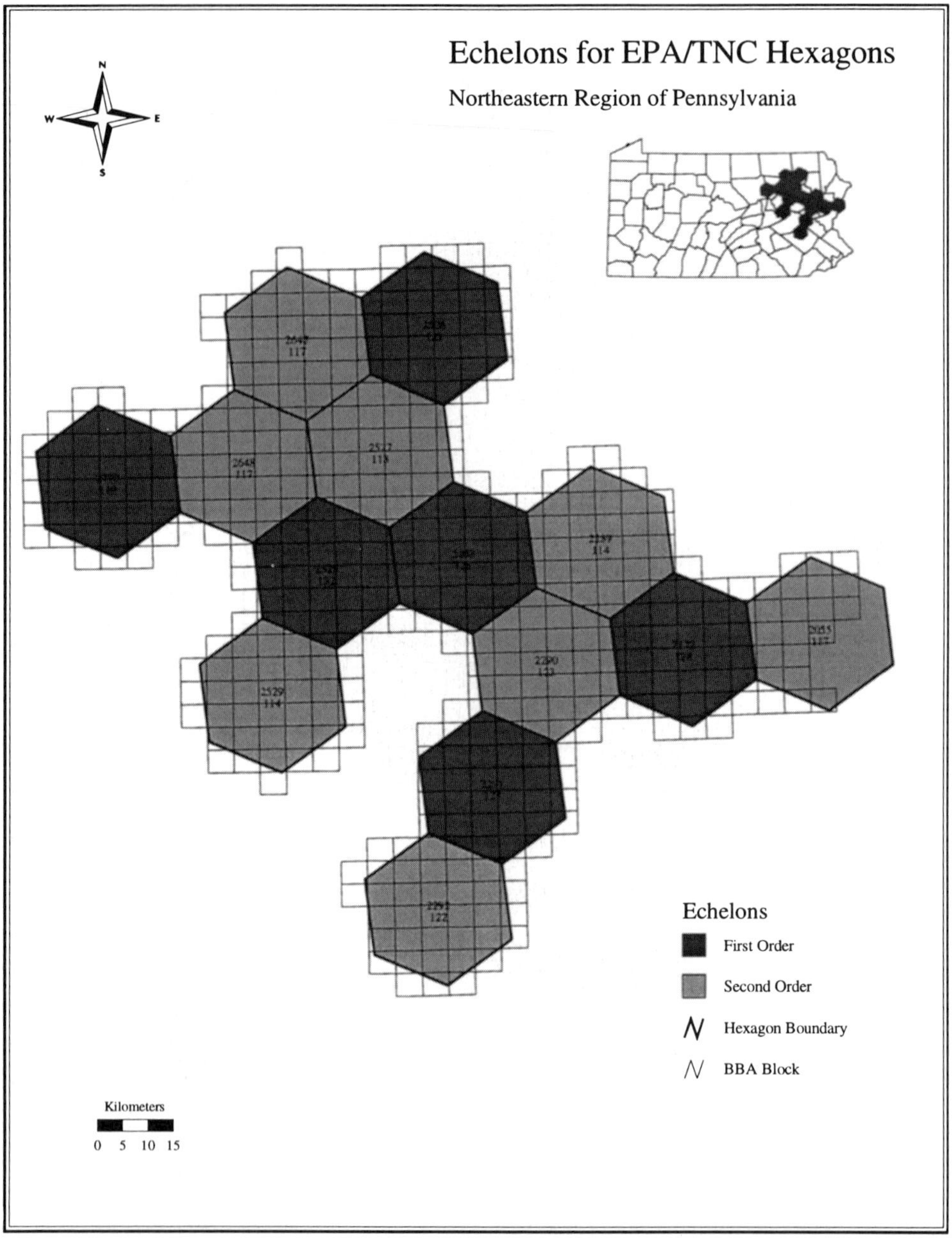

Figure 7. Northeast region EMAP hexagons. The 4-digit number in each hexagon is the EPA-EMAP identifier, while the number below is species richness.

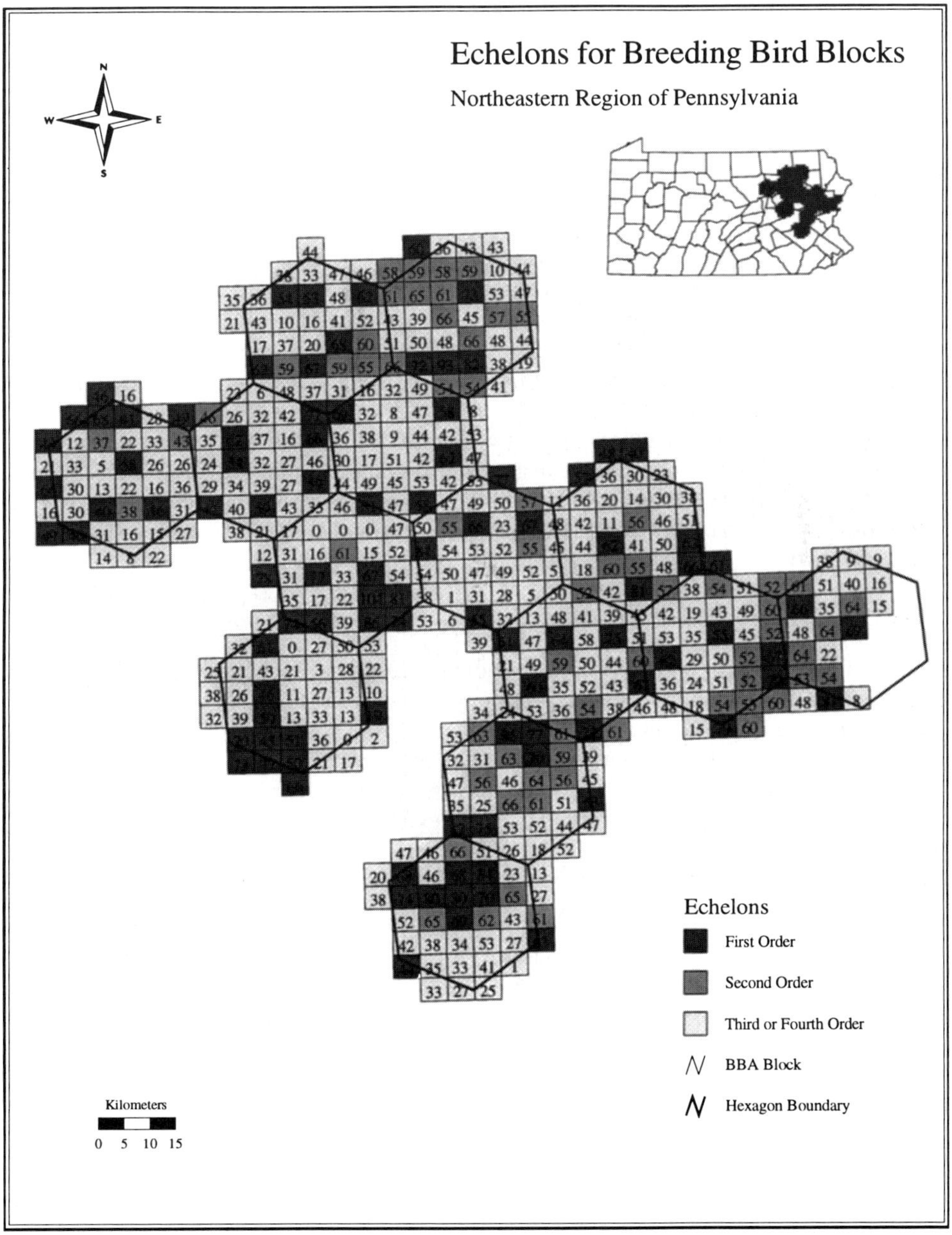

Figure 8. Northeast region Breeding Bird Atlas blocks. The number in each block species richness.

6. Analysis of Select Hexagons

Certain hexagons appeared particularly interesting because they fell into one of two categories. The first category contained members of first order echelons from a statewide perspective that did not appear high within their regions when mapped at the finer scale, as evidenced by containing few blocks that were of first order. We will call these "globally high/locally moderate" hexagons. The second category revealed the opposite phenomenon, where hexagons were of second order from a statewide perspective, although they appeared high within their regions when mapped at the finer scale, as evidenced by containing several blocks of first order. We will call these "globally moderate/locally high" hexagons.

Such observations were not so striking in the northcentral region where underlying land use is more homogeneous than the other two regions. In the northwest and northeast regions, hexagon numbers 4309 and 2408 were of the "globally high/locally moderate" category, while hexagon numbers 3910, 3912 and 2292 were of the "globally moderate/ locally high" category. One could argue that other hexagons fall into one of these categories, but for the purposes of initial exploratory research we focus on those that seem most obvious. Properties of these particular hexagons were further investigated, as discussed next.

ENTROPY MEASURES

In order to assess the distribution of breeding bird atlas blocks amongst the echelon orders within a hexagon, the Shannon entropy was calculated as $-\sum_{i=1}^{k} p_i \ln(p_i)$, where p_i is the proportion of blocks in echelon order i and k is the number of orders. These results were normalized with respect to maximum entropy, $\ln(k)$, which is often termed "evenness" because it reveals the deviation from an even distribution of units amongst the k classes. An evenness of one (1) indicates maximum entropy, and as evenness decreases towards zero (0), this indicates increasing dominance of particular classes. The results are presented in Table 1, where values are also presented for the whole regions.

The "globally high/locally moderate" hexagons, numbers 4309 and 2408, revealed greater evenness of distribution amongst echelon categories, compared to their whole respective regions. Likewise, the "globally moderate/locally high" hexagons revealed

TABLE 1. Spatial entropy and evenness for the three regions and select hexagons, using echelon orders as categories. The northwest and northcentral regions contained three echelon orders, while the northeast region contained 4 orders.

region/hexagon	entropy	evenness
Northwest	.903	.822
hex4309	.911	.829
hex3910	.859	.782
hex3912	.879	.800
Northcentral	.892	.812
Northeast	1.196	.863
hex2408	1.233	.890
hex2292	1.148	.828

lower evenness than their whole respective regions. Further analysis showed that these differences were attributed to a dominance of highest order echelons and, secondarily, first order echelons within all of the "globally moderate/locally high" hexagons; meanwhile, the "globally high/locally moderate" hexagons contained a greater number of mid-order echelons.

Apparently, when per-block species richness is fairly uniformly distributed over a hexagon, the finer scale echelon mapping may not indicate many local first order echelons, although the species add up across the hexagon to reveal a first order echelon from a statewide perspective. Such results may help provide evidence that biodiversity protection efforts may be best directed at the whole of hexagons like 4309 and 2408, while hexagons like 3910, 3912 and 2292 may contain locally species rich areas that deserve high priority for protection although they were members of second order echelons from the statewide perspective.

SPECIES AREA CURVES

In order to investigate the distributions of species, thus revealing the spatial turnover of species throughout a hexagon, we constructed species area curves in a directed manner. Starting with the block that was mostly in the middle of a hexagon, each new block was obtained by spiraling outward in a clockwise direction. The species area curve is thus constructed by adding any additional species that were encountered with each new block that was encountered. These results are presented in Figure 9, where the spiral-directed curves are compared to curves that would be expected under random sampling of blocks within a hexagon, computed as

$$E[S_n] = \sum_{i=1}^{s} \left[1 - \frac{\binom{N - A_i}{n}}{\binom{N}{n}} \right] \tag{1}$$

where $E[S_n]$ is the expected number of species encountered in n blocks sampled at random from a total of N. A_i is the number of blocks, out of the N from which we sample, where the ith species appears and s is the species richness within the hexagon. This expression can also be found in Kobayashi (1974), Engen (1976) and Hurlbert and Archibald (1995). In the following discussion, any references to land use are based on the EPA Multiresolution Land Coverage database (cf. USEPA, 1995) which is not shown here due to loss of clarity through any greyscale renditioning and to economize space.

One would expect a species-rich hexagon that has a fairly uniform distribution of per-block species richness, to reveal a fairly smooth species area curve when blocks are accumulated in a directed manner. This is observed with hexagon 2408 which follows the smooth expected curve fairly closely. The initial "blip" is associated with the early encounter of a first order block, followed immediately by encounter of a fourth order block. Subsequent overshoots of the smooth expected curve are also associated with encountering first order blocks. The majority of hexagon 2408 that is uniformly of third and fourth order echelons is associated with continuous forest interior, spotted with small lakes. Meanwhile, the first and second order echelons in the northwest area of this hexagon are associated with both a large river and urbanization. On one hand, the river, with its associated wetlands, and the diversity of habitat offered by a small urban area surrounded by mostly interior forest may yield increased species richness. On the other hand, some observation bias may be taking place due to ready accessibility around the urban area.

Hexagon 4309 appears to start out fairly smooth, eventually plateauing, but then increases as a pocket of new species are encountered. This second increase is associated with the two blocks in the southern part of hexagon 4309 that are single-block sized first order echelons (see Figure 4). Investigation of land use in this hexagon (not shown here) shows that while habitat is fairly uniformly distributed throughout the hexagon as a patchwork of open land and continuous forest, the two localized first order echelons in the southern part are associated with open water bodies. Indeed, the local second order echelons that extend from these first order echelons are associated with the same water bodies. Thus far, we are seeing how the finer mapping scale of breeding bird atlas blocks reveals echelon structure that is more closely tied with landscape habitat features.

The three hexagons that appeared "globally moderate/locally high" seem to reveal fairly jagged species area curves from spiral-directed sampling, where local increases in slope are generally associated with encountering a block that is of a first order echelon. The uniformly third order blocks in hexagons 3910 and 3912 are associated with mostly continuous forest interior habitat, while the local first order blocks are associated with breaks in such habitat by either a large river, small scale development or transitional land. Land use within hexagon 2292 is quite distinctly segregated into continuous forest in the northwest, while the remainder of the hexagon is mostly agricultural land. Here we see the opposite effect that was seen in the other hexagons because the first order blocks are clustered in the area of continuous forest, while the agricultural area almost uniformly reveals blocks that are of fourth order echelons, regardless of the presence of water bodies. This observation supports the hypothesis that while some forest fragmentation may actually increase species richness from increasing habitat diversity, too much forest loss results in an overall drop in species richness (Noss, 1983).

7. Discussion

Assessing biodiversity variables such as species richness over a large geographic area requires analysis at multiple spatial scales (Turner, 1995; Turner, et al., 1991). Stoms (1994) addressed this issue by rescaling vertebrate species richness maps for two very different areas in Idaho. Although we used non-aligned grids with different shaped cells, Stoms' hierarchically nested scaling range encompassed our two scales. Visual assessment of Stoms' thematic maps indicate some agreement with our findings. Certain species-rich cells that were similar in size to an EMAP hexagon were mapped as a relatively species-poor area when the mapping units were similar in size to the Breeding bird Atlas blocks. More generally, the finer scale maps were necessary to bring out structure that was not predictable from the coarser scale map. One of Stoms' primary objectives was to see if the identification of species-rich locations was scale dependent. Since identification was based on subjective viewing of thematic maps, such an activity should benefit by the increased objectivity provided by echelon analysis.

Structuring a response surface into echelon objects opens up a greater variety of analyses. While all analyses of the original data can still be performed, properties of the echelon objects themselves can also be very informative (Myers, Patil and Joly, 1996). In our study, the combination of entropy analysis of the Breeding Bird blocks with respect to echelon order categories, along with spiral-directed species area curve construction and referencing to known land usage, together proved to be quite informative. After delineating species-rich regions from a statewide perspective, the distributions of species and per-block species richness within particular EMAP hexagons were characterized, along with rational interpretation of these observations with respect to landscape characteristics.

Acknowledments

Prepared with partial support from the United States Department of Interior, the National Biological Service (Gap Analysis Program); the United States Environmental Protection Agency (Environmental Statistics and Information Division, Office of Policy, Planning and Evaluation; and Environmental Monitoring and Assessment Program, EMAP Design and Statistics Group); and the National Science Foundation (NSF/EPA Water and Watersheds Program). The contents have not been subjected to reviews of these agencies and therefore do not necessarily reflect the views of these agencies and no official endorsement should be inferred. We appreciate the comments from two anonymous reviewers.

References

Brauning, D.W. (1992) Atlas of Breeding Birds in Pennsylvania. University of Pittsburgh Press, Pittsburgh. 484 pp.

Brauning, D.W. and Gill, F.B. (1983-1989) Pennsylvania Breeding Bird Atlas data. The Academy of Natural Sciences of Philadelphia, Pennsylvania Game Commission and Wild Resource Conservation Fund, Harrisburg. (as licensed for use in GAP Analysis at the Pennsylvania State University)

Engen, S (1976) A note on the estimation of the species-area curve. *J. Cons. Int. Explor. Mer*, 36(3):286—288.

Hurlbert, S.H. and Archibald, J.D (1995) No statistical support for sudden (or gradual) extinction of dinosaurs. *Geology*, 23(10):881—884, October 1995.

Johnson, G.D., Nussbaum, B.D., Patil, G.P. and Ross, N.P. (1995) Innovative statistical mind sets and novel observational approaches to meet the challenges in the management of hazardous waste sites. *in* Lewis, R.A. and Subklew, G. (eds.), The Proceedings of a Specialty Conference: Challenges and Innovations in the Management of Hazardous Waste, VIP-52. Air and Waste Management Assoc., Pittsburgh, PA. pp. 3-32.

Kobayashi, S. (1974). The species-area relation I. A model for discrete sampling. *Res. Popul. Ecol.*, 15:223—237.

Myers, W.L., Brooks, R., Storm, G. and Bishop, J. (1996) Pennsylvania Gap Analysis Report—1995. ER9602. Environmental Resources Research Institute, Penn State University, University Park, PA.

Myers, W.L., Patil, G.P. and Joly, K. (1996) Echelon Approach to Areas of Concern in Synoptic Regional Monitoring. Technical Report 96-0601, Center for Statistical Ecology and Environmental Statistics, Department of Statistics, Penn State University, University Park, PA. (*to appear in Environmental and Ecological Statistics*)

Myers, W.L., Patil, G.P. and Taillie, C. (1995) Comparative Paradigms for Biodiversity Assessment. *in* Boyle, T.J.B. and Boontawee, B. (eds.), Measuring and Monitoring Biodiversity in Tropical and Temperate Forests. Center for International Forestry Research, Bogor, Indonesia, pp. 67-85.

Noss, R.F. (1983) A regional landscape approach to maintain diversity. Bioscience, 33(11):700-706.

Scott, J.M., Davis, F., Csuti, B., Noss, R., Butterfield, B., Groves, C., Anderson, H., Caicco, S., D'Erchia, F., Edwards, Jr., T.C., Ullimann, J. and Wright, R.G. (1993) Gap analysis: A geographic approach to protection of biological diversity. Wildlife Monographs, 57:1-41.

Stoms, D.M. (1994) Scale dependence of species richness maps. Professional Geographer, 46(3):346-358.

Turner, S.J. (1995) Scale, observation and measurement: critical choices for biodiversity research. *in* Boyle, T.J.B. and Boontawee, B. (eds.), Measuring and Monitoring Biodiversity in Tropical and Temperate Forests, pp. 97-111. Proceedings of a IUFRO Symposium held at Chiang Mai, Thailand 1994.

Turner, S.J., O'Neill, R.V., Conley, W., Conley, M.R. and Humphries, H.C. (1991) Pattern and scale: statistics for landscape ecology. *in* Turner, M.G. and Gardner, R.H. (eds.), Quantitative Methods in Landscape Ecology. Springer-Verlag, New York. pp. 17-49.

USEPA (1995) Mid Atlantic Landscape Indicators Project Plan. EPA/620/R-95-003. 37 pp.

USING COVARIATE-SPECIES COMMUNITY DISSIMILARITY TO GUIDE SAMPLING FOR ESTIMATING BREEDING BIRD SPECIES RICHNESS

GLEN JOHNSON, GANAPATI PATIL AND SONIA RODRÍQUEZ
Center for Statistical Ecology and Environmental Statistics,
Department of Statistics
Penn State University, University Park, PA 16802 USA

Abstract

Estimating species richness over a certain geographic region becomes a problem of sampling enough area to reach the plateau of the true species-area curve. Sampling efficiency is then achieved by minimizing the area that requires measurement. Johnson and Patil (1995) suggested that covariate-directed sampling may help achieve efficiency by choosing sample units that have the greatest chance of containing different habitat and therefore different species. By retrospectively sampling the breeding bird community in Pennsylvania, various sampling protocols are investigated in conjunction with various covariate species. Results indicate which combinations of covariate species and sampling protocols may be more efficient than random sampling and which may be less efficient.

1. Introduction

Estimating species richness over a large geographic area based on a subsample of the area presents a special sampling challenge due to the non-additivity of species richness. The basic problem is one of estimatingthe number of classes in a multinomial population. Various sample-theoretic approaches have been reviewed by Bunge and Fitzpatrick (1993); however, these methods all apply to data that include abundance measurements for each species that is encountered. Data from sampling a large geographic area is most likely to be available as presence/absence recordings for each species encountered within each sample unit.

An approach to this problem has been suggested by Johnson and Patil (1995), which utilizes the species-area relationship. As discussed in Patil, Johnson and Grigoletto (1996), the number of species in an area is expected to grow with increasing area that is sampled, according to a power function, presented as $S = kA^z$, where S is the species richness and A is the area, while z and k are population specific parameters. Since $0 \leq z \leq 1$, this model implies that a "point of diminishing returns" is approached as the area A increases. The objective then becomes to sample enough of the area of interest in order to either encounter all of the species in the area or (more realistically) encounter the plateau region of the true species-area curve. This is desirable for a sample which is intended for estimating the true species richness, whether the estimate is simply the

P. Bachmann et al. (eds.), Assessment of Biodiversity for Improved Forest Planning, 151–158.
European Forest Institute Proceedings no. 18. © 1998 *Kluwer Academic Publishers. Printed in the Netherlands.*

number of species encountered in the sample (Johnson and Patil, 1995), extrapolation of a fitted species-area curve (Patil, Johnson and Grigoletto, 1996) or a bounded monotonically increasing curve (Bunge and Fitzpatrick, 1993), or even when abundance measurements are available so that sample-theoretic methods may apply (Bunge and Fitzpatrick, 1993; Bunge, Fitzpatrick and Handley, 1995).

Fundamentally, we want to encounter as many of the species that occur in the area of interest within a smaller sub-area that can be affordably sampled. This objective therefore provides the criterion for comparing efficiency of various sampling plans.

Postulating that maximizing the habitat heterogeneity within a selected sample will maximize the number of species encountered, Johnson and Patil (1995) investigated covariate-directed sampling. Using a GIS database that provided synoptic coverage of both breeding bird and tree species in Pennsylvania, they retrospectively sampled the statewide breeding bird community, using incremental tree species richness as a covariate to direct sampling. While this approach performed somewhat better than random sampling, it still missed the northern tier of the state which contained some of the highest bird richness. Indeed, the northern tier of Pennsylvania is amongst the least disturbed regions of the state and therefore has more forest interior, but relatively low tree richness. In this paper, we attempt to overcome this problem by investigating some other approaches to maximizing forest community dissimilarity as an investigator moves from one sample unit to the next. We also apply these approaches to other covariate species that are available to us.

2. Data

The database is described in detail by Johnson and Patil (1995). The sampling frame consists of a subset of the national USEPA-EMAP hexagon tessellation that covers Pennsylvania. Each sample unit within this frame is a 635 km^2 hexagon. A thematic map of the breeding bird species richness is presented in Figure 1 for visualizing the known statewide bird community that is retrospectively sampled in this paper.

3. Sampling Protocol

The general protocol for community dissimilarity-directed sampling, using known occurrences of a community of covariate species follows the steps listed below.

1. Select a sample unit with the highest covariate species richness; then enumerate the species in the community of interest (such as birds) that are encountered in this unit.
2. Select the next unit as the one revealing maximum community dissimilarity for the covariate species, compared to either the single most recently sampled unit or to all accumulatively sampled units (ties may be broken by choosing at random).
3. Add the number of newly encountered species from the community of interest to the overall species richness for this community.
4. Repeat until an affordable sample size has been reached.

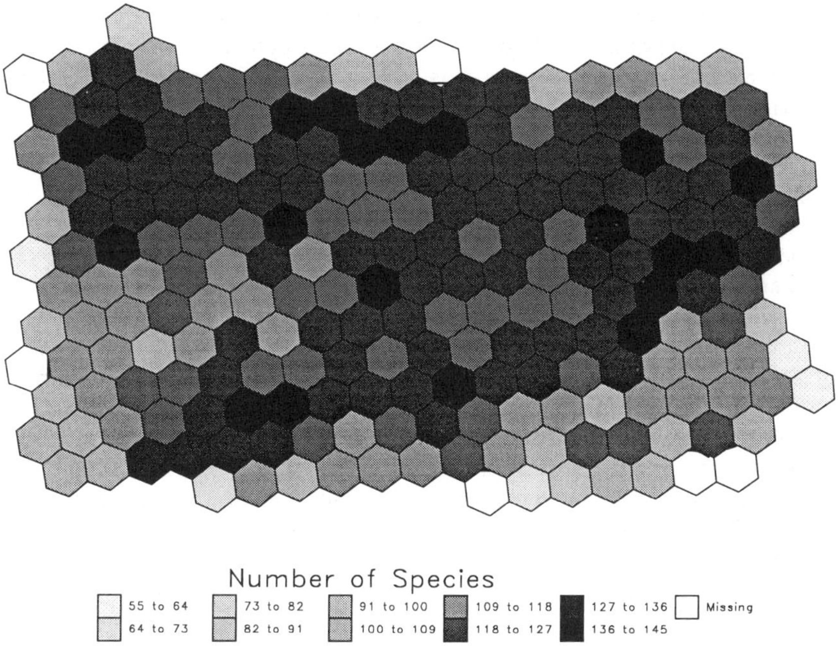

Figure 1. Bird richness in the hexagons

The dissimilarity measurements that were investigated for choosing each new sample unit are summarized below.

- Maximum additional covariate species: Once all covariate species are accounted for, a new sampling cycle may be repeated with all covariate species in the unsampled units.
- Jaccard dissimilarity: For a covariate species that occur in both sample units, b that occur exclusively in the current unit (single or pooled) and c that occur exclusively in the potential unit, dissimilarity is computed as

$$D = \frac{b+c}{a+b+c}$$

Within this protocol, we may pursue two different approaches, as follows.

- Pairwise Jaccard: Maximum dissimilarity is based on pairwise comparisons between the current and each prospective sample unit.
- Accumulative Jaccard: Maximum dissimilarity is based on comparison between all accumulatively sampled units, pooled as one, and each prospective sample unit.

The last method is termed Jaccard dissimilarity because it is the compliment of the classic Jaccard dissimilarity measure (Ludwig and Reynolds, 1988, p. 131), which is designed for presence/absence data. It is simply the compliment of the intersection of two finite sets expressed as a proportion of the full union of these sets.

4. Results and Discussion

Using the directed sampling protocol with the three approaches to measuring covariate-species community dissimilarity, the entire state of Pennsylvania was retrospectively sampled for enumerating breeding bird species. The resulting state-wide species area curves are reported in Figure 2 to 4, along with the expected curve under random sampling, where for each possible sample size (1 to 186), the expected number of species encountered is computed as

$$E=[S_n]=\sum_{i=1}^{s}\left[1-\frac{\binom{N-A_i}{n}}{\binom{N}{n}}\right] \tag{1}$$

where $E[S_n]$ is the expected number of species encountered in n hexagons sampled at random from a total of $N=186$, A_i is the number of hexagons, out of the N from which we sample, where the ith species appears and s is the species richness in Pennsylvania. Of course this expectation could only be computed because we had an exhaustive data set.

No particular covariate species or sampling protocol consistently performed well or poorly. The protocol following " maximum additional covariate species" worked well for trees; the "pairwise Jaccard" protocol worked well for trees as well as reptiles and amphibians; the "accumulative Jaccard" protocol under-performed random sampling for all but mammals which, in turn, yielded the most promising result.

A primary motivation for improvement over the incremental tree richness-directed protocol was to include sample units early in sampling that lied in tree species-poor but bird species-rich areas, such as the northern tier of Pennsylvania. Since the only competitor appeared to be pairwise Jaccard dissimilarity-directed sampling, the location of hexagons that were sampled in this manner were identified. These locations are reported in Figure 5, which can be compared to Figure 6. As seen in these figures, more locations in the bird-rich, tree-poor northern tier are included through Jaccard dissimilarity-directed sampling; however, the clustering effect is still present.

5. Summary

This initial exploratory data analysis allows the use of a synoptic database for a whole state to be used to entertain some new ideas about sampling for the enumeration of a non-additive variable such as species richness. For other datasets we may see very different results; therefore, these protocols should continue to be tested as other datasets become available. A primary limitation of our dataset is the oversized sample units. While a 635 km^2 hexagon may be an appropriate primary sample unit size for continental

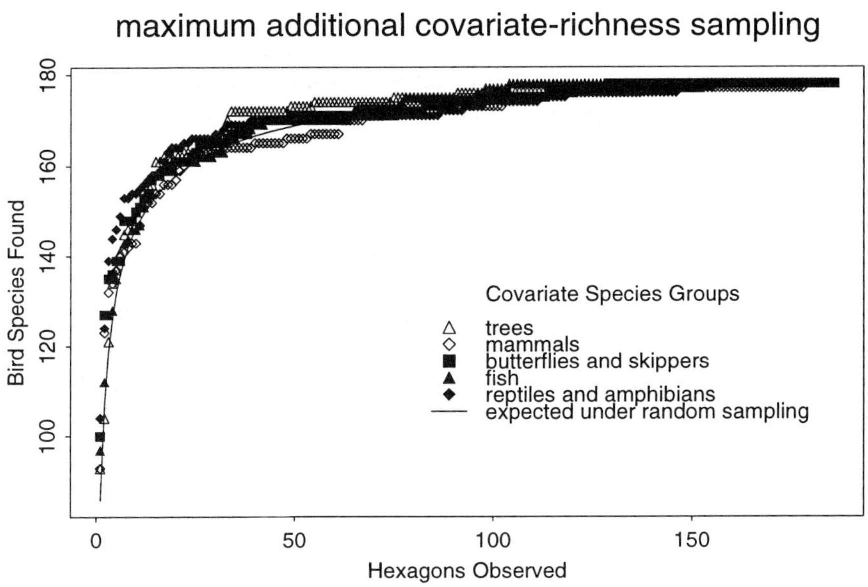

Figure 2. Species area curves from selecting hexagons in order of maximum additional covariate species, until all covariate species are accounted for, then repeating the cycle as needed.

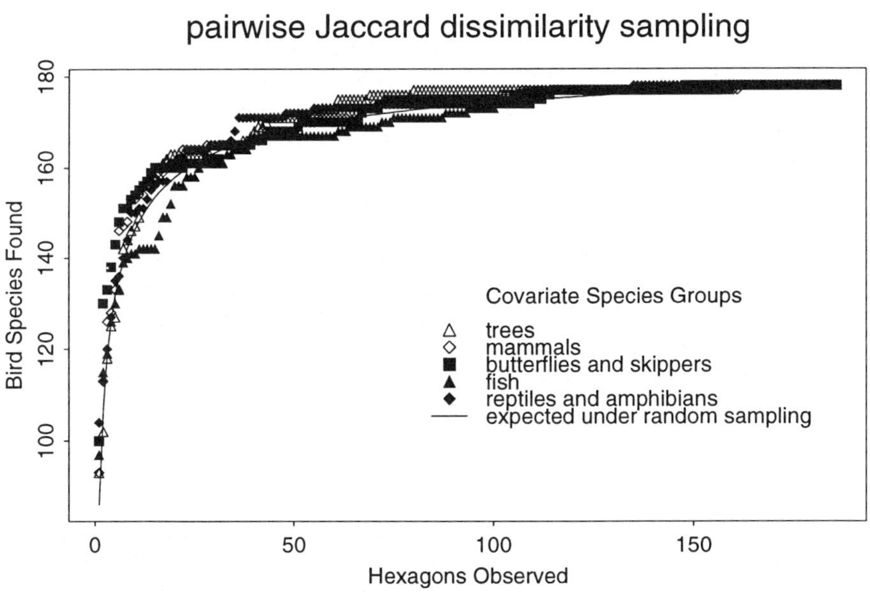

Figure 3. Species area curves from pairwise Jaccard sampling which selects hexagons in order of maximum covariate-species community dissimilarity based on pairwise comparison of the most recently sampled hexagon with each remaining unsampled one.

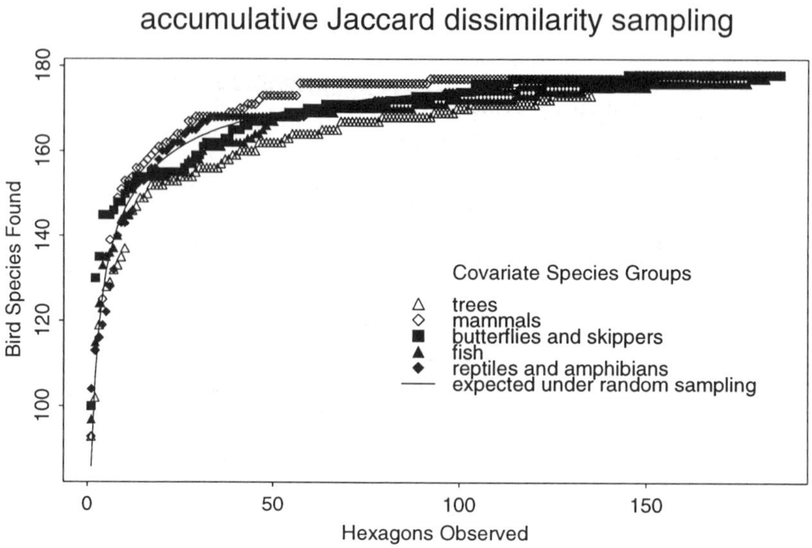

Figure 4. Species area curves from accumulative Jaccard sampling which selects hexagons in order of maximum covariate-species community dissimilarity between all sampled hexagons (pooled) and each remaining unsampled hexagon.

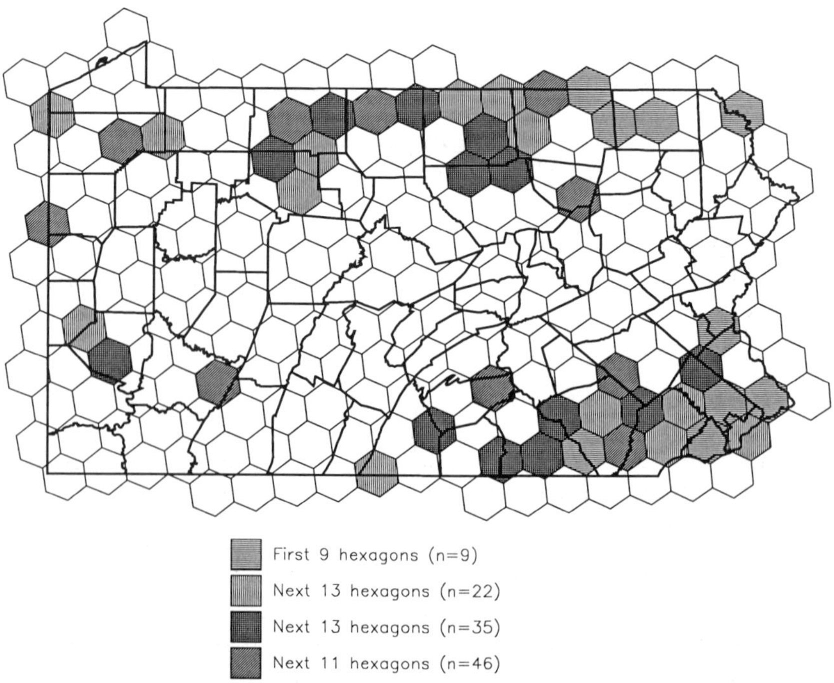

Figure 5. Hexagons included in the pairwise Jaccard dissimilarity-directed sampling, grouped to correspond with the first 4 cycles or incremental tree richness-directed sampling

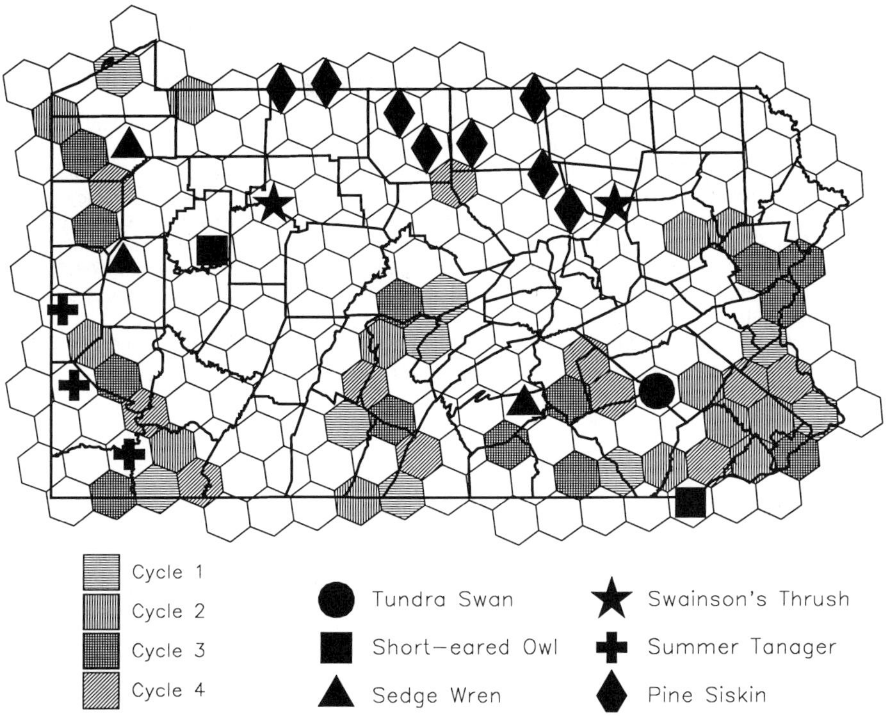

Figure 6. Hexagons included in the first four cycles of incremental tree richness-directed sampling, along with the locations of unencountered bird species (from Patil, Johnson and Grigoletto 1996).

analysis, it appears to be large enough to prevent the observations of any strong correlations between the breeding birds and other covariate species within the state of Pennsylvania.

Acknowledgments

Prepared with partial support from the United States Environmental Protection Agency (Environmental Monitoring and Assessment Program, EMAP Design and Statistics Group under a Cooperative Agreement Number CR-821783). The contents have not been subjected to Agency review and therefore do not necessarily reflect the views of the Agency and no official endorsement should be inferred.

We sincerely appreciate comments on previous work from Woollcott Smith, who convinced us to investigate a dissimilarity measure for covariate-directed sampling. We also benefited greatly from discussion and comments from Charles Taillie and Ted Williams. Finally, we thank two anonymous referees.

References

Bunge, J. and Fitzpatrick, M. (1993) Estimating the number of species: a review. Journal of the American Statistical Assoc., 88:364-373.

Johnson, G.D. and Patil, G.P. (1995) Estimating statewide species richness of breeding birds in Pennsylvania. Coenosis, 10(2-3):81-87.

Patil, G.P., Johnson, G.D. and Grigoletto, M. (1996) Covariate-directed sampling for assessing species richness. {\em in} Mowrer, H.T., Czaplewski, R.L. and Hamre. R.H. (eds.). Spatial Accuracy Assessment in Natural Resources and Environmental Sciences: Second International Symposium. May 21-23, 1996. General Technical Report RM-GTR-277. Fort Collins, CO: U.S. Department of Agriculture, Forest Service, Rocky Mountain Forest and Range Experiment Station. pp. 569-576.

Bunge, J., Fitzpatrick, M. and Handley, J. (1995) Comparison of three estimators of the number of species. Journal of Applied Statistics, 22(1):45-59.

Ludwig, J.A. and Reynolds, J.F. (1988) Statistical Ecology, A Primer on Methods and Computing. John Wiley and Sons, New York. 337 pp.

SOME MATHEMATICAL REMARKS ON FOREST BIODIVERSITY

ALAIN FRANC
ENGREF MAI
Université Paris Sud - CNRS URA 2154
Ecologie des Populations et des Communautés
Bat 362 - 91 405 Orsay Cedex - France

Abstract

Biological diversity, currently and briefly denoted as biodiversity, probably is one of the key features of living world, as it has been wonderfully emphasised by Mayr (Mayr 1982). It is one of the field where a true biological quest, i.e. for the understanding of diversity, meets sophisticated mathematical tools, with a great biological relevance. Diversity studies may be summed up as the research of regular time and space pattern in allocation of individuals between species. The obvious way to study it, and the oldest one from an historical point of view, is just counting the number of different species at given location during a given time period. More sophisticated tools appeared after the Second World War, with the construction of indices which aim at comparing two samples on species diversity and evenness (see Margurran 1988, for a review). After the renewal of theoretical ecology in the early fifties, a great deal of effort was devoted to the understanding of diversity and evenness as a consequence of current ecological processes. In this article, we briefly sketch out some of the keystone ideas, with our own personal bias, which have contributed to the development of this field. Our presentation is divided into two parts: the first one dedicated to the quest for indices, and the second to the quest for understanding.

1. The most common indices

Quantitative indices are means to encapsulate the evaluation of diversity into one figure computed from the distribution of individuals among species. This very simple question may appear as hopeless: if diversity could be encapsulated into one single figure, it would be closer to simplicity and uniqueness of behaviour, rather than to complexity and diversity of behaviour, as it is often implicitly assumed.

The diversity concept is related to the statistical concept of variance: generally speaking, the more diverse a given population, the greater also the variance of the variables describing its individual. But we feel that diversity measurement is a far richer and more subtle task than variance estimation. In fact, we shall see later that one of the commonly used indices, viz. the Simpson index, is closely related to the variance of a distribution function of individual among species. On the other hand, the Shannon index

P. Bachmann et al. (eds.), Assessment of Biodiversity for Improved Forest Planning, 159–169.
European Forest Institute Proceedings no. 18. © 1998 *Kluwer Academic Publishers. Printed in the Netherlands.*

H of diversity is maximum when the variance of the distribution function is minimum. We shall see in a third part that those indices are two items within one unifying family of diversity indicators. Hence, variance is one specific and particular way of measuring diversity.

Suppose we have a community with N individuals where Z species have been recognised. They are labelled 1 to Z, where the order of labelling has no importance. Hence, one species is labelled by one integer:

$$i = 1, \cdots, Z$$

The population is known by the number of individuals per species only. This is a distribution function, and we do not know which precise individual belongs to which species. It is usual to deal with frequencies rather than with numbers. We will denote by n_i the number of individuals belonging to the species i, and by x_i the fraction of the population belonging to this same species, so that:

$$x_i = \frac{n_i}{N}$$

where

$$\sum_i n_i = N$$

and consequently

$$\sum_i x_i = 1$$

Most of the indices are computed from the values x_i or n_i. A more detailed review of different indices, their use and estimation may be found in Pielou (1969), Margurran (1988) and Rozenzweig (1995).

1.1 The Shannon index

The Shannon index, H, is probably the best known and most widely used diversity index in ecological literature. It is defined as:

$$H = -\sum_i x_i \operatorname{Log} x_i$$

The connection between the Shannon index and diversity measurement comes from the connection between information theory and entropy concept within statistical mechanics, which is rather subtle. A precise and simple presentation may be found in Khinchin (1957).

A well known drawback of the Shannon index is that it does not account properly for the importance of rare species. If a frequency is low, say $x_i = 10^{-3}$, then its contribution ($h_i = x_i \operatorname{Log} x_i$) to the Shannon index is low as well (in our example, $h_i = 6.9\ 10^{-3}$). Hence, the existence of rare species does not significantly modify the Shannon index. This may be in contradiction with biological expertise: indeed, the existence of rare species may be a significant component of the uniqueness of a given location as far as diversity is concerned.

This question is far from being innocent, and it is a good example of the difficulty of linking biological expertise with mathematical know-how. We shall see how a slight improvement has been proposed about twenty years ago.

When the Shannon index has to be evaluated from field data, the quantities x_i are not known. The most current technique is to use an estimator of them, say

$$\tilde{x}_i = \frac{n_i}{N}$$

This estimator can be shown to be biased. Moreover, it has been proved that no unbiased estimator of the Shannon index exists. One most current technique to estimate the Shannon index from field data with reduced bias relies on Jack-knife technique.

The Shannon index is an indicator of the amount of information which can be extracted from a theoretical sample of infinite length. When samples of finite length are considered, the information computation is being processed using Brillouin's formula (Brillouin 1962; Pielou 1969):

$$H = \frac{1}{N} \text{Log} \frac{N!}{\prod_i n_i!}$$

It is well known that when the integers n_i are large, a good approximation of $n_i!$ is given by Stirling formula: $n! \approx \sqrt{2\pi n}(n/e)^n$ which leads easily to Shannon's formula when applied to Brillouin's formula.

THE SIMPSON INDEX

The Simpson index (Simpson 1949) was proposed at approximately the same time when Shannon's work was published. The underlying assumption is that interactions constitute the motor which generates diversity: the more interactions, the more diverse is the community. We shall see below that this assumption is far from being obvious, but let us take it for granted since it is the basis of Simpson's index of diversity. If individual meet at random, the probability for an individual of type i to meet an individual of type j is

$$p_{ij} = x_i x_j$$

There will be a 'creative' interaction if both individuals do not belong to the same species, i.e. if $i \neq j$. The probability of this event is then:

$$S = 1 - \sum_i x_i^2$$

which is Simpson's index of diversity. It is sometimes presented as

$$S = \sum_i x_i^2$$

although in the latter form it decreases as diversity increases.

Simpson index may be found equivalent to other indices relevant in other fields, such as Nei index in population genetics. Nei index is the fraction of heterozygous in a population and, if x_i represents the allelic frequencies, its value is $S = \sum x_i^2$, as Simpson index.

Finally, we may note that the Simpson index is related to the variance of distribution law of the population. As

$$v(x) = \frac{1}{N-1}\sum_i \left(x_i - \frac{1}{N}\right)^2$$

it is easy to check that:

$$v(x) = \frac{1}{N-1}\sum_i x_i^2 - \frac{1}{N}$$

1.3 HILL'S SYNTHESIS AND RÉNYI'S INFORMATION

There seems to exist a variety of diversity indices. Which one is it suitable to choose ?

A synthesis between major indices has elegantly been proposed by Hill (Hill 1973), who defines an index N_α within one family such as

$$N_\alpha = \left(\sum_i x_i^\alpha\right)^{\frac{1}{1-\alpha}}$$

It is then easy to check the following relationships between these indices and classical diversity indices (Hill 1973; Margurran 1988):

a	Related index	Relationship
$-\infty$	May $(s_{-\infty})$	$N_{-\infty} = 1/S_{-\infty}$
0	Number of species (s_0)	$N_0 = S_0$
1	Shannon index (s_1)	$N_1 = \exp(S_1)$
2	Simpson index (s_2)	$N_2 = 1/S_2$
$+\infty$	Berger-Parker index $(s_{+\infty})$	$N_\infty = 1/S_\infty$

As noted by Hill himself, the family of indices he proposed is inspired by Rényi's information in statistical physics, defined as follows (Rényi 1970; Beck and Schlögl 1993; Ott et al. 1994): if a distribution function is characterised by its probabilities $x = (x_1, \cdots, x_N)$, then the Rényi's information of order α of the distribution is:

$$I_\alpha = \frac{1}{\alpha - 1}\,\mathrm{Log}\sum_i x_i^\alpha$$

Hence,

$$N_\alpha = \exp I_\alpha$$

It may be noteworthy to understand the connection between Shannon's information and Rényi's information, from slight changes in Khinchin's axiomatic presentation of information (Khintchin 1957). Information is a quantity which should obey the following rules, if it exists:

(i) it is a function of the distribution law of the events only:

$$I(x) = I(x_1, \cdots, x_Z)$$

(ii) the distribution law where the information on the state of the system is minimum is the one with equiprobability between states:

$$I(x) \leq I(\tfrac{1}{Z}, \cdots, \tfrac{1}{Z})$$

(iii) The information of a sample remains unchanged when adding a new item with zero probability:

$$I(x_1, \cdots, x_Z) = I(x_1, \cdots, x_Z, 0)$$

For the fourth axiom, it is possible to choose either *(iv-a)* or *(iv-b)*. Axiom *(iv-a)* may look rather complicated at first glance, but in fact it is quite simple and natural. Khintchin's choice is axiom *(iv-a)*.

(iv-a) Let us suppose we have a system S composed of two subsystems Σ_I and Σ_{II}, with states $i=(1, \cdots, N)$ and $j=(1, \cdots, P)$ respectively. Then, the state of the whole system is given by a couple (i, j). In the general case, both subsystems are not independent, and we may define the probability x_i of subsystem I to be in state i and the probability $y_{j|i}$ of subsystem II to be in state j when subsystem I is in state i. We note by

$$I_i = I\left(y_{1|i}, \cdots, y_{p|i}\right)$$

the information of subsystem II when subsystem I is in state i. Then, the fourth axiom states that information on the whole system is the sum of two parts:

$$I_{I \cup II} = I_I(x) + \sum_i x_i I_{II|i}$$

If systems I and II are independent, then $I_{II|i} = I_{II}$, and

$$I_{I \cup II} = I_I + I_{II}$$

The information of two independent systems is the sum of their respective information. This special case, more restrictive than the general one including conditional probabilities, is the other possibility for the fourth axiom, as follows.

(iv-b) The information is an additive property of independent systems

It is then possible to show (Khinchin 1957) that Shannon's information is the only function $I(x)$ which satisfies axioms *(i)* to *(iii)* and *(iv-a)*, whereas (Beck and Schlögl 1993) Rényi's information is the only family which satisfies axioms *(i)* to *(iii)* and *(iv-b)*. As axiom *(iv-b)* is a special case of axiom *(iv-a)*, it shows that Shannon's information is a special case of Rényi's information.

1.4 LINKS WITH THERMODYNAMICS AND CHAOTIC SYSTEMS

The striking feature of the relationship between Hill's family for measuring diversity on a sample and Rényi's information is that relevant figures for level a are equivalent in both areas. The "go-between" is the concept of "Rényi's dimension"(Beck and Schlögl 1993), or "Rényi's entropy" (Rényi 1970; Grassberger and Procaccia 1983; Ott et al. 1994).

a	Diversity index	Rényi's dimension
$-\infty$	May's index	Negative limit dimension
0	Number of species	Number of items
1	Shannon index	Information dimension
2	Simpson index	Correlation dimension
$-\infty$	Berger-Parker index	Positive limit dimension

Another striking feature is that similar notions have been used for more than ten years to decipher the strangeness of attractors of chaotic systems (Grassberger and Procaccia 1983; Beck and Schlögl 1993; Ott et al. 1994). The analogy is as follows: as the use of analytic tools to study long term behaviour of chaotic systems leads quickly to untractable equations, most efforts have focused on statistical properties of their attractors, which are usually called strange attractors. The statistical problem is the study of the allocation of points within the phase space through iteration of a chaotic map, or density patterns. This is done by dividing the phase space into boxes, labelled from 1 to Z, and computing the fraction of the points which are in a given box. The equivalent in ecology is the allocation of individuals to species, and computing the fraction of individuals belonging to a given species. It is rather striking that these two problems independently lead to quite equivalent and homologous indices, thought to be relevant in both respective fields.

As far as we know, such an equivalence had not as yet been noted in the literature. It can be worked out far further and deeper with classical tools of statistical physics, such as "escort distribution", to provide a clear understanding of relationships between diversity indices on one hand, and thermodynamics quantities on the other. We feel that such an understanding is valuable for diversity analysis.

For example, it is well known that information dimension, related to the Shannon index and the usual fractal dimension of the attractor, is small if the entire attractor may be properly covered with a few boxes only. This means that boxes with scarce points within the attractor do not contribute to the dimension of the attractor. Hence, the ecological equivalent to this well known property is that the Shannon index does not account for rare species which may play a key role in diversity biological richness. It is even relevant to assert that the Shannon index is an evaluation of the possibility to ignore rare species, (resp. scarce points) in the description of the whole population (resp. the whole attractor). The way to account for the existence of scarce points within the attractor, or rare species within a population, is the use of multifractal analysis, which relies on Rényi dimensions, and the ecological equivalent of which is the Hill's family of diversity indexes.

On these grounds, it seems sensible to encourage the use and further study of Hill's index family for studying diversity, in a way parallel to multifractal analysis of strange attractors of chaotic maps.

Moreover, beyond this formal analogy through similar formulas, there is a sort of duality between thermodynamics on one side, and diversity indices and chaotic maps on the other, which has been stressed for chaotic maps by Beck and Schlögl (1993): in the first case, physical systems are known by the means of macroscopic features, such as energy, etc. ..., and assumptions are made on microstates which could lead to this macroscopic structure, whereas in statistical properties of strange attractors of chaotic maps and diversity indices, the whole complexity of the microscopic pattern is known, and some relevant macroscopic features which should encapsulate this complexity are looked for. Hence, we feel that those fields might feed each other, as both aim at relating diversity of microscopic behaviours to simplicity of macroscopic indicators.

2. Is diversity a consequence of ecological processes?

The use of diversity indices allows a description of encountered and observed diversity. But the question of explaining diversity remains quite open and virgin after that, whatever the quality of the mathematical tools which have been used. In this section, we stress some hints which have been proposed in current ecological literature, and are or may be relevant for explaining forest biological diversity.

At first glance, it could be temptable to relate diversity to some key components of ecological processes. Since the pioneering work of MacArthur and Wilson (MacArthur and Wilson 1963) on island biogeography, relating observed diversity to observed ecological processes has been a trend in both theoretical and applied ecological research. As this field is still in full development, a comprehensive synthesis is currently out of reach (see Ricklefs and Schluter 1993 for an up to date development on this topic). It seems more sensible to propose some quick glances at prominent approaches, which give an indication of current trends and open questions. The choice of the topics is biased by our own personal views, and we explicitly do not attempt at giving an objective and up to date review of this subject.

The main limitation to deciphering relationships between diversity and ecological processes probably is that current ecological processes may not be the main reason for current biological diversity: the evolutionary history of a phylum certainly is a keystone for understanding its current diversity.

2.1 DIVERSITY AND PRODUCTIVITY AS AN EXAMPLE

A much debated issue is the experimental relationship between site productivity and community diversity (Rozenzweig and Abramsky 1993). The relationship is often humped, with a maximum at medium productivity, and minimum at both extremes, either high or low productivity. Such a humped curve has been reported both in vegetation communities and in animal communities. Up to now, there is no general model which could explain such a pattern, and this open question has been thoroughly studied in the late 80's (see Rozenzweig and Abramsky 1993 for more details).

In the hump shape of the relationship, if it exists, the increase of diversity with productivity is not puzzling. It can easily be related to the well known 'limiting factor law'. For any species in any region, there is a productivity threshold, such as below it, the site factors are too limiting for the species to survive. As the threshold decreases, more and more species cross this fatal boundary, and less and less species can survive. But the decrease in diversity with increased productivity beyond a given hump is more puzzling, and researches did focus on this very point. Many explanations have been suggested, but none gave rise to a consensus. Among the numerous hypotheses which have been reviewed in Rozenzweig and Abramsky (1993), which underpin if necessary the diversity of processes involved in diversity production within communities, we hereafter retain two of them, as prominent examples.

One hypothesis relates species diversity to microsite diversity. It is a well known result in theoretical ecology, even if still debatable in empirical ecology, that one ecological niche can feed only one species. This theory is called the 'niche theory'. The hypothesis asserts that mesic sites have more niches than either very poor or very rich sites. Hence they also have a higher species diversity. We will illustrate this using three common site factors: heat, water and nutrient. In poor sites (cool, dry and nutrient poor), all factors have low values. Hence, there is a unique combination of factors which leads to a unique site with very low productivity. In rich sites (warm, moist and nutrient rich), as well there is a unique combination of site factor values which leads to one site. By contrast, in mesic site many combinations may exist: cool, moist and acidic, or warm, dry and calcareous, and so on. There are many combinations of site factor values which can lead synthetically to mesic sites. Each combination is a specific ecological niche, which can sustain a given community. The assumption states that the diversity of communities reflects the diversity of sites, which is at its maximum for mesic sites. This hypothesis has been favoured in Tilman and Pacala (1993).

The second hypothesis is a consequence of the limiting factor law. At high productivity, almost all species may potentially survive on the site. Hence, competition is fierce, and low diversity might be a consequence of the survival of the fittest. This hypothesis is linked to the previous one, but its origin is more dynamic.

2.2 DIVERSITY AND STABILITY AS A MODEL

There was a well-established consensus among ecologists that species diversity is a good assurance for the long term survival of a community: a community with very only a few species is more liable to extinction than one with many more species. Hence, diversity is implicitly related to stability. This principle, which was based on common sense more than on properly argued scientific reasoning, received a severe shock when May (May 1973) showed by using large Lotka-Volterra systems that stability decreases when the complexity of interactions increases, i.e. when the Simpson index of diversity increases. May's argument was based on a stability analysis through linearisation in the vicinity of an equilibrium. The stability of an equilibrium, i.e. its ability to resist to small perturbations, is guaranteed if every eigenvalue of the Jacobian matrix has a strictly negative real part. The Jacobian matrix is called the community matrix in population dynamics models. May constructed a random community matrix as follows: *(i)* All coefficients but a proportion equal to c, called 'connectance', are equal to zero *(ii)* Every non zero coefficient is the realisation of a Gaussian law, with mean zero and standard deviation s *(iii)* There are n species in the community. What May showed, using the semi-circular law which describes the distribution of the eigenvalues of a random matrix in the complex plane, is that the probability of the permanence, i.e. of a stable equilibrium with all species being present, is close to zero as soon as

$$\sigma\sqrt{cn} > 1$$

and close to one when

$$\sigma\sqrt{cn} < 1$$

with a transition gap of the order of $n^{-2/3}$. Hence, at a given intensity of interaction between species (s), and number of species (n), the stability suddenly drops when connectance (c) crosses the value $1/n\sigma^2$. Pointedly stated, a random Lotka-Volterra system is stable either if there are only a few connected species, or if the intensity of connections is low. Many important connections lead to instability. Whether this property remains true for general systems, i.e. whether it is generic to systems, is still debatable.

This simple model was something like a shock to the ecologist community. The quick answer was to stress that connections within an ecosystem are not distributed randomly, hence the reasoning might be biased. The idea was that proper structuration of interactions might lead to stability with an important number of related species. This structuration within models, i.e. the departure from the hypothesis that the connections within a community matrix are randomly distributed, can take two ways: *(i)* the concept of food web, in animal communities, and *(ii)* spatial interactions, within vegetal communities. Since the pioneering work of May, the study of permanence has been an active field in ecological research, labelled 'community assembly'. A review of recent developments may be found in Sigmund (1995).

2.3 DIVERSITY AND MIGRATIONS AS A COUNTEREXAMPLE

There is a natural tendency among field ecologists to relate observed data in an ecologically well structured framework. The previous paragraph shows such an example, where community diversity might be related to site factors and/or to microsite diversity. But evolution at geological time scale might be as relevant as current factors in understanding diversity patterns, as emphasised in the following example on tree species diversity in angiosperms in the Northern Hemisphere.

It has been known for a long time (Humboldt and Bonpland 1807; Wallace 1876) that diversity within many taxa follows a clinal pattern with a decrease from the Equator to the Poles. For example, there exist about 1,500 tree species in the Flora of French Guyana, at Northern shore of the Amazonian Basin, and the average number of different species encountered within one ha is close to 180, from about 600 to 700 stems. These figures may be compared with those from similar situations in temperate world, presented in Latham and Ricklefs (1993):

Two facts are noticeable: *(i)* while there is a general climatic similarity between those biomes, the variety of diversity between them is high *(ii)* figures for the warm temperate East Asia are close to those from French Guyana, although the former region definitely belongs to the temperate world and the latter to Equatorial Moist Forest Belt.

	Europe	Asia	North America, East	North America, West	Total
Subclasses	5	9	6	9	10
Orders	16	37	14	26	39
Families	21	67	19	46	74
Genera	43	177	37	90	213
Species	124	729	68	253	1166

Diversity in Tree Species, Temperate Northern Hemisphere
Source: Latham and Ricklefs 1993

Hence, a close causal relationship between climatic factors and forest tree diversity is not supported by the diversity of the tree flora in Northern Hemisphere. The usual explanation of the discrepancy between the richness of Asia or Eastern North America and the one in Europe is the existence of glaciations: in the Quaternary Ice Ages, flora could migrate from North to South back and forth in Eastern Asia or Eastern North America, whereas in Europe, flora were trapped between the periglacial climatic conditions off the ice sheet and the Mediterranean Sea. This explanation is supported by the abundance of a very rich Tertiary Flora within the European continent, which has been destroyed.

But, and we mention it to remind the reader that things seldom are that simple, Latham and Ricklefs (1993) showed that this explanation only accounts for a fraction of the discrepancy between floras, using rarefaction techniques. To complete the explanation, speciation processes both close to the supposed autochtonous birth place of angiosperms in South Eastern Asia and during migration across Siberia and Continental Asia have been suggested by Latham and Ricklefs (see Latham and Ricklefs 1993 for more details). Hence, a complex intrication of speciation, migration and selection is necessary to understand the pattern of diversity within the tree flora between the temperate biomes of the Northern hemisphere.

DIVERSITY AND POPULATION GENETICS

Let us now remark that *(i)* the limiting factor rule is a specific case of a much broader selection process active at any scale in any community, *(ii)* speciation has been evoked in tree diversity in the Northern hemisphere, and *(iii)* also migration are important processes for understanding diversity patterns.

We did not emphasise those three driving forces just by chance: there are deeply related to the four classical processes in population genetics recognised for the evolution of gene pools within and among populations (selection, mutation, drift and migration; see for example Hartl (1988) for a concise and easily accessible presentation of population genetics).

Hence, population genetics should play a major role in deciphering diversity patterns, in addition to ecological processes. Moreover, ecological processes and population genetics are naturally connected when tuning time scale, with emphasis on ecological processes at fine time scale, comparable with a generation length, and emphasis on population genetics processes at time scales involving the succession of many generations.

3. As a conclusion ...

Our conclusion will be rather short and, maybe, tautological. While presenting different sections, we emphasised the role of taxonomy, mathematics, statistical physics, ecology and population genetics for a better understanding of observed diversity. Far from being stagnant and rooted in well established paradigms within each of these fields, the study of evaluation and understanding of biological diversity on the whole, and forest biological diversity more precisely, is rapidly evolving at the cross-roads between those well established fields.

References

Beck, C. and Schlögl, F. 1993. Thermodynamics of chaotic systems. Cambridge University Press. Cambridge. 286 p.

Brillouin, L. 1962. Science and information theory. Academic Press. New York.

Grassberger, P. and Procaccia, I. 1983. Measuring the strangeness of strange attractors. Physica. 9D: 189.

Hartl, D.L. 1988. Population genetics. Sinauer. Edition française: génétique des populations. Flammarion. 1994.

Hill, M.O. 1973. Diversity and evenness: a unifying notation and its consequences. Ecology. 54: 427-431.

Humboldt, A. and Bonpland, A. 1807. Essai sur la géographie des plantes accompagnées d'un tableau physique des régions équinoxales. Paris. Cited in Latham and Ricklefs 1993. Op. Cit.

Khinchin, A.I. 1957. Mathematical foundations of information theory. Dover. New York.

Latham, R.E. and Ricklefs R.E. 1993. Continental comparisons of temperate-zone tree species diversity. In: Ricklefs, E. and Schluter, D. (eds.). Species diversity in ecological communities. Historical and geographical perspectives. Pp. 294-314.

MacArthur, R.H. and Wilson E.O. 1963. An equilibrium theory of insular zoogeography. Evolution. 17: 373-387.

Margurran, A. 1988. Ecological diversity and its measurement. Princeton University Press.

May, R.M. 1973. Stability and Complexity in Model Ecosystems. Princeton University Press.

Mayr, E. 1982. The growth of biological thought. Diversity, Evolution and inheritance. Harvard University Press. Edition française: Histoire de la biologie. Diversité, Evolution et Hérédité. 1989. Livre de Poche. 1205 p.

Ott, E., Sauer, T. and Yorke, J.A. 1994. Coping with chaos. Wiley Interscience. New York.

Pielou, E.C. 1969. An introduction to mathematical ecology. Wiley Interscience. New York.

Ricklefs, E. and Schluter, D. (eds.) 1993. Species diversity in ecological communities. Historical and geographical perspectives. Chicago University Press. Chicago. 414 p.

Rozenzweig, M.L. 1995. Species diversity in space and time. Cambridge University Press. Cambridge.

Rozenzweig, M.L. and Abramsky, Z. 1993. How are diversity and productivity related? In: Ricklefs, E. and Schluter, D. (eds.). Species diversity in ecological communities. Historical and geographical perspectives. Pp. 52-65.

Schluter, D. and Ricklefs, E. 1993. Species diversity: an introduction to the problem. In: Ricklefs, E. and Schluter, D. (eds.). Species diversity in ecological communities. Historical and geographical perspectives.

Sigmund, K. 1995. Darwin's "circle of complexity": assembling ecological communities. Complexity. 1: 40-44.

Simpson, E.H. 1949. Measurement of diversity. Nature. 163: 688.

Tilman, D. and Pacala, S. 1993. The maintenance of species richness in plant communities. In: Ricklefs, E. and Schluter, D. (eds.). Species diversity in ecological communities. Historical and geographical perspectives. Pp. 13-25.

Wallace, A.R. 1876. The geographical distribution of animals. Cited in Ricklefs, E. and Schluter, D. (eds.). Species diversity in ecological communities. Historical and geographical perspectives.

PUTTING DIVERSITY INDICES INTO PRACTICE

– Some Considerations for Forest Management in the Netherlands

M.J.J. HEUSÈRR
Dept. of Forestry, Wageningen Agricultural University
P.O. Box 342; 6700 AH Wageningen
The Netherlands

Abstract

Uneven-aged stands are diverse, both in spatial distribution of trees and in tree characteristics. The structures and natural processes in these stands are hard to describe, as are the planned and executed silvicultural operations. For clear and objective management planning in uneven-aged stands it has often been suggested to use diversity indices to describe aspects of stand structure such as diversity in species and size. However, an understanding of the meaning of these one-dimensional measures is complicated by i) the ambiguity of the concept of diversity, ii) the great diversity in structures, dynamics and silvicultural operations in uneven-aged stands, and iii) the indices' dependence on sampling design. Most diversity indices require the user to have sound theoretical and empirical knowledge.

The meaning given to an index is a particular viewpoint of reality unique to each individual, depending on personal theoretical and empirical knowledge of the index, and the variable context in which the index is understood. From this viewpoint the objectivity of indices rests solely at the debate among the manager and others involved, converging to a consensus on the meaning of an index.

Introducing diversity indices into forest management in the Netherlands requires care and thought, as well as the involvement of managers and other potential users. A number of selection criteria are recommended for choosing useful indices to describe aspects of structure. Given that few data are available in the Netherlands and that there is little empirical knowledge, special attention should be given to learn using the selected diversity indices from experience.

1. Introduction

Attention for nature conservation in forest policy has increased rapidly in the Netherlands and its neighbouring countries in the last decade, mainly in response to public pressure. Many forest enterprises desire more ecological and economic stability, since storms, and other disasters have destroyed many single-species plantations, and financial returns have been low (Hekhuis 1991). The changing demands of the general public and of forest owners have lead to the adoption of uneven-aged management strategies, abandoning clearcuts in favour of selection cut. Uneven-aged, mostly mixed-

P. Bachmann et al. (eds.), Assessment of Biodiversity for Improved Forest Planning, 171–180.
European Forest Institute Proceedings no. 18. © 1998 *Kluwer Academic Publishers. Printed in the Netherlands.*

species stands are created, as these are considered to be more stable than even-aged monocultures in economic and ecological terms, and to fulfil the forest's timber production function and nature conservation function better (Jagt et al. 1996; Klingen and Sevenster 1991).

In uneven-aged stands, trees of various sizes are mixed on a relatively small scale in various patterns. In the Netherlands there is yet little understanding of how and to what extent uneven-aged stand structures can meet the desired forest functions, and which silvicultural operations are required where and when. Planning and control in such stands require other knowledge and skills on the part of the manager than in even-aged stands (Jagt et al. 1996; Klingen and Sevenster 1991; Knuchel 1953; Leibundtgut 1977), as stand structure and natural processes are more complex. This increased complexity makes it difficult to describe type and intensity of silvicultural operations, stand structure, and stand dynamics. According to Gadow (1993), there are too many possible ways of treating an uneven-aged stand, and simple guiding images are not always sufficient to prevent foresters from becoming confused.

Bos and Hekhuis (1991) argue that in order to improve decision making in forest management, stand structures need to be formulated by a set of measurable entities, and relationships between these measures and the fulfilment of forest functions have to be defined clearly. Measures should change in response to natural causes and to management activities. Many diversity indices from ecological studies have been introduced and applied in forest management research in recent years to describe diversity (i.e. variety) in size, species, spatial distribution and other aspects of stand structure (e.g. Buongiorno 1995; Gadow and Füldner 1994; Höher 1994; Meer et al. 1994; Pretzsch 1993; Swindel et al. 1984). The expected complications for management practice entailed in using these one-dimensional measures appear to be rather serious (e.g. Köhl and Zingg 1995), but have hardly been mentioned.

The main goal of this paper is to clarify, why the choice of diversity indices for use by forest managers in the Netherlands ought to be well considered. First, the complications of using diversity indices to describe aspects of stand structure and dynamics and the type and intensity of silvicultural operations in uneven-aged stands are explained. It is then argued that in practice it is usually difficult to achieve an understanding of many indices. The paper concludes with recommendations to assure rapid appraisal of suitable diversity indices in practice, and to increase their objectivity in use.

2. Modelling aspects of the structure

In the first place, the structure of a forest stand is characterised by a spatial distribution of trees*. Secondly, each individual tree in this stand has several characteristics, such as a species name and stem volume. Aspects of stand structure, such as stand density and species diversity, may be described by combining a number of trees in terms of spatial distribution and tree characteristics in a model.

Indices now are models that typically represent one or more aspects of the stand structure in a single value, e.g. stem number and basal area. Stand dynamics, and type and intensity of silvicultural operations can be expressed in terms of the measured or desired change in these.

* Other organisms, both plants and animals, are of great importance to the stand structure too. The argument made in this paper largely may also be made for these non-tree elements in a forest stand. However, for simplicity's sake they have been omitted in this paper.

Density indices such as relative spacing and basal area have long been applied in forest management (von Gadow 1994). They indicate how completely a site is occupied by trees (von Gadow 1994) and have often been used to represent competition in growth and yield modelling (Vanclay 1994).

Generalising one could say that trees in even-aged monocultures normally have a regular spatial distribution, and there is little diversity in tree characteristics. Often a density index and its standard deviation suffice to describe stand structure, dynamics and competition when management activities are being planned or controlled.

Uneven-aged stands here are assumed to have a much less regular spatial distribution and a lot of diversity in tree characteristics. When planning management strategies here, additional indices, or even a completely new set of indices, are required to describe the increased diversity.

Diversity indices are often used in ecological research, to characterise species diversity in relation to nature conservation and environmental monitoring (Grassle et al. 1979; Magurran 1988; Spellerberg 1991). These indices have attracted attention in forest management research in recent years, and have been applied to express diversity in spatial distribution, tree characteristics and competition in forest stands. Shannon's and Simpson's indices are well known (Köhl and Zingg 1995; Magnussen and Boyle 1995; Swindel et al. 1984), less known are e.g. Poisson's index (Cox 1971; de Vries 1986) and the aggregation index (von Gadow 1994; Pretzsch 1993).

3. Increasing complexity accompanying increasing diversity

The obstacles for the practical use of diversity indices revealed in ecological and mathematical studies have hardly been discussed in forestry literature. Three of the most important complications are outlined below.

The first complication is related to the definition of the concept of diversity. In ecology diversity shows to be an ambiguous concept, including both richness and evenness (Hurlbert 1971; Magurran 1988; Peet 1974). Richness* is defined as the number of classes per number of individuals, evenness* as the extent to which individuals are equally distributed over these classes. Many popular indices, e.g. Shannon's and Simpson's, take richness and evenness into account in one single value, and are therefore difficult to interpret. Other ways of measuring diversity have been suggested, but, partly due to this ambiguity, only lead away from any uniformity in defining diversity (Hurlbert 1971; Peet 1974).

Richness and evenness are often positively correlated, but such correlation is neither biologically nor mathematically necessary (Hurlbert 1971). To overcome this inconsistency the concept of intrinsic diversity ordering has been suggested (Patil and Taillie 1982), but its use seems restricted (Köhl and Zingg 1995). Choosing of one of the many diversity indices implicitly involves a decision regarding the diversity ranking of communities that are not intrinsically comparable (Patil and Taillie 1982). For example Shannon's index is sensitive to changes in the importance of the rarest classes, whereas Simpson's index is largely influenced by the most abundant classes (Peet 1974).

The complications in discriminating between various structures, processes and activities as their diversity increases form the second obstacle. A forest stand consists of one or more eco-units within natural or man-made limits, e.g. creeks or roads. An eco-

* Richness also is referred to as variety, evenness also as equitability

unit then is every surface on which at one moment in time vegetation has begun to develop, and whose architecture, eco-physiological functioning and species composition are ordained by one set of trees until the end (Oldeman 1990).

In even-aged plantations trees are normally artificially regenerated as a whole group (Oldeman 1990), that is as one single eco-unit in the total stand area. If samples are only taken within the plantation there is little difference between plots, because there is a simple, more or less regular structure within this one eco-unit (Figure 1). The variance in spatial distribution, species and other tree characteristics within plots and between plots, is relatively small.

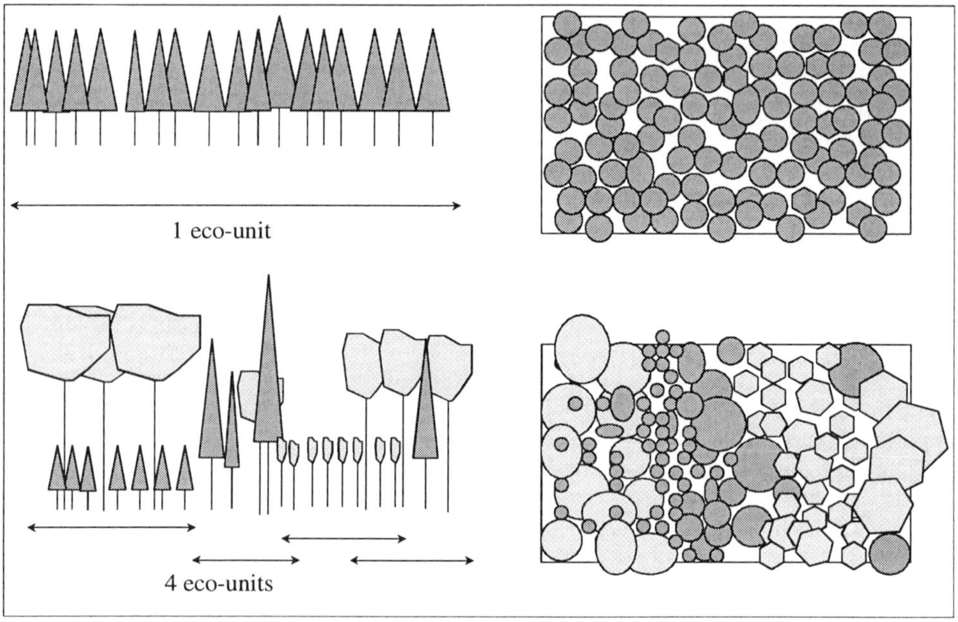

Figure 1. Plan view of a sample plot in a mono-specific even-aged stand (above) and in a mixed-species uneven-aged stand (below).

On the other hand, in managed uneven-aged stands the eco-units are small due to single-tree or group-wise harvesting and regeneration (Oldeman 1990), and a number of eco-units occur within a stand. The size of these eco-units and their boundaries may vary considerably. Samples are often taken over boundaries between eco-units of different size and shape (Figure 1), and therefore there is a relatively large variance in the spatial distribution and tree characteristics within plots.

Moreover, in an uneven-aged stand the composition of an eco-unit will not be not as regular as in a plantation. Trees here are often naturally regenerated in all kinds of irregular spatial distributions, and may be of different size and species (Oliver and Larson 1990). This leads to a high variance in the spatial distribution and tree characteristics within plots.

Clearly, for uneven-aged forest types estimates of indices will show higher variance than those for even-aged types. Complicating is that for most indices it is still largely unknown how diversity in structure influences their estimates and variances.

Thirdly, the use of most diversity indices is complicated, because they are affected by the sampling design, often in a non-linear way (Bongers et al. 1988;Magurran 1988; Magnussen and Boyle 1995; Peet 1974; Peet 1975; Spellerberg 1991). Species richness tends to increase with the sample size, while evenness tends to decrease (Hurlbert 1971). For a number of stands, Köhl and Zingg (1995) showed Simpson's and Shannon's index to be more sensitive to the use of stem number or basal area in calculating species diversity than to the number of species.

The number of individuals per plot and the number and type of classes included in the sample may affect an estimate, and lead to difficulties in comparing two estimates of an index (see also Hendrickson (1979)). Plot size and threshold diameter influence the number of trees as well as the number of eco-units within a sample plot. Classes of tree characteristics can be composed in many different ways, e.g. according to species, genera, or families, leading to a different number of classes, and a different number of individuals per class.

In even-aged plantations the relationship between plot size, threshold diameter, classes and indices is normally reasonably clear, so comparisons from different studies can be made fairly easily. In uneven-aged stands this relationship is rather complex and it is more difficult to make such comparisons. Measurement of plots with a variable size and a constant number of trees do not simplify this problem.

The dependence of a number of popular diversity indices on sampling design has been studied in detail. For example, an unbiased estimator was found for Simpson's index (Smith et al. 1979), whereas there is no unbiased estimator for Shannon's index (Blyth 1958 in: Smith et al. 1979). For Poisson's index the influence of plot size has been nullified using a 'zero-plot' diagram (Wenk et al. 1990). Nevertheless, only part of the relationship between sampling design and indices has been clarified, seldom improving user-friendliness.

If trees are incorporated into a system of co-ordinates by measuring distances between trees, any plot size smaller than the original plot may be simulated. This facilitates comparisons between different studies. However, distance-dependent indices and models still appear to be rather expensive for application in the Netherlands, and a thorough study has to be done to demonstrate how they can be incorporated into practice.

4. Understanding diversity indices

Diversity indices could be used by a manager personally, e.g. to control the results of his silvicultural activities, or simply to store information for further calculations. However, they are generally intended to have a communicative function, describing the actual or desired situation, stand dynamics, planned operations and their results.

For example, when forest managers have difficulties in justifying to the owner or the public the extent to which a desired forest function has been fulfilled, in making sure that loggers do what they are supposed to do, or to inform a newly appointed forest manager about the forest resources. Use of indices in the latter case seems valuable to ensure continuity in management, as managers in the Netherlands today seem to be employed merely for their organisational skills rather than for their silvicultural knowledge.

Such communication is often considered to become more objective by use of indices. This is because indices are often thought to be impersonal, to exist independently, as they can be checked by reference. Anyone can control the index value and see whether it has changed in accordance with the strategy pursued, but the objectivity of diversity indices is restricted by the meaning attributed to them. Indices may have different meanings to different persons, as each individual has a personal scale for interpreting these, depending on the individual's knowledge of an index.

Knowledge, the information that is stored in memory, ranges from facts, truths, principles associated with professional practice to an item that a person validates by applying personal criteria (Boud et al. 1993; Glaser et al. 1983). Much of it is essentially knowing how, experiential knowledge or practical wisdom from observing, undergoing or encountering, that cannot always be specified.

Considering the obstacles outlined above in light of this epistemological viewpoint, the meanings given to and index value will be a particular viewpoint of reality unique to an individual, and specific to time and place (cf. Hinrichs 1995). The scale used to interpret diversity indices has to be partly developed by knowledge of the theory underlying the index, partly by experience of various kinds of stand types, sometimes by measuring with various sampling designs, and partly by a bank of pre-existing yet changing signification and cultural meanings. Due to this the meaning assigned to an index by different persons may be analogue, but not univocal.

Similarly to semantic-categorical variables, diversity indices are not understood in isolation, but need to be put into a context (Albert et al. 1996; Spellerberg 1991). This context may consist of comment, of the estimate's variance, of its actual or theoretical minimum and maximum, or of additional information on the same or related forest characteristics from other indices and models.

For example, consider a forest manager analysing the stand structure and dynamics of two different stands. The estimates of Shannon's index for species diversity in both stands are 0.9. If the manager is merely shown the figures, both estimates appear univocal. From the estimates' variance or a short walk through the forest the manager will learn, that the two identical estimates do not refer to an identical forest structure; they are not univocal, but analogue. The meaning of a diversity index to an individual is clarified in the wider context that describes the structure.

Now communication is neither a transfer nor sharing of information, but a process in which knowledge is created by the joint involvement of all parties encompassing among others scientific and aesthetic judgements. When using an index in forest management, the person doing the describing should make sure that the recipient of the description attributes the same meaning to the index, as there is no guarantee that they will attribute the same meaning to it. Objectivity by using a diversity index can therefore only rest upon consensus being reached after open debate among managers, public, forest workers and others, creating the context in which the index is to be placed.

Index values can be converted into plain text to simplify communication. Gadow and Füldner (von Gadow and Füldner 1994) suggest that descriptions of structures, processes or management activities will be better understood as numeric indices than as discrete semantic-categorical variables. Numbers may be continuous whereas semantic-categorical variables are merely discrete, e.g. a reduction of the Shannon-Wiener index by 0.4 versus 'felling 75% of the undesired tree species throughout the stand'. However, considering the argument made here it is clear that there is no reason to believe that numeric indices in themselves will be understood better than semantic-categorical variables.

5. Assuring objectivity in management use

From the above it is clear that the understanding of diversity indices as well as other indices describing aspects of stand structure in general depends on personal knowledge of these indices, acquired by studying the sometimes complex mathematical and ecological theories about these indices, personal observations relating to the indices in the forest, and the broad context in which the indices are understood. Ideally, the empirical and theoretical knowledge on indices is first broadened by research on the complex relationships between structure, processes, activities, sampling design, indices and variances before being applied in actual management.

However, forest managers in the Netherlands are transforming plantations into uneven-aged forest and they need tools to guide this transformation process now. Diversity indices new to management practice need to be selected and introduced as soon as possible. This selection should be well considered, so that objectivity in describing aspects of stand structure by diversity indices is assured as much as possible.

Before beginning with the selection of indices it is advisable to clearly and univocally define some concepts and indices. Not only the concept of diversity is unclear, but also, for example, the definition of eco-unit boundaries (Meer et al. 1994) and of Shannon's index. In the formula of the latter a natural logarithm (Meer et al. 1994; Spellerberg 1991), a $^{10}\log$ (Peet 1974) and a $^{2}\log$ (Bongers et al.1988) have been used.

To avert serious problems in use, indices and sampling design must be standardised. Suitable diversity indices may be selected using the six following criteria (partly based on Magurran 1988).

i) *the aspect of diversity being measured, and its direct relation to function fulfilment*

ii) *the index's ability to discriminate between different structures, processes and treatments*
For a better understanding an index must be able to discriminate between various situations and changes. The variance or measure of variation should be given with the index to improve this understanding (see also Spellerberg 1991). The relationships between various stand types differing in composition and size of eco-units, the index-values, and their variances need to be investigated. Indices with relationships that are easy to interpret should be selected, e.g. linear ones.

iii) *the index's dependence on sampling design*
In a similar way the relationship between sampling design and an index should be studied. For some indices it is recommended to standardise the sampling design to enable comparisons of different studies (Peet 1974, Spellerberg 1991).

iv) *other indices and models selected to describe structure, processes and treatments*
Other indices and models selected make up large part of the context in which a selected index is understood. They are required to understand what is going on in a stand. Here it should be noted that the argument outlined holds to a greater or lesser degree for models in general.

v) *costs and returns of estimating a set of suitable indices*
From indices and models based on well understood theory, a suitable set of indices and models that complement each other may be selected if it is known what aspects

of stand structure need to be described and in what detail. As sampling is laborious, it is costly. Therefore it is very important to weigh up costs and returns of using indices or models instead of global observations in the first place.

Then when indices are to be selected, it has to be remembered that it is more favourable to assess all desired indices from one set of plots, all of the same size, or another well-designed sample. Estimating indices from various plot types may be very accurate, but the high costs will be unacceptable for practice.

Nowadays it may be relatively easy to choose new sampling methods for practical use in uneven-aged management. The sampling techniques and indices currently used in Dutch forest management are unsatisfactory (Schoonderwoerd and Klein 1993), and therefore managers may be willing to change these to learn more about uneven-aged forests and their management.

vi) *whether the index is already widely 'understood' and measured*
In the light of the previous criterion, diversity indices new to forest management can only be understood by managers and others investing a considerable amount of time and research effort. Several observation techniques, sampling designs, indices, and other models are already in use, and are fairly well understood by managers. Though they may not perform optimally, the costs of introducing new indices and another sampling design may not justify abandoning the methods already in use. Especially if these methods are already well understood, it may be worthwhile merely slightly modifying these.

The potential users should be involved in selecting a suitable set of indices by these criteria. They need simple, efficient indices, whose strengths and drawbacks are understood. Learning from experience plays an essential role in understanding the selected indices, next to the study of their theoretical aspects. Understanding becomes internalised by using the indices; learning is enhanced. Therefore the complexity of the indices selected need not be constrained by the current knowledge of the potential users.

6. References

Albert, M., Gadow, K. von, and Kramer, H. 1996. Zur Strukturbeschreibung in Douglasien-Jungbeständen am Beispiel der Versuchsflächen Manderscheid und Uslar. Allg. For. Jag. Zeit. (166): 205-210.

Bongers, F., Popma, J., Meave del Castillo, J., and Carabias, J. 1988. Structure and floristic composition of the lowland rain forest of Los Tuxtlas, Mexico. Vegetatio. (74): 55-80.

Bos, J. and Hekhuis, H.J. 1991. Beheersplanning bos met meervoudige functievervulling: doeloperationalisering en technische produktiefuncties. Report 671. Institute for Forest and Nature Research. Wageningen. The Netherlands.

Boud, D., Cohen, R., and Walker, D. (eds.). 1993. Using Experience for Learning, SHRE and Open University Press. Buckingham. United Kingdom.

Buongiorno, J., Peyron, J., Houllier, F. and Bruciamacchie, M. 1995. Growth and Management of Mixed Species, Uneven-Aged Forests in the French Jura: Implications for Economic Returns and tree Diversity. For. Sc. (41): 397-429.

Cox, F. 1971. Dichtebestimmungen und Strukturanalyse von Pflanzenpopulationen mit Hilfe von Abstandmessungen. PhD-thesis, Forestry Faculty. University of Göttingen. Germany.

Davis, K.P., and Johnson, K.N. 1987. Forest Management. McGraw-Hill. USA.

Gadow, K. von. 1993. Management-base Control Inventory. In: Vanclay, J.K., Skovsgaard, J.P. and Gertner, G.Z. (eds.). Growth and Yield Estimation from Successive Forest Inventories. Proceedings IUFRO Conference. Forskningsserien 3. Forest and Landscape Research Institute and Royal Veterinary and Agricultural University. Denmark. Pp. 25-30.

Gadow, K, von. 1994. Modelling Growth in Managed Forests – Realism and limits of lumping. Paper for the NATO Advanced Workshop on Ecosystem Modelling. 21st February 1994. Bischofsgrün. 16 p.
Gadow, K. von, and Füldner, K. 1994. How to define a thinning in a mixed deciduous beech forest. In: Pinto da Costa, M.E. and Preuhsler, T. (eds.). Mixed Stands. Proceedings IUFRO working groups S4.01-03 and S4.01-04. Universida Tecnica de Lisboa. Lisbon. Portugal. Pp. 31-42.
Gadow, K. von, and Füldner, K. 1995. Zur Beschreibung forstlicher Eingriffe. Forstw. Cbl. (114): 151-159.
Glaser, E.M., Abelson, H.H. and Garrison, K.N. 1983. Putting knowledge into use – facilitating the diffusion of knowledge and the implementation of planned change. Jossey-Bass Inc. San Francisco. USA.
Grassle, J.F., Patil, G.P., Smith, W., and Taillie, C. 1979. (eds.). Ecological Diversity in Theory and Practice. Statistical ecology. Volume 6. International Co-operative Publishing House. Maryland. USA.
Hekhuis, H.J. 1991. Oorzaken van verschillen in exploitatieresultaat tussen particuliere bosbedrijven. Nederlands Bosbouwtijdschrift. (63): 186-193.
Hendrickson Jr., J.A. 1979. Estimates of species diversity from multiple samples. In: Grassle, J.F., Patil, G.P., Smith, W. and Taillie, C. (eds.). Ecological Diversity in Theory and Practice. Statistical ecology. Volume 6. International Co-operative Publishing House. Maryland. USA. Pp. 145-158.
Hinrichs, A. 1995. Assessment of Information Systems from the management viewpoint: a contribution to the discussion about the usefulness of Geographic Information Systems in forest enterprises. Allg. For. Jag. Zeit. (166): 225-229.
Höher, G.C. 1994. Waldbauliche Untersuchungen zur Enstehung, Struktur und Weiterentwicklung der Erdmannbestände im Forstamt Erdmannshausen. Erdmannshausen. Germany.
Hurlbert, S.H. 1971. The non-concept of species diversity: a critique and alternative parameters. Ecology. (52): 577-586.
Jagt, J.L., Jans, M.F.F.W., and Paasman, J.M. 1996. Geintegreerd bosbeheer ingekleurd. Volume B-17. IKC-Natuurbeheer. Wageningen. The Netherlands.
Kelty, M.J., Larson, B.C. and Oliver, C.D. (eds.) 1992. The Ecology and Silviculture of Mixed Species Forest. Forestry Sciences. Volume 40. Kluwer Academic Publishers. Dordrecht. The Netherlands.
Klingen, S. and Sevenster, J. 1991. Geïntegreerd Bosbeheer. Natuur en Milieu Overijssel. Zwolle. The Netherlands.
Knuchel, H. 1953. Planning and control in the managed forest. Oliver and Boyd. Edinburgh. United Kingdom.
Köhl, M., and Zingg, A. 1995. Eignung von Diversitätsindizes bei Langzeituntersuchungen zur Biodiversität in Waldbeständen. Allg. For. Jag. Zeit. (167): 76-85.
Leibundgut, H. 1977. Über die Notwendigkeit naturnahen Waldbaues. Schw. Z. f. Fw. (128): 490-501.
Magurran, A.E. 1988. Ecological Diversity and Its Measurement. Croom Helm. London. United Kingdom.
Magnussen, S., and Boyle, T.J.B. 1995. Estimating sample size for inference about the Shannon-Weaver and the Simpson indices of species diversity. For. Ecol. Man. (78): 71-84.
Meer, P.J. van de, Bongers, F., Chatrou, L., and Riéra, B. 1994. Defining gaps in tropical forests. Acta Oecologia. (15): 701-714.
Nagel, J. 1994. Structural analyses of the forest reserve Saubrink on the basis of a permanent sample plot data. In Pinto da Costa, M.E. and Preuhsler, T. (eds.). Mixed Stands. Proceedings IUFRO working groups S4.01-03 and S4.01-04. Universida Tecnica de Lisboa. Lisbon. Portugal. Pp. 81-96.
Oldeman, R.A.A. 1990. Forests: Elements of Silvology. Springer Verlag. Berlin. Germany.
Oliver, C.D. and Larson, B.C. 1990. Forest Stand Dynamics. Biological Resource Management Series. McGraw-Hill. New York. USA.
Patil, G.P., and Taillie, C. 1982. Diversity as a concept and its measurement. J. Am. Stat. Assoc. (77): 548-561.
Peet, R.K. 1974. The measurement of species diversity. Ann. Rev. Ec. Sys. (5): 285-307.
Peet, R.K. 1975. Relative diversity indices. Ecology. (56): 496-498.
Pretzsch, H. 1993. Analyse und Reproduktion räumlicher Bestandesstrukturen. Schriften aus der Forstlichen Fakultät der Universität Göttingen und der Niedersächsischen Forstlichen Versuchsanstalt. Göttingen. Germany. Volume 114.
Schoonderwoerd, H. and Klein, J.P.G. de. 1993. Inventarisatie van kleinschalig bos. Nederlands Bosbouwtijdschrift. (65): 100-109.
Smith, W., Grassle, J.F. and Kravitz, D. 1979. Measures of diversity with unbiased estimates. In: Grassle, J.F., Patil, G.P., Smith, W. and Taillie, C. (eds.). Ecological Diversity in Theory and Practice. Statistical ecology.. International Co-operative Publishing House. Maryland. USA. Volume 6. Pp. 177-191.
Spellerberg, I.F. 1991. Monitoring Ecological Change. Cambridge University Press. Cambridge. United Kingdom.
Swindel, B.F., Conde, L.F. and Smith, J.E. 1984. Species diversity: concept, measurement, and response to clearcutting and site-preparation. For. Ecol. Man. (8): 11-22.

Vanclay, J.K. 1994. Modelling forest growth and yield, applications to mixed tropical forests. CAB International. Wallingford. United Kingdom.

Vries, P.G. de. 1986. Sampling Theory for Forest Inventory. Springer Verlag. Berlin. Germany.

Wenk, G., Anatanaitis, V. and Smelko, S. 1990. Waldertragslehre. Deutscher Wirtschaftsverlag Berlin. Berlin. Germany.

TECHNIQUES FOR ADDRESSING SPATIAL DETAIL IN FOREST PLANNING

HOWARD HOGANSON
Department of Forest Resources, University of Minnesota,
1861 Hwy 169 East, Grand Rapids MN 55744; USA

JOSE BORGES
Departamento de Engenharia Florestal, Instituto Superior de Agronomia,
Tapada da Ajuda, 1399 Lisboa Codex, Portugal

DENNIS BRADLEY
North Central Forest Experiment Station, USDA Forest Service,
1992 Folwell Avenue, St Paul MN 55108; USA

Abstract

Specialized model solution approaches can be designed for forest management scheduling problems by utilizing an understanding of the problem. A specialized decomposition approach has made it possible to address larger problems. It has proven successful in applications. Concepts of moving windows from geographic information systems can be combined with dynamic programming (DP) techniques to address adjacency considerations in large problems. This DP approach and the specialized decomposition approach can likely be combined to help identify ways of sustaining both timber production and forest-wide spatial conditions such as the amount of forest edge or interior space. The specialized decomposition approach has been expanded to address spatial interactions between timber markets. Similar expansions seem plausible to address broader spatial environmental concerns related to forest biodiversity.

1. Introduction

Forest managers are being asked to sustain healthy ecosystems and provide timber for an increasing world population. Biodiversity, an important element of healthy ecosystems, must be addressed by considering the spatial arrangement of management activities. Spatial factors are difficult to address in forest planning models because of interdependencies between individual management units. This difficulty at least partially explains why spatial factors have received relatively little attention in scheduling models used in practice. Yet computer technology has potential to help substantially. Geographic information systems also offer enormous potential to help formulate analyses and interpret results.

Spatial aspects of forest management are complicated by the many objectives of management, each with its own spatial facets. Several key measures have been identified

P. Bachmann et al. (eds.), Assessment of Biodiversity for Improved Forest Planning, 181–188.
European Forest Institute Proceedings no. 18. © 1998 *Kluwer Academic Publishers. Printed in the Netherlands.*

regarding spatial arrangement. Forest edge has long been considered an important spatial measure impacting wildlife habitat. Another is interior space of mature forests. Other important measures are being identified as the importance of forest biodiversity becomes better understood. Clearly, it would be desirable if better modelling tools were available to help identify effective strategies for sequencing harvests across large landscapes. It is important to understand trade-offs between timber production opportunities and spatial dynamics of the forest. Such understanding is essential for policymaking. In this paper we review some potentially promising modelling techniques for better addressing spatial aspects of management. Key to the techniques are the ability to integrate an understanding of the forestry problem into the mathematical solution processes used.

2. A Specialized Decomposition Approach for Large Problems

Hoganson and Rose (1984) developed an approach for forest management scheduling that decomposes the problem into subproblems and uses an economic interpretation to help determine an optimal solution in terms of an equivalent linear programming (LP) formulation. The DUALPLAN model implements this approach (Hoganson and Rose 1989). The approach is a form of lagrangian relaxation (Fisher 1981) with specialized techniques to search for the values of the dual variables. Key to the process is the similarity between a forest-wide scheduling model and basic economic analyses of individual projects. This similarity has also been emphasized by others (Paredes and Brodie 1989) and is key in interpreting model results. Essentially each DUALPLAN subproblem is an economic cash flow analysis for a specific forest management unit; in each analysis, shadow prices (dual variables) associated with the forest-wide constraints are used to measure the impact of management unit decisions on forest-wide constraints. An iterative process is used to search for shadow prices that make the overall schedule within tolerable limits of the forest-wide constraints.

A recent forest management text (Davis and Johnson, 1987, p. 672) gives a positive review of the specialized decomposition approach. It works well for several reasons. First, impacts of changes in forest values are fairly predictable as harvest timings can usually be adjusted to take advantage of periods of high product prices and avoid periods of low prices. Shadow prices on timber products are thus likely correlated over time unless harvest levels are free to fluctuate greatly over time. Second, the forestry problem itself is not as precisely defined as one might interpret from a strict mathematical formulation; for most situations some minor violations of the forest-wide constraints are acceptable. More common primal solution methods emphasize the need to maintain strict feasibility, and this comes at some computation cost.

Like linear programming models, DUALPLAN provides estimates of the value of all management options for each unit. Such information is valuable in implementation when managers consider whether factors not recognized in the model are reason to select another alternative. DUALPLAN solutions also do not split management units by assigning more than one alternative to each unit. With more traditional methods this requires the use of mixed integer programming techniques. Recent work has shown that DUALPLAN can address additional forest-wide constraints involving management inputs or outputs other than harvest amounts (Hoganson and McDill 1993, Hoganson and Kapple 1995).

3. Addressing Adjacency with Dynamic Programming

With the large and increased interest in ecosystem management, methods for addressing spatial relationships within forest management scheduling models have received considerable attention in the forestry literature. Emphasis in the literature has been on adjacency concerns. Examples include: Jones et al. (1991), Murray and Church (1994), Nelson and Errico (1993), Weintraub et al. (1994) and Snyder and Revelle (1996). As noted by Weintraub et al. (1994) most studies are theoretical with little consideration as to how such methods would be implemented. Methods generally involve either integer programming formulations or simulation approaches. With simulation, stands are scheduled sequentially with spatial implications associated with stands scheduled earlier eliminating alternatives for stands scheduled later. The sequencing of stands for analysis is thus important. With integer programming, model size is limiting because of numerous additional constraints needed to represent spatial limitations. Simulation approaches leave open the question of performance in terms of optimality and the need for considering other forest-wide constraints is generally ignored or greatly simplified. The USDA Forest Service uses a revised version of SNAP (Sessions and Sessions, 1988), a simulation model, to help address spatial concerns. SNAP emphasizes site-level constraints but is not tied to forest-wide objectives. LANDIS (Mladenoff et al. 1996), an ecosystem model for the Lake States is also a simulation model that links directly with a GIS and has routines for tracking various indices of landscape pattern. It does not have a scheduling routine for harvests and depends on user input for such decisions.

Dynamic programming (DP) techniques, when applicable, are usually far more efficient than linear or integer programming. Model efficiency is critical when there is desire to expand models to recognize additional details. Hoganson et al. (1994) proposed a DP approach for addressing adjacency concerns. In the proposed DP formulation, each stage of the problem corresponds directly to the management decision for a specific stand (management unit). State variables relate to the schedule implied for adjacent stands addressed earlier in the DP network. Each arc represents a management alternative for the associated stage (stand). Objective function values associated with each arc are the net returns for with the corresponding alternative. For arcs that violate adjacency constraints either a large cost can be assumed or the arc can be eliminated. The recursive nature of DP links the value of all stands represented in the network.

Key to the DP formulation is the definition of the state variables. For each stage, there is an associated "front" which separates the "scheduled" stands from the "unscheduled" stands. "Scheduled" stands are those stands represented in earlier stages of the network (the schedules selected for them are determined until the optimal solution is traced through the network after all stages are addressed). Each "scheduled" stand along this front represents one state dimension for the corresponding stage. The condition tracked with each dimension is the status of the stand over time in terms of adjacency concerns. By tracking conditions for all stands along the front, it is possible to identify which alternatives for the stand corresponding with each stage would violate adjacency constraints. Key to the formulation is the fact that the formulation does not grow in size like a decision tree because state descriptors are to longer needed for those stands represented earlier in the network and no longer along the front.

From the notion of a front, it is clear that the order in which stands are sequenced in the DP network will influence model size. A case in which model size would be smallest is one where stands in the forest are all in a single row with no stand adjacent to more than two others. With this arrangement, one could start the DP formulation at one end of the forest and move sequentially across it with the "front" always only one stand wide. The number of stands in the forest would not be critical as only the number of stages would increase with increasing stand numbers, not the number of state dimensions at each stage. Expanding the concept to a forest of two stands wide (stands of similar size and shape), the DP formulation would have two to three stands along the front depending on how stands are arranged. In general, assuming that the number of alternative types does not vary substantially between stands, DP formulations will be smaller for narrow forests such that the stands can be sequenced "lengthwise" to keep the "front" narrow.

To solve large, realistic-size problems, an approach proposed by Hoganson et al. (1994) links DP formulations of overlapping subsets of the forest. The approach is defined by five steps:

1) Start at one edge of the forest and select a strip of the forest that can be solved using DP. Make the strip long, generally running the entire length of the forest and narrow enough to be manageable in terms of a DP formulation for stands within the strip (window).
2) Formulate and solve the DP problem assuming the only stands in the forest are the stands within the strip.
3) If the entire forest has been included in a DP window, then all stands yet to be scheduled can be scheduled according to the optimal DP solution found in step 2. Otherwise, for stands along the outside edge of the strip (subproblem) in step 2, schedule those stands according to optimal DP solution determined.
4) Re-define the strip by: (a) eliminating from the latest strip, all stands scheduled in step 3, and (b) adding to the strip, a set of stands adjacent to the strip that have not yet been analyzed.
5) Return to step 2.

The approach is comparable to the use of moving windows in geographic information systems. Windows in this case are long in one direction. Window widths influence the number of states in each stage of the DP and must be matched with model size limitations. In effect, each pass of the window schedules stands along the outside edge of the strip (window). It is assumed that the optimal schedule for these stands are not influenced by adjacency constraints involving stands not yet considered in a window. The ability of the approach to find near-optimal solutions thus depends on what is lost by this simplifying assumption.

A hypothetical 1000 stand forest was used to examine performance of the "moving windows" approach for large problems (Hoganson et al. 1994). Each stand was one cell of a rectangular grid of stands. Up to 7 management alternatives were considered for each stand. Information for each stand was generated randomly to mimic age and site quality conditions of the aspen forest type in the Lake States. A one-period exclusion period was assumed for adjacency constraints. For each of three test cases, the approach was applied varying both the side (edge) of the forest on which to start the process and the window size. For the case of an older forest with more stands financially mature, like those typical of Minnesota, the lowest-cost solution had adjacency constraints costing $45.70 per acre. This cost was approximately 9.5 percent of the net present value of the forest under the unconstrained case. For each of the three test cases the approach was compared to the approach proposed by Weintraub et al. (1994). A window size only two stands wide outperformed the Weintraub algorithm in all cases. Solutions showed no sensitivity to the side of the forest used to start the process. All formulations involving less than 16,000 states at any stage (window size of 4 to 5) were solved in less than 3 minutes using a 50 MHz 486DX2 computer.

Considering forests with irregular shaped stands complicates the implementation as formulating the DP efficiently is not straightforward. This has been the subject of additional research (Borges et al. 1994). Recent unpublished work has implemented the DP approach to 20 test cases involving forests with 800 to 3200 stands of irregular shapes and sizes. Formulations with as many as 8 periods and 16 management alternative types have been considered. Model performance has been especially encouraging. Tests suggest the approach can come very close to optimal solutions. With irregular shaped stands the approach still easily outperformed the Weintraub approach and greatly outperformed simpler heuristics. Over large areas savings from a model like this could be enormous. Case studies mimicked relatively low-valued, pulp-producing, forest lands in northern Minnesota. One would expect greater gains in areas involving higher-valued forest products.

Initial reviews of recent results also suggest some potentially important implications for management. Although still preliminary, results suggest that returns from timber production under adjacency constraints are not overly sensitive to the method used to design management unit boundaries. If true, this could have substantial implications regarding appropriate management strategy. It would suggest it is appropriate to focus primarily on environmental concerns when designing management unit boundaries. Not as surprising, results also suggest that costs of adjacency constraints will vary tremendously depending on whether sustainable flow constraints are also imposed. Results also add insight to the impact of adjacency constraints on overall spatial conditions. Cases show that even with adjacency constraints imposed, conditions like the amount and type of forest edge and interior space will vary tremendously over time. This clearly suggests that improved problem formulations are needed to better address spatial concerns. The DP approach may offer some additional appeal for improved formulations as described below.

4. Integrating the Decomposition Approach and the DP Adjacency Approach

Forms of hierarchial planning have been suggested for linking detailed subproblems (Weintraub 1994, Church et al. 1994). Consensus is lacking as to how to make the linkages. A concern with hierarchal approaches is the potential of ignoring important spatial interactions between areas that are assigned to different subproblems. Ideally, one would like to define the subproblems such that boundaries follow natural boundaries which result in relatively independent subproblems. However, with such subdivisions the size subproblems may be too large.

Integer programming approaches usually treat adjacency interrelationships as constraints that must be satisfied at all costs. The DP approach offers potential to view adjacency constraints differently. With the DP approach, violations of adjacency constraints are prevented by assigning a high cost to all arcs corresponding with combinations of management alternatives that violate any adjacency constraint. Modifications in adjacency cost assumptions make it possible to consider the impact of allowing some violations. Benefits in fact might be assigned. Such might be the case for combinations of alternatives for neighboring stands that sustain existing interior space. In effect, one could consider trade-offs between spatial objectives and timber production. One could thus focus on the intent of spatial constraints rather than on the constraints themselves. For example, rather than focus on violations of a forty acre clearcut rule, one could shift emphasis to the amount and type of forest edge and interior space produced over time under various assumptions concerning the value of these spatial "products."

The decomposition approach used in DUALPLAN emphasizes the linkage between the forest-wide problem and economic analyses at the stand level. The DP approach has the potential to work well within this decomposition approach. The DP approach could serve as the basic method for addressing the stand-level subproblems during each iteration of the larger forest-wide model. It would be the link that recognizes interdependencies between the subproblems. Although many DUALPLAN iterations may be needed, tests on the DP approach to date suggest that very small DP formulations lead to good solutions; thus the time required for each iteration of the forest-wide model might still be minimal as small window sizes may be adequate for intermediate iterations of the DUALPLAN process. Shadow prices would incorporate into the DP analyses the effects of forest-wide constraints such as allowable cut levels or market-level production targets. For the final few iterations it may be desirable to consider larger more detailed DP formulations. Research is planned at the University of Minnesota to focus on this modelling approach.

5. Beyond Adjacency: A Possible Approach for Broader Spatial Concerns

The DUALPLAN modelling approach was expanded in the software package DTRAN (Hoganson and Kapple 1991) to recognize market interactions when multiple timber markets share a common timber production area. Of central focus in DTRAN are transport costs and stand-level decisions involving timber market destinations. DTRAN overcomes the combinatorial nature of integrating harvest timing decisions and wood shipping decisions by utilizing maps of wood procurement zones to make shipping decisions. An optimal shipping schedule is identified for each management unit by simply referencing its location on maps which define procurement zones for each market and product combination. In terms of an LP formulation of the problem, it is the marginal values for the different products in each period at each market location that define the zone boundaries. Boundaries change on the maps in the solution process as estimates of the marginal value of each product changes for each market. The model thus overcomes the need to utilize separate decision variables to enumerate combinations of possible shipping options and harvest timing options.

DTRAN served as the forest management scheduling model for the $900,000 Minnesota Generic Environmental Impact Statement (GEIS) on Timber Harvesting and Management (Jaakko Pöyry Consulting 1994). Statewide analyses for the GEIS involved over 10,000 analysis areas and up to four timber product groups for each of six major market locations. Results demonstrated that this specialized technique can be used to help model an enormous amount of detail at even a statewide level. Furthermore, these techniques offer "room" to incorporate substantially more detail. A logical next step in model improvement is the explicit recognition of additional spatial interrelationships.

The DTRAN solution process searches for market prices that move the system to equilibrium. Maps are used in the process to reflect the impact of location on forest product values. Maps might also be used directly in a solution process to address spatial variation in landscape conditions and values related to ecosystem management objectives. These objectives might be much broader than adjacency-like considerations. As in DTRAN, the model could adjust value estimates to move the forest-wide system towards equilibrium. Equilibrium in this case would relate to how landscape-level spatial measures compare to landscape-level objectives.

To describe this approach further, consider a simplified example. Assume an objective of management is to maintain a site-specific environmental index at some average value for the forest as a whole. Several specific spatial measures might go into this index. For a specific management schedule and its summary of local conditions, one could determine the impact that changes in these conditions would have on the index. For example, in one location the index might change by +.1 for each additional land unit maintained as old aspen, and in another location, the change might be -.2 per each additional land unit of old aspen. Thus spatially one could differentiate the impact of producing additional units of the forest conditions important for the index. This information could be mapped. Using the map and assuming a single marginal value for a targeted forest-wide environmental index average, one could create a map of marginal values for the key stand conditions locally. These values could be used in the DUALPLAN valuing process. The maps of site-specific values for spatial factors would also be valuable for understanding management trade-offs between potential objectives. Examples of specific spatial measures that might be addressed include habitat suitability indices and local measures of forest biodiversity. Exploring the potential strengths and weaknesses of this approach seems desirable.

6. Conclusions

Computer models offer enormous potential to explore an almost limitless set of potential management schedules. Yet forest management problems are so complex that when spatial concerns like those related to biodiversity are recognized, a mathematical formulation of the problem get so enormously large and complex that it is impractical to solve it using standard operations research techniques. Yet we have enormous potential to combine our understanding of the forestry problem and operations research to design specific modelling approaches to make practical application possible. And much of what can be integrated need not be all that complicated. Major gains can likely be made simply by utilizing maps in solution processes. Mathematical modelling and common sense can mix to help lead us to better forest management decisions.

References

Borges, J. 1994. A modelling approach to spatial management constraints in forest management. PhD dissertation. University of Minnesota. 117 pp.

Church, R., A. Murray and K Barber. 1994. Designing a hierarchical planning model for USDA Forest Service planning. pp 401-409 *in* Proceedings of the 1994 Symposium on Systems Analysis in Forest Resources, Sept. 6-9, 1994, Pacific Grove, CA 482 p.

Davis, L. S. and K. N. Johnson. 1987. *Forest management*. McGraw-Hill, New York. 790 p.

Fisher, M. 1981. The Lagrangian relaxation method for solving integer programming problems. *Management Science* 27(1):1-18.

Hoganson, H. J. Borges and D. Rose. 1994. A dynamic programming approach to solving adjacency problems. in forest management. Paper presented at the 1994 Symposium on Systems Analysis in Forest Resources, September 6-9, 1994, Pacific Grove, California.

Hoganson. H. and D. Kapple. 1995. Estimating impacts of extended rotation forestry. Final research report submitted to the Minnesota Department of Natural Resources. 170 pp.

Hoganson, H. M. and D. C. Kapple. 1991. DTRAN 1.0: A multi-market timber supply model. College of Natural Resources and Agricultural Experiment Station, Department of Forest Resources Staff Paper Series Report No. 82, University of Minnesota, St. Paul. 66 pp.

Hoganson, H., and M. McDill. 1993. Relating reforestation investments in northern Minnesota with forest industry needs, nontimber values and mitigation strategies. research report submitted to the Charles K. Blandin Foundation. 232 p.

Hoganson, H. M. and D. W. Rose. 1984. A simulation approach for optimal timber management scheduling. *Forest Science*. 30(1):220-238.

Hoganson, H. M. and D. W. Rose. 1989. DUALPLAN version 1.0 users manual. College of Natural Resources and Agricultural Experiment Station, Department of Forest Resources Staff Paper Series Report No. 73, University of Minnesota, St. Paul. 48 pp.

Jaakko Pöyry Consulting, Inc. 1994. *Generic Environmental Impact Statement on Timber Harvesting and Forest Management in Minnesota.* Tarrytown, NY: Jaakko Pöyry Consulting, Inc. 813 p.

Jones, J., J. Meneghin, and M. Kirby. 1991. Formulating adjacency constraints in optimization models for scheduling projects in tactical planning. *Forest Science*. 37(5):1283-1297.

Mladenoff, D., G. Host, J. Boeder and T. Crow. 1996. A spatial model of forest landscape disturbance, succession and management. in: Goodchild, M. et al. eds. GIS and Environmental Modeling: progress and research issues. Fort Collins CO: GIS World Books: 175-179.

Murray, A. T. and R. L. Church. 1994. Adjacency constraint aggregation. pp 131-138 in Proceedings of the International Symposium on Systems Analysis and Management Decisions in Forestry. March 9-12, 193, Valdivia, Chile. 482 p.

Nelson, J., and D. Errico. 1993. Multiple pass harvesting and spatial constraints: an old technique applied to a new problem. *Forest Science*. 39(1):1-15.

Paredes V., G., and J. Brodie. 1989. Land value and the linkage between stand and forest level analyses. *Land Economics*. 65(2):158-166.

Sessions J. and J. Sessions. 1988. Scheduling and network analysis program (SNAP). User's Guide. Dept. of Forest Management. Oregon State University. Corvallis OR.

Snyder, S. and C. ReVelle. 1996. The grid packing problem: selecting a harvest pattern in an area with forbidden regions. *Forest Science*. 42(1):27-34.

Weintraub, A., F. Barahona and R. Epstein. 1994. A column generation algorithm for solving general forest planning problems with adjacency constraints. *Forest Science*. 40(1):142-161.

DESIGN-BASED APPROACHES FOR INFERENCE ON DIVERSITY

L. BARABESI
Dipartimento di Metodi Quantitativi, Università di Siena
P.zza S.Francesco 17, 53100, Siena, Italy

L. FATTORINI
Dipartimento di Metodi Quantitativi, Università di Chieti
V.le Pindaro 42, 65127, Pescara, Italy

Introduction

A number of measures have been proposed to quantify species diversity in a community from theoretical, mathematical or ecological point of views. For an updated review of these proposals see Gove et al. (1994). Obviously, all of these indexes are functions, say Δ (N), of the unknown abundance vector N $=\left[N_1, N_2, \ldots, N_k\right]^T$, where N_l represents the number of units belonging to each species $(N_l > 0, l = 1,2,\ldots,k)$ when there is a total of k species in the entire population. Accordingly, the problem consists in the estimation of Δ by means of suitable statistics on the basis of a sampling design.

A number of results are available in literature in the case of random sampling, i.e., when n units are randomly and independently drawn from the population. Blyth (1958) is the first one to deny the existence of any unbiased estimator of the Shannon measure. Subsequently, Bowman et al. (1971) provide a bias expansion for the sample Shannon's measure, while Adams and McCune (1979) investigate the performances of a generalized jackknife procedure to reduce this bias. Moreover, Smith et al. (1979) give very general results on the estimation of dichotomous diversity measures and obtain the minimum variance unbiased estimators for the expected species diversity measures. Finally, Patil and Taillie (1979) perform the estimation of Δ_β indexes and expected species diversity measures by using the ordinary jackknife.

However, very few ecological communities have a frame, so that it is not usually feasible or practical to obtain a random sample of their members. Pielou (1966) is the first one to account for the actual field conditions by using a cumulative pooled quadrat sampling plan. Given a random sample of quadrats the object index is computed on the accumulated quadrats by adding quadrats one at a time until the index, on visual inspection, stops varying. However, no theoretical results on the statistical properties of the resulting estimates are available, except for those provided by Heyer and Berven (1973), who extend the Pielou procedure to different random orderings, so as to obtain improved estimates as well as standard error evaluations. The aim of this paper is to take further steps in the Pielou's direction by using alternative sampling designs as well as functions of the sample data which are "suitable", in the sense that the sampling distribution of the resulting estimators is simply obtainable by the standard results of statistics.

P. Bachmann et al. (eds.), Assessment of Biodiversity for Improved Forest Planning, 189–195.
European Forest Institute Proceedings no. 18. © 1998 *Kluwer Academic Publishers. Printed in the Netherlands.*

2. Preliminaries and notations

Consider a delineated study area of size A and a community of N individuals on the area. At first suppose k to be known, even if this is not the case in most instances, and let $e_1, e_2, \ldots, e_k$ be the standard basis of R^k. Now, denote by $y_1, y_2, \ldots, y_N$ the marks associated with each individual, where $y_j = e_l$ if the j-th individual belongs to the l-th species $(j = 1,2,\ldots,N, l = 1,2,\ldots,k)$. Moreover, denote by S a sample of individuals selected from the community according to a design. Referring to units by their indexes, S is a subset of the population indexes $\{1,2,\ldots N\}$. Obviously, each sampling design induces a probability of selection $\Pr(S)$ for each $S \in N$, where N denotes the family of subsets for the basic set $\{1,2,\ldots,N\}$. Throughout the paper it is supposed $\Pr(\phi) = 0$.

As is well known, in a design-based approach locations, as well as marks, are considered as fixed characteristics and no attempt is made to model the mechanism generating them. In this case the object parameter is the abundance vector

$$N = \sum_{j=1}^{N} y_j$$

where obviously $N = 1^T N$.

Now, suppose that N be estimated by means of a function of sample data, say $\hat{N} = t(S)$. In a design-based approach the uncertainty stems from the fact that only some individuals of the community are observed, i.e., the accuracy of the resulting estimate depends on the selected sample. Thus, through a suitable choice of both the sampling design and t, a design-unbiased estimator such that

$$E\left(\hat{N} \mid y_1, y_2, \ldots, y_N\right) = \sum_{S \in N} t(S) \Pr(S) = N$$

can be achieved without making any assumption about the population. Moreover, a further goal of a design-based approach is to ensure unbiased or, at least, conservative estimators for the variability of $\hat{N}$ on the basis of sample data.

On the other hand, in a model-based approach the locations and the marks in the community are viewed as a realization of a marked point process in the plane. However, a suitable realistic model for marks may have a very complex structure. For example, marks may depend on each other as well as on the point locations which, in turn, may exhibit various kinds of interaction (e.g., inhibition or attraction) or spatial trend. In these cases, simplifying assumptions on point processes (e.g., independence, stationarity, isotropy) are likely to produce models that are very far from being adequate. Thus, when dealing with natural populations, a design-based approach seems to be preferable.

3. Some suitable sampling plans

Plot sampling designs are frequently used in forest studies. The sample plots can be laid out in a regular way or at random. Obviously, the use of randomly selected plots, which are referred to as *floating plots* (see, e.g., De Vries 1986), allows for an easier statistical treatment of sample data.

When rectangular plots of length L and width $2b$ are used, the sampling design is as follows: i) select a point at random over the study area and let it be the center of the plot; ii) choose or select at random a direction θ on $(0,\pi)$ and construct a rectangular plot with orientation θ, iii) repeat the procedure n times. Disregarding irregularities near the study area margin, the j-th unit is included in a plot if the plot center falls in a rectangle of length L and width $2b$ with orientation θ, having the j-th unit as its center. Boundary effects can be removed by using the well-known reflection method (see Gregoire, 1982). Accordingly, the probability of selecting the j-th unit by the i-th plot turns out to be

$$\gamma_j = \frac{2Lb}{A}, j = 1,2,\dots,N, i = 1,2,\dots,n.$$

Now, let S_i be the set of units in the i-th plot. For each plot, the Horvitz-Thompson estimator of the abundance vector is given by

$$Z_i = \sum_{j\in S_i} \frac{1}{\gamma_j} = \frac{A}{2Lb} \sum_{j\in S_i} y_j, i = 1,2,..,n.$$

Obviously, each z_i constitutes an unbiased estimator for N, with a variance-covariance matrix, say Σ, which is structure dependent, in the sense that it depends on the relative locations of units. Moreover, $z_1, z_2, \dots, z_n$ are n *iid* random vectors.

Very similar results are obtained when circular plots of radius b are used. In this case, the sampling design reduces to: i) select a point at random over the study area and let it be the center of a circle of radius b; ii) repeat the procedure n times. Thus, disregarding boundary effects (which can be removed by using the reflection method), the probability of selecting the j-th unit by the i-th plot is given by

$$\gamma_j = \frac{\pi b^2}{A}, j = 1,2,\dots,N, i = 1,2,\dots,n,$$

so that the Horvitz-Thompson estimator of the abundance vector turns out to be

$$Z_i = \sum_{j\in S_i} \frac{1}{\gamma_j} y_j = \frac{A}{\pi b^2} \sum_{j\in S_i} y_j, i = 1,2,\dots,n.$$

Also in this case $z_1, z_2, \dots, z_n$ constitute n *iid* random vectors with expectations N and a structure-dependent variance-covariance Σ.

The strip sampling may be viewed as a modification of a rectangular plot sampling, obtained by using strips of fixed direction and width $2b$ running across the whole study area. Consider a straight line thrown arbitrarily outside the study area and define as baseline the shadow cast by the study area onto the line. Let ω be the baseline length. The following steps have to be performed: i) select a point at random onto the baseline and throw a transect across the study area, perpendicular to the baseline at the selected position; ii) include in the sample all the individuals at distance less than b from the transect; iii) repeat the procedure n times. Disregarding irregularities near the edges of the baseline, the j-th unit is at a distance less than b from the transect if the selected point falls in an interval of width $2b$ around the projection of the j-th unit onto the baseline. If the transect is selected at a distance $d < b$ from an edge of the baseline, boundary effects can be removed by running an additional transect at distance d from the opposite edge outside the study area (see Thompson 1992). Accordingly, the probability of selecting the j-th unit by the i-th transect turns out to be

$$\gamma_j = \frac{2b}{\omega}, j = 1,2,\ldots,N, i = 1,2,\ldots,n,$$

so that

$$Z_i = \sum_{j \in S_i} \frac{1}{\gamma_j} y_j = \frac{\omega}{2b} \sum_{j \in S_i} y_j, i = 1,2,\ldots,n,$$

is the Horvitz-Thompson estimator for N, with a structure dependent variance-covariance Σ. Obviously, the n replicated transects give rise to n *iid* random vectors $z_1, z_2, \ldots, z_n$.

In some instances, mapping the individuals to carry out plot sampling may be time-consuming or impractical. In these cases, distance methods are preferred to plot count methods. The simplest distance-based method, which is referred to as *point-to-plant* sampling, involves the following steps: i) draw a point at random on the study area; ii) select the nearest unit; iii) repeat the procedure n times. As Overton and Stehman (1995) point out, at each drawing the probability of selecting the j-th unit is proportional to the size υ_j of its Voronoi polygon, i.e.,

$$\beta_j = \frac{\upsilon_j}{A}, j = 1,2,\ldots,N.$$

Accordingly, the random vector Z_i, where $Z_i = \beta_j^{-1} y_j$ if the j-th unit is selected $(i = 1,2,\ldots,n, j = 1,2,\ldots,N)$, turns out to be an unbiased estimator for N, with unknown variance-covariance Σ. Moreover, constitute $z_1, z_2, \ldots, z_n$ n *iid* random vectors.

Practically speaking, each sampling plan provides n *iid* unbiased estimators for N with variance-covariance matrix Σ. Thus, an improved estimator for N is given by

$$\bar{Z} = \frac{1}{n} \sum_{i=1}^{n} Z_i,$$

which turns out to be unbiased with variance-covariance Var $(\bar{Z}) = n^{-1} \sum$. Moreover, an unbiased and consistent $(n \to \infty)$ estimator for $Var(\bar{Z})$ is obtained by using $\mathrm{Var}(\bar{Z}) = n^{-1} S$, where

$$S = \frac{1}{n-1} \sum_{i=1}^{n} [Z_i - \bar{Z}][Z_i - \bar{Z}]^T.$$

Finally, the straightforward use of the Central Limit Theorem ensures that

$$\sqrt{n}(\bar{Z} - N) \xrightarrow{d} N_k\left(0, \Sigma\right).$$

As to the point-to-plant sampling, note that $\bar{Z}$ represents the Hansen-Hurwitz estimator for N.

4. Sampling distribution

In the field, the species richness k is rarely known. Accordingly, a population diversity index Δ (N) cannot be estimated by $\Delta(\bar{Z})$ - since Δ is a function of the k population abundances - but by $d(\bar{Z})$, where d is the same function computed on the non-null abundance estimates. From a mathematical point of view, d may be seen as an extension of Δ from Q_k to Q_k^*, where $Q_k = \{z_l > 0, l = 1,2,...,k\}$ and $Q_k^* = \{z_l \geq 0, l = 1,2,...,k, z \neq 0\}$, i.e., the two functions coincide on Q_k. Accordingly, since in determining the asymptotic distribution of $d(\bar{Z})$ only the properties of d in a neighborhood of N are of interest, one can refer to d or Δ indifferently.

More precisely, if $\partial\Delta(z)/\partial z$ is defined in a neighborhood of N, is non-null and continuous at N, a straightforward application of the delta method ensures that

$$\sqrt{n}(d - \Delta) \xrightarrow{d} N\left(0, \sigma^2\right),$$

where

$$\sigma^2 = g^T \Sigma g = \sum_{l=1}^{k} \sum_{h=1}^{k} g_l g_h \sigma_{lh}, \; g = \frac{\partial\Delta(z)}{\partial z}\Big|_{Z=N}$$

Moreover

$$\hat{\sigma}^2 = \sum_{\bar{z}_l \neq 0} \sum_{\bar{z}_h \neq 0} \hat{g}_l \hat{g}_h s_{lh},$$

where

$$\hat{g}_l = \frac{\partial\Delta(z)}{\partial z_l}\Big|_{Z_l = \hat{Z}_l}, \hat{z}_l \neq 0,$$

provides a consistent estimator of σ^2, in such a way that

$$\frac{\sqrt{n}}{\hat{\sigma}}(d - \Delta) \xrightarrow{d} N(0,1).$$

Even if d is asymptotically unbiased, a bias occurs for finite samples. It is a stated result that most diversity indexes, when evaluated from the sample, consistently underestimate the population counterparts. Thus, a jackknife procedure may be used to reduce bias in order to generate better centered confidence intervals. In addition, jackknife also provides variance estimates. In this framework, the jackknife estimator of a diversity index is given by

$$d_{jack} = \frac{1}{n}\sum_{i=1}^{n} d^{(i)},$$

where

$$d^{(i)} = nd(\bar{z}) - (n-1)d(\bar{z}_{-1}), i = 1,2,\ldots,n,$$

are the n pseudovalues and

$$\bar{z}_{n-i} = \frac{1}{n-1}\sum_{h=1,h\neq i}^{n} z_h, i = 1,2,\ldots,n.$$

Moreover, the jackknife estimate of the sampling variance σ^2 turns out to be

$$\hat{\sigma}^2_{jack} = \frac{1}{n-1}\sum_{i=1}^{n}\left(d^{(i)} - d_{jack}\right)^2.$$

Zahl (1977) proposes the use of jackknife in a plot sampling in order to estimate Simpson's measure. Other jackknifing procedures in plot sampling are due to Heltshe and Bitz (1979) to estimate Brillouin's measure, Heltshe and Forrester (1983), Smith and van Belle (1984) and Palmer (1990, 1991) to estimate species richness, and Gove et al. (1994) to estimate Δ_β curves. However, the findings of these papers are usually obtained by model assumptions, simulation studies or field works. From a more general point of view, since this is the simplest case in which a function of the sample mean, $d(\bar{z})$ is used to estimate a function of the population mean, d (N) = Δ, (N) one has at disposal standard results (see, e.g., Shao and Tu 1995). More precisely, it is well-known that if $\partial\Delta(z)/\partial z$ is defined in a neighborhood of N, is non-null and continuous at N, then

$$E(d) - \Delta = \frac{1}{2n} tr(H\Sigma) + o(n^{-1}), H = \frac{\partial^2\Delta(z)}{\partial_z\partial_z^T}\Big|_{z=N},$$

providing that Δ is second order differentiable and the Hessian is continuous at N. On the other hand, d_{jack} has a $o(n^{-1})$ bias, $\hat{\sigma}^2_{jack}$ is a consistent estimator for σ^2 and

$$\frac{\sqrt{n}}{\hat{\sigma}_{jack}}\left(d_{jack} - \Delta\right) \xrightarrow{d} N(0,1).$$

Note however that these results hold for well-behaved (sufficiently smooth) functions Δ. For example, they do not hold when estimating species richness or right-tail sum diversity indexes if $N_l = N_h$ for some $l \neq h = 1,2,...,k$.. In these cases further investigations are needed.

5. Further developments

We feel that some other commonly applied sampling designs give rise to diversity index estimates sharing the above mentioned properties. For example, it can be straightforwardly proved that this is true for the well-known Bitterlich sampling. Moreover, the multivariate extension of the jackknife results, may lead to the construction of conservative confidence envelopes for some diversity curves, such as Δ_β curves.

References

Adams, J.R. and McCune, E.D. 1979. Application of the generalized jackknife to Shannon's measure of information used as an index of diversity. In: J.F. Grassle, G.P. Patil, W. Smith, and C. Taillie (eds.). Ecological Diversity in Theory and Practice. International Co-operative Publishing House, Fairland (MD). pp. 117-131.

Blyth, C.R. 1958. Note on estimating information. Annals of Mathematical Statistics 30, 71-79.

Bowman, K.O., Hutcheson, K., Odum, E.P., and Shenton, L.R. 1971. Comment on the distribution of indices of diversity. In: G.P. Patil, E.C. Pielou and W.E. Waters (eds.). Statistical Ecology, Vol. 3. Pennsylvania State University Press, University Park (PA). pp. 315-359.

De Vries, P.G. 1986. Sampling theory for forest inventory. Springer-Verlag, Berlin.

Gove, J.H., Patil, G.P., Swindel, B.F., and Taillie, C. 1994. Ecological diversity and forest management. In: G.P. Patil and C.R. Rao (eds.). Handbook of Statistics, Vol. 12 (Environmental Statistics). Elsevier Science B.V., Amsterdam. pp. 409-462.

Gregoire, T.G. 1982. The unbiasedness of the mirage correction procedure for boundary overlap. Forest Science 28, 504-508.

Heltshe, J.F. and Bitz, D.V. 1979. Comparing diversity measures in sampled communities. In: J.F. Grassle, G.P. Patil, W. Smith, and C. Taillie (eds.).Ecological Diversity in Theory and Practice. International Co-operative Publishing House, Fairland (MD). pp. 133-144.

Heltshe, J.F. and Forrester, N.E. 1983. Estimating species richness using the jackknife procedure. Biometrics 39, 1-12.

Heyer, R.V., and Berven, K.A. 1973. Species diversity of herpetofaunal samples from similar microhabitats at two tropical sites. Ecology 54, 642-645.

Overton, W. S. and Stehman, S. V. 1995. The Horvitz-Thompson theorem as a unifying perspective for probability sampling: with examples from natural resource sampling, The American Statistician 49, 261-268.

Palmer, M.W. 1990. The estimation of species richness by extrapolation. Ecology 71, 1195-1198.

Palmer, M.W. 1991. Estimating species richness: the second-order jackknife reconsidered. Ecology 72, 1512-1513.

Patil, G.P., and Taillie, C. 1979. A study on diversity profiles and orderings for a bird community in the vicinity of Colstrip, Montana. In: G.P. Patil and M. Rosenzweig (eds.). Contemporary Quantitative Ecology and Related Econometrics. International Co-operative Publishing House, Fairland (MD). pp. 23-48.

Pielou, E.C. 1966. The measurement of diversity in different types of biological collections. Journal of Theoretical Biology 13, 131-144.

Shao, J. and Tu, D. 1995. The jackknife and the bootstrap. Springer-Verlag, Berlin.

Smith, E.P. and van Belle, G. 1984. Nonparametric estimation of species richness. Biometrics 40, 119-129.

Smith, W., Grassle, J.F. and Kravitz, D. 1979. Measures of diversity with unbiased estimates. In: J.F. Grassle, G.P. Patil, W. Smith, and C. Taillie (eds.). Ecological Diversity in Theory and Practice. International Co-operative Publishing House, Fairland (MD), pp. 177-191.

Thompson, S. K. 1992. Sampling. Wiley, New York.

Zahl, S. 1977. Jackknifing an index of diversity. Ecology 58, 907-913.

Assessment and Monitoring Techniques for Forest Biodiversity

ESTIMATION OF FOREST DIVERSITY WITH MISCLASSIFICATION

GEORGE GERTNER
Department of Natural Resources and Environmental Sciences
University of Illinois at Urbana-Champaign
Urbana, IL 61801
U. S. A.

XIANGCHI CAO
National Center for Super Computer Applications (NCSA)
University of Illinois at Urbana-Champaign
Urbana, IL 61801
U. S. A.

DIETER PELZ
Universität Freiburg i. Br.
Abteilung für Forstliche Biometrie
Werderring 6
D-79085 Freiburg
Germany

Abstract

In this paper, the consequences are assessed of species identification errors when estimating species diversity with the Shannon-Weaver index. Species misclassification can be due to recording error, editing error, poor field crew training, etc. In certain situations, misclassification can lead to biased estimates of the biodiversity index as well as inflated variance estimates. Different approaches are presented for assessing the consequences of misclassification. The results of a control study are presented. The work presented is part of an ongoing study to develop error budgets for different types of comprehensive stochastic dynamic modelling systems for both plant and forest communities.

1. Introduction

Diversity is an average property of a community and that property is identified as the species variety and abundance. The study of biodiversity is so important because it is one the central themes in ecology, and the diversity of a system is often seen as an indicator of the well-being of the system. Ecosystems with more diversity are potentially more productive and more stable (Culotta 1996).

P. Bachmann et al. (eds.), Assessment of Biodiversity for Improved Forest Planning, 199–210.
European Forest Institute Proceedings no. 18. © 1998 *Kluwer Academic Publishers. Printed in the Netherlands.*

There are many indices that have been developed for measuring diversity (Magurran 1988). The Shannon index is one of the most commonly used index for evaluating forest diversity (Boyle and Boontawee 1995). Although there are many problems with the Shannon index (Pielou 1975; Pielou 1995), the appeal of the Shannon index is that it summarises both species richness and species evenness into a single number.

The following is the formula of the Shannon index (Shannon and Weaver, 1949):

$$H = -\sum_{i=1}^{s} p_i \times \log(p_i), \tag{1}$$

where H is the Shannon index, p_i is the relative abundance of species i, and s is the total number of species.

When using the Shannon index, it is assumed that species are properly classified in order to correctly estimate relative abundance and total number of species. In practice, misclassification of species can occur for many reasons in forest surveys for biodiversity. The most likely cause of species misclassification is the mis-identification of species because of lack of knowledge or poor training of field crews (Boyle and Boontawee 1995). Particularly in forest communities that are species rich, misclassification due to mis-identification can be extremely high. With the guidance of taxonomic experts, this error can be kept small, however this is not always possible in practice. Mis-identifying species can be related to weather, seasonal and human factors, etc. Mis-coding of species, using poor quality specimens can be other sources of errors (Pielou 1975). Recording errors can occur when there are many similar species codes, such as ABCD1 and ABCD2. When there are many species with very similar species codes, the recording errors can potentially be very large.

In this paper, the consequences of misclassification are assessed when estimating biodiversity with the Shannon index. First, the results of a control study that was designed to assess misclassification is presented. The results from this control study are specific to the particular forest community monitored. Next, to generalise the results for other forest community types, a sensitivity analysis of the Shannon index for different forest community types is conducted. Finally, some approaches to adjust for misclassification are given.

2. Control Study

A control study was conducted to assess the consequences of misclassification on the estimate of biodiversity using the Shannon index for one community type. It was the explicit intention of this control study to have misclassification of species. A forest stand 200 by 200 meters in size was monitored for biodiversity. The forest is located in south-western Germany in the federal state of Baden-Württemberg near Freiburg. The forest is a temperate broadleaved forest that is administered by the Baden-Württemberg Forest Service. For the particular forest community, the maximum number of species that can be expected in this type of forest community is 30 species.

In late winter, when trees were dormant and without their leaves, twenty circular plots 15 meters in radius were monitored. These plots were independently visited by seven graduate students from the Department of Forest Biometrics and the Landscape Institute at the University of Freiburg. All seven students had a course in dendrology during their undergraduate studies at the university. The students were requested to identify the species of trees without the use of any twig keys or other references. Communication between the different students was prohibited. Prior to the survey, all species of the trees were properly identified when the leaves were visible by a well-trained forester who was very knowledgeable about tree species in the region. It was assumed that the forester identified all species without errors.

The total number of trees classified by each student was 920. Table 1 shows the species codes, numbers, and proportional abundance, and percent misclassification by species. "New species" refers to mis-identified species that were classified with species names that did not actually exist on the plots. Across the seven replications (students), the number of trees that were misclassified was 5.48%: 1.41% of trees were misclassified as a more abundant species, 3.55% as a less abundant species, and 0.52% as a "new species". There was a slight tendency of more common trees being misclassified as less abundant species.

TABLE 1. Summary of number of species, true relative abundance and percent of species correctly identified near Freiburg, Germany.

Rank	Species Code	True Number	True Relative Abundance	Percent Classified Correctly (Misclassified)	
1	hbu	556	0.604348	98.76%	1.24%
2	es	106	0.115217	94.93%	5.07%
3	erl	79	0.08587	92.56%	7.44%
4	bah	64	0.069565	80.27%	19.73%
5	ei	33	0.03587	94.7%	5.30%
6	hasel	29	0.031522	86.21%	13.79%
7	bi	15	0.016304	78.33%	21.67%
8	rei	13	0.01413	84.62%	15.38%
9	bu	8	0.008696	89.06%	10.94%
9	tkir	8	0.008696	29.69%	70.31%
10	li	6	0.006522	62.5%	37.50%
11	rob	2	0.002174	100%	0%
12	pap	1	0.001087	62.5%	37.50%
"New species"	faulb	0	0	0%	100%
"New species"	i	0	0	0%	100%
"New species"	mehlb	0	0	0%	100%
"New species"	ulme	0	0	0%	100%
"New species"	wei	0	0	0%	100%
Total		920			

TABLE 2. Jack-knife estimates of the Shannon index and variance by student.

Replication (student)	Shannon Index	Variance
1	1.4633	0.0055
2	1.4802	0.0052
3	1.4546	0.0073
4	1.4083	0.0041
5	1.4334	0.0053
6	1.4399	0.0062
7	1.4059	0.0067
Control (forester)	1.4624	0.0057

Table 2 shows the Shannon estimates of biodiversity for the control and for the seven replications (students). Jack-knifing (Zahl 1977) was used to obtain estimates of the Shannon indexes and variances. The average Shannon index across replicates was 1.4408. The estimated bias was -0.0216 (-1.48%) and the increase in variance due to misclassification was 0.0008 (increase of variance due to misclassification of 11.82%). It can be seen from these results, that the bias due to misclassification was fairly small and there was an increase in variance due to misclassification.

The results of the control study are limited to the particular forest community monitored. The community type can be very important when misclassification is considered. This will be shown in the next section with a sensitivity analysis.

3. Community Type and its Sensitivity to Misclassification

It is the purpose here to show that sensitivity to misclassification can be different for different community types. When measuring biodiversity, not all community types are as sensitive to misclassification as others. The methodology to assess this sensitivity is based on an extension of an approach developed by Magnussen and Boyle (1995) for determining the expected Shannon index and its variance. Their approach does not consider misclassification error. In the extension presented here, their approach is extended to considered this type of error.

3.2. EXPECTATION AND VARIANCE OF SHANNON WITHOUT MISCLASSIFICATION

In some systems, objects are distributed with well defined patterns. The distribution of objects in these systems can be described precisely in a mathematical fashion. Also, when data or information is not sufficient to make statistical conclusions about the distribution of the objects studied, it is sometimes useful to make assumptions based on different theoretical distributions. One theoretical distribution function that has been widely used because of its high flexibility is the beta function.

The beta function was used as a prior likelihood function by Magnussen and Boyle (1995) to represent the most probable species abundance distribution for different forest communities. The Shannon index was calculated based on the likelihood function of the most probable species abundance.

Suppose the following beta function, also called a likelihood function, can be used to describe a forest community species abundance distribution:

$$L(p_s|\alpha,\beta) = p_s^{\alpha-1} \times (1-p_s)^{\beta-1} / B(\alpha,\beta) \;,\; 0 < p_s < 1,\; \alpha,\beta > 0 \tag{2}$$

where $L(p_s|\alpha,\beta)$ is the likelihood of species s having a relative abundance p_s; α and β are two parameters of the beta function; and $B(\alpha,\beta)$ is the complete beta function of $\Gamma(\alpha)\Gamma(\beta)/\Gamma(\alpha+\beta)$. For different combinations of the α and β values, the likelihood function $L(p_s|\alpha,\beta)$ gives different types of likelihood curves. Figure 1 displays the most probable species distributions for four "typical" forest communities. They are the inverse J-shaped, J-shaped, bell-shaped and flat-shaped distributions (Magnussen and Boyle 1995). The inverse J-shaped distribution usually represents communities in which there are many rare species and very few dominant species. In contrast, the J-shaped

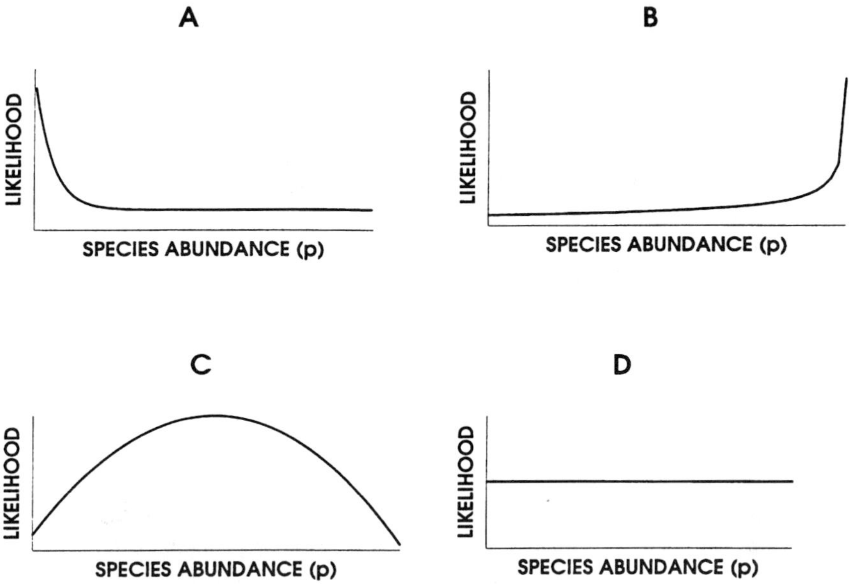

Figure 1. Four typical types of species abundance distributions: (A) inverse J-shaped $L(p_s|0.5,\ 6.0)$, (B) J-shaped $L(p_s|3.0,\ 0.5)$, (C) bell-shaped $L(p_s|2.0,\ 2.0)$ and (D) flat-shaped $L(p_s|1.0,\ 1.0)$ (from Magnussen and Boyle).

distribution depicts the situation in which the communities are dominated by a few dominant species and few rare species. The bell-shaped species distribution appears in many Montana-temperate forests in which both extremes (rare species and dominant species) are not common. The flat-shaped distribution represents the situation of evenly distributed species.

If the beta distribution can be used to represent the most probable species abundance distribution for a forest community, then the expected Shannon index for the forest community can be found by summing over all possible relative species abundance $(0 < p_s \leq 1)$, with each sum given a weight equal to its likelihood $L(p_s|\alpha,\beta)$. The conditional expectation of the Shannon index is given by Magnussen and Boyle (1995) as:

$$E(H|\alpha,\beta) = \frac{\int_0^1 - p_s \times \log(p_s) \times L(p_s|\alpha,\beta) dp_s}{\int_0^1 p_s \times L(p_s|\alpha,\beta) dp_s} \tag{3}$$

$$= \frac{\int_0^1 - p_s^{\alpha} \times (1-p_s)^{\beta-1} \times \log(p_s) dp_s}{B(\alpha,\beta) \times \alpha / (\alpha+\beta)}$$

where $\int_0^1 p_s \times L(p_s|\alpha,\beta)dp_s = \alpha/(\alpha+\beta)$ is a normalising constant.

Based on (3), the conditional variance of $E(H|\alpha,\beta)$ can be shown to be:

$$Var(H|\alpha,\beta) = \int_0^1 [-p_s \times \log(p_s) - \frac{\alpha}{\alpha+\beta} E(H|\alpha,\beta)]^2 \times L(p_s|\alpha,\beta)dp_s \,. \tag{4}$$

Due to a typographical error in their article, this conditional variance is slightly different from the variance function given by Magnussen and Boyle. This conditional variance represents the natural variation of a forest.

Without misclassification error, equations (3) and (4) give the expectation and variance of the Shannon index (1). Figure 2 shows the expected Shannon index when α and β ranged from 0.5 to 10.

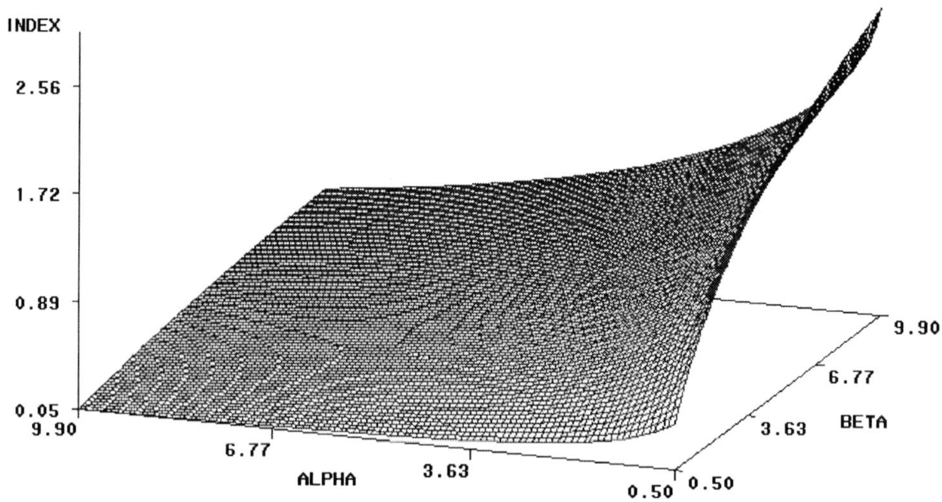

Figure 2. The expected Shannon index with α and β ranged from 0.5 to 10.

3.2. EXPECTATION AND VARIANCE OF SHANNON WITH MISCLASSIFICATION

In general, classification is the process of dividing objects or items into mutually exclusive groups such that the members of each group are as "close" as possible to one another, and different groups are as "far" as possible from one another, where distance is measured with respect to specific variable(s) or properties. In many applications, objects are characterised by their properties. These properties are usually measured by discrete variables or categorised continuous variables. Based on their measures, objects are classified into classes. In an error-free world, every object belongs to its correct class. However, in practice, objects may be measured with error and placed into wrong classes. Misclassification of this type does not change the total number of objects or items, but it changes the distribution of objects among the classes. For example, let $O = (o_1, \ldots, o_{10})$ be the set of 10 objects, and $C = (c_1, c_2, c_3)$ be the set of 3 classes. Without error, the class set is $C = \{ \{o_1, o_5\}, \{o_2, o_3, o_9, o_{10}\}, \{o_4, o_6, o_7, o_8\} \}$. When the objects are measured with error, the class set might be $C = \{ \{o_5, o_7, o_{10}\}, \{o_1, o_2, o_3\}, \{o_4, o_6, o_8, o_9\} \}$.

Similarly, it is assumed here that misclassification does not change the size of the estimated population when monitoring biodiversity, but it only changes the species distribution of the sampled population. Misclassification of species causes the potential reallocation of species to other species that already exist in the forest community, but also to "new species". "New species", as already mentioned, refers to mis-identified species that do not actually exist in the sample, but artificially occurs due to mis-identification.

If only misclassification is considered when using the Shannon index, and it is assumed that the error in the size of the total population due to misclassification is zero, then errors in absolute species abundance is linearly related to the errors in relative species abundance. Let p_s be the relative abundance of species s without misclassification, p_s' be the observed relative abundance of species s with misclassification and e_s be the misclassification error in relative species abundance. The following relationship can then be written:

$$p_s' = p_s + e_s \tag{5}$$

Assuming that the population is a closed system and p_s has a beta distribution, p_s' must then also have a beta distribution. This property makes it difficult to find the distribution for the error term e_s.

As an alternative to finding the distribution for the error term e_s, a functional relationship between the error and the distribution shift of likelihood function was formulated. Misclassification actually just shifts the beta distribution from L($p_s|\alpha,\beta$) to L($p_s'|\alpha',\beta'$), where α and β are the parameters of the beta function without misclassification, and α' and β' are the parameters of the beta function with misclassification. The form of this functional relationship can be described as:

$$\frac{100 \bullet \int_0^1 | L(p_s|\alpha,\beta) - L(p_s'|\alpha',\beta') | p_s dp_s}{\int_0^1 p_s L(p_s|\alpha,\beta) dp_s} = d \tag{6}$$

where d is the percent absolute difference in the area between L($p_s|\alpha,\beta$) and L($p_s'|\alpha',\beta'$). It can be shown that when there is d percent misclassification of species, the difference in absolute difference in the area between L($p_s|\alpha,\beta$) and L($p_s'|\alpha',\beta'$) is d percent misclassification (Cao 1997).

With this relationship, the conditional expectation of the Shannon index with misclassification can be numerically calculated based on a sample function of $(\alpha',\beta'|d)$. The bias and variance due to misclassification is then given by

$$\begin{aligned} Bias(H) &= E(H|\alpha,\beta) - E(H|d) \\ Var(H) &= Var(H|\alpha,\beta) + Var(H|d) \end{aligned} \tag{7}$$

where $E(H|d)$ and $Var(H|d)$ are the expectation and the variance increase due misclassification of the Shannon index, respectively, when there is d percent misclassification. The appeal of this formulation is that both change in community richness and relative abundance due to misclassification can be simultaneously accounted for when calculating the Shannon index.

3.3. SENSITIVITY ANALYSIS FOR SOME TYPICAL FOREST COMMUNITIES

A program written in the language of C was written to numerically determine the expectations and variances when there are errors due to misclassifications. Due to computational intensity of the algorithm needed to determine the expectations and variances, the program had to be run on a Convex Exemplar Super Computer operated by the National Center for Supercomputing Applications (NCSA) in Champaign, Illinois, USA.

For the sensitivity analysis, (α,β) in the likelihood function were set as (1.0, 1.0), (2.0, 2.0), (3.0, 0.5) and (0.5, 6.0) for the flat-shaped, bell-shaped, J-shaped and inverse J-shaped functions, respectively. In Table 3 the estimates of the Shannon index are based on equations (3) and (4) when there are no misclassifications. The variances shown in Table 3 can be considered to be due to the natural variation of a forest community assuming that the most probable relative species abundance distribution of the community follows a beta distribution. Table 4 shows the bias and the variance of

TABLE 3. Shannon index without misclassification error.

Likelihood distribution shape	Flat	Bell	J	Inverse J
Expected Shannon index[1]	0.5000	0.5833	0.1388	2.0365
Variance without misclassification error[2]	0.0115	0.0064	0.0121	0.0126
Percent coefficient of variation[3]	21.47	13.74	79.32	5.51

[1] $E(H|\alpha,\beta)$, [2] $V(H|\alpha,\beta)$, [3] $100\sqrt{V(H|\alpha,\beta)}\,/\,E(H|\alpha,\beta)$

TABLE 4. Bias and variance of the Shannon index due to misclassification.

Likelihood distribution shape	Flat		Bell		J		Inverse J	
Percent species misclassification (d)	10%	20%	10%	20%	10%	20%	10%	20%
Expected Shannon index[1]	0.5078	0.5381	0.5882	0.6039	0.1487	0.1556	1.9643	1.7664
Bias[2]	0.0078	0.0381	0.0049	0.0206	0.0099	0.0168	-0.0722	-0.2701
Percent bias[3]	1.56%	7.63%	0.84%	3.53%	7.13%	12.10%	-3.55%	-13.26%
Total variance with misclassification error[4]	0.0138	0.0205	0.0084	0.0155	0.0122	0.0124	0.0866	0.0956
Variance due to misclassification error[5]	0.0023	0.0090	0.0020	0.0091	0.0001	0.0003	0.0740	0.0830
Percent of total variance due to misclassification error[6]	16.64%	43.85%	23.74%	58.62%	0.82%	2.41%	85.48%	86.85%

[1] $E(H|d)$, [2] $E(H|\alpha,\beta)-E(H|d)$, [3] $100\bullet Bias(H)\,/\,E(H|\alpha,\beta)$, [4] $Var(H|\alpha,\beta)+Var(H|d)$,
[5] $Var(H|d)$, [6] $100\bullet Var(H|d)/(Var(H|\alpha,\beta)+Var(H|d))$

Shannon index for when the percent misclassification is 10% and 20%. It can be seen from the table that the different species distributions have different sensitivities to misclassifications. The flat-shaped and bell-shaped distributions have the most resistance to misclassifications. With d=10%, the bias of the Shannon index is about 2%, and with d=20%, the Shannon index is biased 8%. One of the reasons for the lower sensitivity to misclassification for the flat-shaped and bell-shaped distributions is that the chances of misclassifying of a rare species into a more common species and misclassifying a common species into a rare species are approximately equal. In other words, the effects of misclassification are cancelled out. In terms of variance increase due to misclassification, both the flat-shaped and bell-shaped distributions are fairly sensitive to misclassification.

In the J-shaped case, the chances of misclassifying a rare species into a common species are less than that of misclassifying a common species into a rare species because there are fewer rare species and the population is dominated by common species. Misclassification may create “new species” or increase the evenness of species distribution. This causes a systematic increase of the Shannon index. With d=10% the Shannon index increases about 4%, and with d= 20% the increase is 13%. In contrast, with the inverse J-shaped case, the Shannon index decreases with misclassification. The Shannon index decreases 6% with d=10%, and 21% with d=20%. In the inverse J-shaped case, because there are many rare species and very few dominating species, it is more likely that rare species might be overlooked. This will reduce the number of species and causes a decrease of the Shannon index. In terms of variance increase due to misclassification, the J-shaped distribution is extremely robust and insensitive to misclassification, while the inverse J-distribution is extremely sensitive to misclassification.

3.3. ANALYSIS OF FREIBURG CONTROL STUDY WITH LIKELIHOOD METHOD

The likelihood approach was applied to a forest community near Freiburg. A method similar to a Gibbs sampler (Tanner 1991) was employed to estimate the parameters of the beta distribution for the forest community. The parameter estimates were $\hat{\alpha} = 2.152$ and $\hat{\beta} = 8.853$. Figure 3 shows the likelihood distribution for the forest community to be log-normal like in form. The distribution falls between the bell-shaped and inverse J-shaped functions described in the above sensitivity analysis. Without misclassification error, the expected Shannon index was 1.46 and variance was 0.0056. Listed in Table 5 is the bias and variance corresponding to a misclassification error of 5.48% based on Equation (5).

TABLE 5. Bias and variance of the Shannon index due misclassification for the control study near Freiburg Germany.

Type of Misclassification	Unweighted	Weighted more common to less common
Expected Shannon index with misclassification error	1.4910	1.4935
Bias	0.0287	0.0312
Percent bias	1.97 %	2.14 %
Total variance with misclassification error	0.0065	0.0064
Variance due to misclassification error	0.0009	0.0008
Percent of total variance due to misclassification error	13.82 %	12.48 %

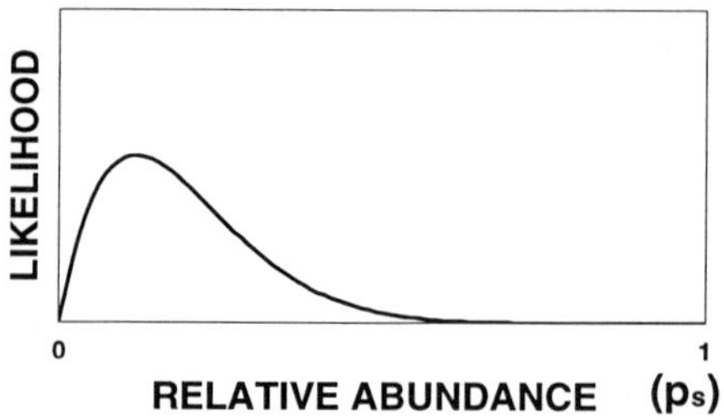

Figure 3. Species abundance distribution for broadleaved forest community near Freiburg, Germany.

A weighted analysis was also conducted, where there was a systematic misclassification of more common species to less common species. Results for both the unweighted and weighted analyses are very similar and agree quite closely to the results of the control study.

4. Methods for Adjusting for Misclassifications

Currently an on-going study of misclassification errors in biodiversity surveys of rare and endangered plant communities around the United States is being conducted. For certain plant communities the biases and variances of the Shannon index have been found to be extremely large. This has been particularly true for plant communities where community species richness is high.

In the analyses of these plant surveys, the likelihood method has been used, but other techniques for correcting or adjusting for misclassification are being developed. The basic assumption for the method of the likelihood function is that species abundance distribution follows a beta distribution. However, this assumption might not always be appropriate for all forest and plant communities. For this reason, methods that are not based on a theoretical distributions for probable species abundance distribution are being developed.

The general problem of misclassification has been considered from different viewpoints by many investigators (for example, Tenenbein 1970; Chen 1979 and 1989; Geng 1989; York 1992; Viana 1994). Chen (1979 and 1989) gave a review of methods for misclassified categorical data and the maximum likelihood estimation for loglinear models. Geng (1989), York (1992) and Viana (1994) applied Bayesian estimation methods to the problems of misclassification and incomplete data.

Both a Bayesian approach and double sampling approach are being developed to estimate and adjust the Shannon index for misclassification. The approaches are extensions of schemes developed by Viana (1994) and Geng (1989). With the Bayesian approach, it is assumed that the misclassification probabilities are known a priori for a community. Based on these misclassification probabilities, adjustments can then be made. With the double sampling approach, a random subsample of observed plants or trees that have been classified using standard field methods are then classified again using more precise methods (for example, a taxonomic expert). Based on the subsample, the misclassification probabilities can be determined which can then be used for making adjustments to the Shannon index.

5. Conclusion

The research presented here describes part of a larger research project to develop error budgets (Cao 1997; Gelb 1974; Gertner et al. 1995) for a forest and plant community stochastic dynamics model. These models are being designed to provide standard diversity measures, monitor the development of the communities in terms of species diversity and structure diversity, testing the significance of the influence of human activities on communities, and to estimate the rehabilitation time of a disturbed community.

The error budgets being developed can be used for many purposes such as data correction, model evaluation, quality control assessment, and decision-making by management The benefit of an error budget is substantial. First, error budgets can be used to evaluate the statements made from a survey or a model. Given all sources of errors, the error budgets can be used to determine if biodiversity estimates and projections are valid or invalid. Estimates and projections are valid only if they are within certain error limits. Second, error budgets can be a guide in designing surveys and models. With error sensitivity analysis, all types of error sources can be tested to find their contribution to final estimates and predictions. Effort can be then directed towards controlling the important sources of error. Third, an error budget provides information for error correction. To correct errors, knowledge of all sources of error are needed. Errors from different sources may need different procedures for error correction. Finally, an error budget can be used as a basis for "virtual experimentation" when using models for testing different environmental hypotheses and scenarios.

6. Acknowledgments

The authors would like to thank Mr. Karl-Heinz Rudishauser for helping to design and install the control study, and the seven graduate students who took time from their busy research schedules to take part in the study. We would like to also thank the National Center for Supercomputing Applications (NCSA) at the University of Illinois at Urbana-Champaign for the computer time.

References

Boyle, T. and Boontawee, B. (eds.) 1995. IUFRO Symposium on Measuring and Monitoring Biodiversity in Tropical and Temperate Forests. Sponsored by Center for International Forestry Research. Chiang Mai. Thailand. Pp. 5-17.

Cao, X. 1997. Stochastic Models of Plant Diversity. Ph.D. Thesis. University of Illinois. (Expected completion date: spring 1997.)

Chen, T. 1979. Log-linear models for categorical data with misclassification and double sampling. JASA. 74: 481-488.

Chen, T. 1989. A review of methods for misclassified categorical data in epidemiology. Statist. Medicine. 8: 1095-1106.

Culotta, E. 1996. Exploring Biodiversity Benefits. Science. 273: 1045-1046.

Gelb, A. 1974. Applied Optimal Estimation. The MIT Press. Cambridge MA. 374 p.

Geng, Z. 1989. Bayesian estimation methods for categorical data with misclassifications. Commun. Statist. – Theory Meth. 18(8): 2935-2954.

Gertner, G. and Köhl, M. 1995. Correlated observer errors and their effects on survey estimates of needle-leaf loss in Switzerland. Forest Science. 41(4): 758-776.

Gertner, G., Cao, X. and Zhu, H. 1995. A quality assessment of a Weibull based growth projection system. Forest Ecology and Management. 71: 235-250.

Gertner, G. 1987. Approximating precision in simulation projections: an efficient alternative to Monte Carlo methods. Forest Science. 33: 230-239.

Magurran, A. 1988. Ecological Diversity and its Measurement. Princeton Paperbacks. Princeton. New Jersey.

Magnussen, S. and Boyle, T. 1995. Estimating sample size for inference about the Shannon-Weaver and the Simpson indices of species diversity. Forest Ecology and Management. 78: 71-84.

Pielou, E. 1975. Ecological Diversity. Wiley. New York.

Pielou, E. 1995. Biodiversity versus old-style diversity measuring biodiversity for conservation. In: Boyle, T. and Boontawee, B. (eds.). Proceedings of IUFRO Symposium on Measuring and Monitoring Biodiversity in Tropical and Temperate Forests. Sponsored by Center for International Forestry Research. Change Mia. Thailand. Pp. 5-17.

Shannon, C. and Weaver, W. 1949. The Mathematical Theory of Communication. University of Illinois Press. Urbana. Illinois.

Tanner, M. 1991. Tools for Statistical Inference. Wiley. New York.

Tenenbein, A. 1979. A double sampling scheme for estimating binomial data with misclassification. JASA. 65: 1350-1361.

Viana, M. 1994. Bayesian small-sample estimation of misclassified multinomial Data. Biometrics. 50: 237-243.

York, J.C. 1992. Bayesian methods for the analysis of misclassified or incomplete multivariate discrete data. Ph.D. dissertation. Department of Statistics. University of Washington.

Zahl, S. 1977. Jackknifing an index of diversity. Ecology. 58: 907-913.

TOWARDS MORE RIGOROUS ASSESSMENT OF BIODIVERSITY

JEROME K. VANCLAY
Center for International Forestry Research
PO Box 6596 JKPWB, Jakarta 10065, Indonesia

Abstract

Biodiversity is often ill-defined and subjectively surveyed, resulting in inefficient and ambiguous estimates. Strengths and deficiencies of prevailing survey techniques are appraised through a review of selected literature. Analogies with forest inventory are used to suggest options for more efficient and rigorous biodiversity assessment. Techniques such as variable-probability and model-based sampling, especially when used in conjunction with generalised linear modelling, offer efficient alternatives to more traditional assessments based on quadrats and nested plots. Bayesian methods offer scope to combine expert and local knowledge with formal samples, and warrant further investigation. Suggestions for further research are given.

1. Introduction

Since Wilson (1988) popularised the contraction "biodiversity" (for biological diversity), it has become a popular buzzword, both in the scientific literature and the popular press. However, the term is often used casually and imprecisely, and clear definitions are hard to find (e.g. Taylor 1978; Solomon 1979). For example, Article 2 of the Convention on Biological Diversity (UNEP 1992) offers one definition, namely

> *"... the variability among living organisms from all sources including, inter alia, terrestrial, marine and other aquatic ecosystems and the ecological complexes of which they are part; this includes diversity within species, between species and of ecosystems",*

which leaves considerable scope for interpretation and provides no unambiguous basis for assessment. This ambiguity precludes clear guidelines and optimal inventory designs, so I begin with an overview of some concepts and a re-examination of some indices. Fortunately, there is a common basis underlying most biodiversity indices, namely "the study of the number of its species (the community's species-richness) and their relative abundance (called variously evenness, equitability or dominance). Diversity increases with richness and evenness. The term species is to be interpreted broadly..." (Solomon 1979).

Since no two species are equal in every sense, it is difficult to avoid value judgements concerning the relative worth of genes, species and habitats, when compiling aggregate

P. Bachmann et al. (eds.), Assessment of Biodiversity for Improved Forest Planning, 211–232.
European Forest Institute Proceedings no. 18. © 1998 *Kluwer Academic Publishers. Printed in the Netherlands.*

statistics on biodiversity. I avoid these rather subjective aspects, merely noting that the various indices may be weighted according to the rarity, uniqueness or other "value" of a trait, organism or assemblage, in the same way that indices can be weighted for numbers, biomass or energy flows (e.g. MacArthur 1955).

Despite some important differences between forest inventory and biodiversity surveys (the former seeks good estimates of the mean within homogeneous strata, whereas the latter often involves extreme values within heterogeneous strata, e.g. Valencia et al. 1994), I use analogies to suggest more rigorous and efficient methodologies. Since the literature on this topic is extensive, my review is highly selective.

2. Why assess biodiversity?

One of the most important steps in planning an inventory is to clearly define the objectives, purpose and expected outcome, and the inventory of biodiversity is no exception in this regard. Efficiency may be compromised by appraising biodiversity if it is merely a surrogate for other objectives. Thus I begin by exploring some situations where recognition of the real issues may reveal a more efficient methodology than the biodiversity surrogate.

2.1 BIODIVERSITY AS A SURROGATE

If the real question is one of the following, it may be more efficient to design the inventory to serve that goal directly, rather than to attempt to quantify biodiversity. Questions sometimes posed to justify biodiversity surveys are contrasted with alternative formulations that may offer better insights for inventory design:

- Should we conserve this tract of forest? Is this the most important representative example of this type? Will this complement other reserves and ensure reproductive success of target species? "Importance" need not embrace biodiversity, but may include area, fragmentation, disturbance, placement (e.g. outlier), protection, etc. (see e.g. Margules and Usher 1981; Prendergast et al. 1995).
- Is this place "natural" and "environmentally healthy"? What tangible signs of anthropogenic disturbance can be found in this vicinity? What can we infer from the status of selected indicator species (see e.g. Spellerberg 1991)?
- Are there "unique" or locally endemic species here? What is known about the distribution and status of species thought to be confined to this locality? Endemic records may "simply reflect taxonomists predilections to visit certain areas" (Burley and Gauld 1995).
- What is the potential for ecotourism? How do scenic quality and charismatic species rate on a regional and international scale? (see e.g. Brown et al. 1990).
- What is the risk that a species found here will become extinct? Is there sufficient information to build a model of population dynamics for species thought to be at risk (e.g. Burgman et al. 1993)?
- Where am I most likely to find a new species? Where are existing surveys inadequate or incomplete? (In an empirical study, Namkoong (1995), observed that rare species were not well correlated with either general species diversity or any single environmental factor).

- Where can I find plants of potential commercial interest (e.g. so-called wonder drugs)? Where is the required adaptation expected to have evolved? (Beattie 1995).
- Where are regional "hot spots" of biodiversity? Hot spots for one group of organisms may not coincide with those for other groups (even for groups as similar as terrestrial ants, beetles and spiders, e.g. Oliver and Beattie 1996), so this question cannot be resolved without defining the groups of interest, and surveys conducted for other purposes may have little relevance.

2.2 BIODIVERSITY PER SE

There are also many bona fide reasons for assessing biodiversity. Among these are

- Checklists and other indicators of species or genetic richness: Species checklists are of interest to naturalists and resource managers, and are usually compiled from historical records and casual surveys within the defined area. They are usually confined to particular groups of organisms such as birds and vascular plants. To date, there are no complete assessments of forest diversity anywhere in the world, although partial assessments based on small subsets of species exist.

  Issue: record identities of all species (including vagrants) within a defined area.
  An analogue of this issue is the monitoring of germplasm of progenitors of commercial crop plants. Despite widespread concerns, the extent of genetic erosion of crops and their progenitors has rarely been estimated (Brush 1995). Serendipity may play a role in prospecting for new sources (e.g. Iltis 1988), but in general, crop breeders and botanists who have evaluated germplasm know more about its variability than collectors who can only observe phenotypes (Williams 1988).

  Issue: understanding rather than sampling the genotype.
  A related issue is the number of species globally, an intriguing question, not yet to be answered satisfactorily, despite an extensive literature (see e.g. Stork 1993).

  Issue: an approximate estimate of richness, extrapolated from samples.

- Map biodiversity and species distributions: Many planners would like maps with isolines of diversity (cf. contour map) or systematic point estimates enabling species distributions and richness surfaces (cf. digital elevation models) to be constructed and used in geographic information systems for land use planning (e.g. Scott et al. 1993). Depending on the extent of prior information (see below), systematic surveys may be an efficient way to provide such data, but it is very difficult to ensure uniformly good coverage.

  Issue: a consistent series of point estimates suitable for interpolation or correlation studies (also to integrate diverse geographical information in a compatible way).

- Quantify baselines: The establishment of baseline data is an investment in the future, and offers the potential to detect and monitor changes. The uncertainty of future uses of such data make it impossible to give explicit prescriptions for such work, but it is apparent from diverse inventory experience that when resources are limited, quality should not be sacrificed for quantity. A few detailed sites, carefully documented and maintained over long periods, offer greater utility than a larger number of inferior samples.

Issue: detailed "benchmark" samples for comparison with other studies (at different times and in different places); the question is not so much how to sample, but rather where to put them.

- Monitor changes: Monitoring poses many challenges, especially when the objective cannot be determined in advance. One of the challenges in the context of this paper is the design and placement of sample units, and this too depends on objectives, notably if the object is to detect, map or quantify changes. If an early-warning system is envisaged, the placement of plots is critical, and the concept of triage is relevant: intervention may be futile, unnecessary, or effective, and the challenge is to recognise the latter.

  Issue: where to sample; what to measure; how to anticipate and detect impacts.

All of these objectives, whether bona fide or surrogate, impinge on the design of the inventory, and on the extent and circumstances under which inferences be made (Table 1). For example, despite superficial similarities, the compilation of a species checklist (by casual surveys) requires markedly different inventory and data management techniques than compilation and testing of a species distribution map.

TABLE 1. Data needed to address some questions about biodiversity. Notice that as the task becomes more complex, data needs become more demanding.

Issue	Method and data requirements
How many species?	Extrapolate from localised samples of richness (e.g. Erwin 1982).
Progenitor diversity	Interpolate from presence/absence data. May need social (use) data.
Map distributions	Extrapolate by individual species from environmental surfaces and presence/ absence data.
Monitor (early warn.)	Extrapolate from localised relative abundance data.

3. Quantifying Diversity

3.1 DIMENSIONS OF DIVERSITY

Diversity is a complex concept, spanning several dimensions which may be characterised in part by:

1. Units, ranging from
   - Genes (or traits),
   - Species (or morpho-species),
   - Structural aspects (e.g. size classes, morphological groups, etc.), to
   - Landscape components.

For convenience, I refer mainly to species, but the concepts can be applied equally to genes, structural groups and landscape components.

2. Grain, ranging from (terms in parentheses after Whittaker 1977)
   - Individual samples and micro-habitats (point diversity),
   - Habitats (alpha diversity),
   - Regions (gamma diversity), to
   - Large biogeographic areas (epsilon diversity).

   Whittaker's (1977) definitions, e.g. of gamma diversity are not universally accepted (cf. Primack 1993), so care must be taken with the use and interpretation of these terms.

3. Patterns, embracing (Figure 1)
   - Richness (number of species or assemblages, either numerical, e.g. per thousand individuals, or areal, e.g. per hectare),
   - Evenness (relative abundance, or presence/absence if on a binary scale),
   - Contagion (whether clustered or regularly spaced; encompasses species turnover), and
   - Fractal dimension (i.e. the edge:area ratio; whether clumps tend to form compact circles or elongated dendritic features).

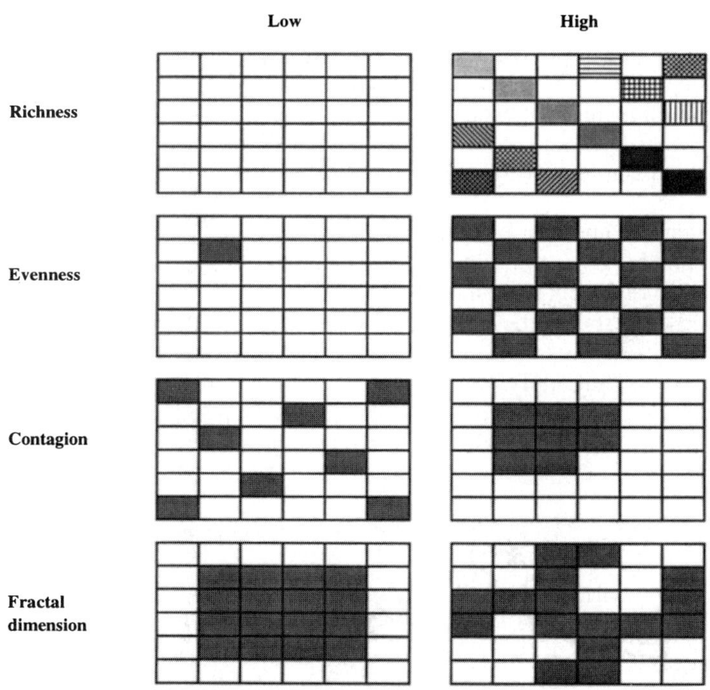

Figure 1. Patterns of diversity: richness, evenness, contagion and fractal dimension.

Biodiversity indices commonly encompass richness, evenness and contagion (implicitly as species turnover, the inverse of similarity), but usually neglect fractal dimension. Fractal dimension may be more obvious at the landscape level, but need not be irrelevant at the species and genetic levels. Note that if contagion is high, samples will differ in composition, so species turnover will also be high and similarity will be low (consider Figure 1 with a 2×2 sampling frame).

However, note that there are other ways to interpret the concept of biodiversity (e.g. Boyle and Sayer 1995; Faith and Walker 1996). Note also that these spatial concepts have a temporal analogue (e.g. 1: timescale, seconds to centuries; 2: rate of succession, slow or fast; 3: nature of disturbances over time, including seasons) which has rarely been explored in the biodiversity literature, despite its relevance, e.g. in systems driven by gap dynamics (but see e.g. Grubb 1977).

The most prevalent concepts of diversity, alpha, beta and gamma diversity are closely interrelated, and warrant further explanation. Whittaker (1977) defined alpha diversity as the species richness (with or without weighting for evenness) at the habitat scale, and identified analogues at the micro-habitat, regional and biogeographic scales. While alpha diversity refers to within-sample diversity, beta diversity refers to between-sample diversity, indicating how species diversity changes along a gradient or in different parts of a habitat. Whittaker (1960 and 1977) coined the term beta diversity to describe the variation in alpha diversity within a region. There are also analogues at other scales: so that "pattern diversity" relates to the variation between point samples within a habitat, and delta diversity relates to the relationship between gamma and epsilon diversity (Note that others use pattern diversity in a very different context, e.g. Scheiner 1992). Alpha, beta and gamma diversity, and their analogues, are closely interrelated (Table 2):

- ⇒ alpha cannot exceed gamma,
- ⇒ if either alpha or beta is high, gamma must be high,
- ⇒ if gamma is low, both alpha and beta must be low,
- ⇒ thus, gamma ≈ alpha × beta.

Notice that these parameters (alpha, beta, etc.) also relate to the species-area relationship (Figure 2):

- ⇒ alpha, gamma and epsilon correspond to points on the curve at the habitat, region and biogeographic scales (thus estimates will depend on the area selected to represent a habitat, etc.),
- ⇒ beta and delta reflect the slope of the curve between these points.

Most of the caveats that apply to the species-area relationship (e.g. Palmer and White 1994) also apply to biodiversity estimates. Notice that if the slope (i.e. turnover, whether at the "pattern", beta or delta scales) is steep, the variance associated with estimates of

TABLE 2. Feasible combinations of, and interrelationships between alpha, beta and gamma diversity.

Alpha (Habitat richness)	Beta (Species turnover)	Gamma (Regional richness)
High	-	High
-	High	High
Low	Low	Low

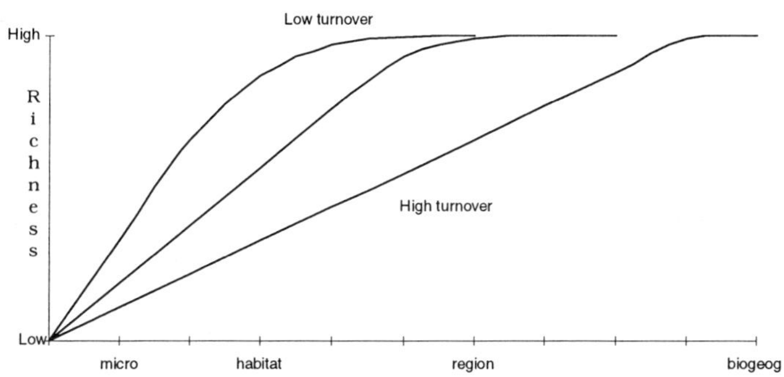

Figure 2. Indicative species-area curves showing the relationship between grain, richness and turnover.

alpha and gamma is likely to be high. The difficulty of finding asymptotes in tropical forests (e.g. Hubbell and Foster 1983) serves as a note of caution regarding the likely accuracy of the various estimates of diversity. If, however, an asymptote is evident, estimates are likely to be more robust.

One further caveat is required to warn about possible ambiguities in estimates of gamma diversity. While there seems to be reasonable agreement about the general concepts of alpha and beta diversity, conflicting definitions of gamma diversity exist (cf. Whittaker 1977; Primack 1993). And even if it is agreed that gamma diversity should represent species diversity at the regional scale, estimates will depend upon whether relative abundance is computed from habitat-level or micro-habitat-level data (i.e. if it is computed using the proportion of individuals or the proportion of plots with the target species).

3.2 DIVERSITY INDICES

Meaningful comparisons of diversity require that these concepts be quantified in standard ways, and many indices have been proposed during the past hundred years (i.e. since Jaccard 1902; e.g. see Magurran 1988). One of the first, the alpha diversity index of Fisher et al. (1943) probably contributed to the current alpha-beta terminology.

Most of the popular indices relating to alpha and gamma diversity belong to a family of indices denoted Db (Patil and Taillie 1979; N.B. b is not related to beta diversity, but merely indicates a parameter to be provided; see also Hurlbert 1971 and Hill 1973 for similar series):

$$\Delta_\beta = \left(1 - \sum_{i=1}^{s} \pi_i^{\beta+1}\right)\beta^{-1}, \quad -1 \leq \beta < \infty \tag{1}$$

where s is the number of species and p_i is the relative abundance of the ith species. This family includes species richness (less one, b = –1), the Shannon index (b → 0), and the Simpson index (b = 1; in the form appropriate for an infinite community). Notice that the influence of relative abundance on $\Delta\beta$ increases as β increases. Thus the Shannon form is sensitive to changes in rare species, whereas the Simpson form is sensitive to changes in common species (Peet 1974; May 1975). The Shannon form is more efficient at statistically discriminating samples (Magnussen and Boyle 1995), but the Simpson form is more meaningful, representing the probability of an interspecific encounter (e.g. predation, etc.), and thus presumably reflecting community stability (Hurlbert 1971). Users may find it informative to plot the value of $\Delta\beta$ for a range of β (cf. transformations for stabilising variance in statistical analyses, Box and Cox 1964). One problem with this family and many other diversity indices is the implicit assumption that all species have been sampled. Since this is rarely satisfied, most indices will be biased towards underestimates of diversity. A further concern is the lack of asymptotic behaviour as sample size increases (Peet 1975).

There are also many indices of similarity and turnover (cf. beta diversity; see Magurran 1988; Spellerberg 1991). The information-rich indices require pairwise comparisons, but one simple index of overall species turnover is Whittaker's (1960):

$$\beta = \frac{s}{\alpha} - 1 \tag{2}$$

where s is the number of species in the system (gamma diversity) and β is the average sample richness (alpha diversity). However, it is not necessarily the most revealing index, and it may be useful to compute a range of indices for comparison. Morista's (1959) index of community similarity is one of the preferred alternatives, being relatively unaffected by sample size and species richness (Wolda 1981):

$$C = \frac{2\sum_{i=1}^{s} n_{1i} n_{2i}}{(\lambda_1 + \lambda_2) N_1 N_2}, \quad \text{where} \quad \lambda_j = \frac{\sum_{i=1}^{s} n_{ji}(n_{ji} - 1)}{N_j(N_j - 1)} \tag{3}$$

n_{ji} is the number of individuals of species i in sample j, and N_j is the number of individuals in sample j (in some variants, the -1 are omitted, e.g. Magurran 1988). Notice that this is a measure of similarity, so reflects the inverse of beta diversity. Like most indices of similarity, it provides a basis for pairwise comparisons only, and classification or ordination may be need to examine the relationships between many samples.

The analogy between the assessment of species turnover and contagion in spatial tree distributions does not seem to have been explored in biodiversity research. Ripley (1977) and Moeur (1993) considered contagion in forest stands, and their work appears to have implications for species diversity.

In all these cases, the problem of assessing biodiversity is reduced, in part, to assessing the numbers of species and their relative abundance (by number, biomass, etc.). In most cases, the number of species is critical, and indices will be biased towards underestimates if all species are not sampled. Thus, I dwell mainly on the assessment of species presence, rather than on the estimation or statistical properties of any given set of indices.

3.3 SAMPLING DIVERSITY

Despite the intricacy of the biodiversity concept, the issue for inventory remains one of assessing the presence and relative abundance of species within well defined "patches", at the point, habitat and regional scales. Notice that the definition of patches and the corresponding sample size are critical. Take a chequerboard situation, and consider the implication of increasing the sample size from single squares to groups of squares, or of using a sample not aligned with the chequerboard's grid (Figure 3, Table 3). Clearly, samples contained entirely within a patch (e.g. 1 & 2) assess something very different than larger samples (5 & 6) and samples straddling patch boundaries (3 & 4).

Obviously, patch size depends on the organism under consideration. For example, a homogeneous patch for trees may be quite large, while the tree itself may look like a chequerboard to a smaller organism such as an insect (e.g. Southwood and Kennedy 1983). The numbers of organisms per unit area may vary by 25 orders of magnitude (10^{21} for soil bacteria to $10^{-5}m^2$ for deer, Odum 1968), and even within a single species, the mass of individual plants may vary 50,000-fold (Harper 1977). This poses real challenges for efficient inventory. In this context, the micro-habitat-region-biogeography scales identified above should be interpreted loosely (e.g. Lawton 1976, who used frond-patch-country-continent scales for insects on bracken).

The nature of the patches impinges on inventory design, since samples should be large enough to characterise a patch, yet small enough to be contained entirely within a patch (Figure 3). This interdependence between organism, patch and sample makes it difficult to offer anything more than general guidelines for defining patches, and means that both people with local knowledge and researchers specialising in the organisms of interest should be intimately involved in the inventory design.

One of the basic assumptions underlying sampling and statistical theory is that of randomness. Many factors (predators, competition, habitat, etc.) lead organisms to aggregate, making it unlikely that individuals will be randomly sampled (Pielou 1975; Southwood 1978; Magurran 1988). Furthermore, recognition and description of organisms is non-random (Gaston et al. 1995). Thoughtful design and careful conduct of planned inventories can help to minimise the consequences, but possible implications of these and other deficiencies should be considered when existing data are used.

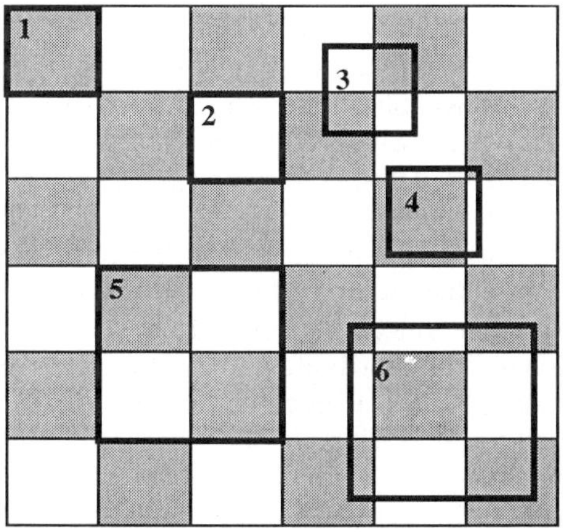

Figure 3. The size and alignment of sample units has a critical effect on estimates of richness, evenness and turnover (see Table 3).

TABLE 3. Implications of different sample placement in a chequerboard environment (Figure 3).

Sample	Richness	Evenness	Turnover
1	Low	High	High
2	Low	High	Medium
3	High	High	Medium
4	High	Low	Medium
5	High	High	Low
6	High	High	
1-4	High	High	Medium
5-6	High	High	Low
True			
Point	Low	High	High
Habitat	High	High	Low

3.4 SOURCES OF DATA

Biodiversity surveys typically rely on a combination of desk studies based on existing data, and specially commissioned surveys. Both sources offer ample scope for bias and confusion. All surveys need careful design and conduct to provide uniform results, but compilation of existing data poses additional hazards that warrant careful scrutiny. The first difficulty is that different sources of data typically involve different sampling intensities and methods, making comparisons and compilations difficult, and sometimes impossible. Both species richness and evenness depend on sample size, so can only be validly compared if samples are of the same size or if subsampling is simulated (Hurlbert 1971).

When attempts are made to combine data from different sources to make inferences about species richness, the accuracy of locational details may be critical. For instance, herbarium records may lead to spurious "hot spots" rich in species where records refer locality names or have been recorded to the nearest degree (recall that a minute of longitude is about 2 km at the equator), or apparently rich in rare individuals at sites which are favourite and accessible places for forays (see Vanclay 1993, for further discussion). This problem becomes evident when attempts are made to use herbarium records at the local scale (e.g. to delineate reserve boundaries), even though they may still be useful at the continental scale (e.g. Chippendale 1981). This does not imply that existing data should not be used, nor that one should defer action until "perfect" data are available, but demands judicious use of all data, and careful consideration of possible implications. One important practical use for existing data is to help plan supplementary surveys by providing a basis for stratifying and estimating variances. However, not that specimens can only provide "presence" records and only careful surveys can provide the "absence" records needed for modelling and more sophisticated inference.

Several other limitations of desk studies were discussed by Burley and Gauld (1995), who summarised three major concerns with existing data:

1. The actual geographic range of many organisms is likely to be seriously under-represented, so that high levels of endemicity may simply reflect taxonomists predilections to visit certain areas.

2. Specimens are not usually a random sample of the biota of a region. Taxonomists "sampling" is highly selective, with a general tendency to maximise the species-richness obtained in a collecting foray, favouring rarity or scarcity, rather than sampling in any ecologically meaningful way.
3. There is a strong bias towards collecting and describing large and showy organisms.

These deficiencies are common to most data sets, and indicate the degree of caution required in the compilation of desk studies.

4. Designing an Inventory

Inventories involve many questions like

- Why?
- What to measure? (e.g. presence or absence of vascular plant species);
- What is a species?
- What constitutes presence?
- How to sample? (e.g. point or plot-based surveys);
- How many samples?
- When and Where? and finally
- Will this satisfy needs (accuracy, cost, level of detail, time constraints)?

Fieldwork should not be commenced until the overall objectives can be clearly stated and satisfactory answers can be given to each of these questions. Because it may be difficult to answer some of these questions in advance, an inventory should be an iterative process, with a preliminary survey preceding the main inventory effort. However, some compromise may be necessary to avoid infinite loops! It is important to establish at the outset that resources are sufficient, or the logical decision may be not to conduct the inventory (Hamilton 1979).

4.1 SPECIES PRESENCE/ABSENCE

It is critical to the success of an inventory that the object of sampling can be reliably determined. In the case of biodiversity assessment, this means that species (genetic or landscape formations, etc.) must be able to be identified in a repeatable way, and that presence or absence must be able to be determined with certainty: two demanding requirements.

Reliable determination of species may be difficult, particularly if mature and fertile specimens are not available. Identification of smaller and less-studied organisms may be equally difficult. Even amongst the better-known groups of organisms, nomenclature may not be entirely rigorous and objective (Stork 1993). One solution is to use "morpho-species" (e.g. Stork 1995; Oliver and Beattie 1996); another is to use structural groups. Many studies have shown that structural diversity is a good indicator of species diversity, for both direct (i.e. plant species richness predicted from plant structural diversity, e.g. Gillison 1996) and indirect relationships (e.g. birds: MacArthur and MacArthur 1961; Pearson 1993; insects: Southwood et al. 1979). Forest structure may be characterised in terms of age classes, size classes (e.g. MacArthur and MacArthur 1961) or morphological attributes (e.g. Gillison 1996; Williams and Humphries 1996).

Assessing presence or absence of a species also involves complications. The presence of a mature individual with viable propagules (viz. plant with fertile seeds; animal raising offspring) may be confidently taken as a confirmed "presence", but sightings of solitary immature individuals are more ambiguous. Are they vagrants with a merely transient presence (see e.g. Burley and Gauld 1995), or are they indicative of a more substantial involvement? Do viable but dormant seeds constitute a presence, and if not, how should one accommodate an influx of pioneers following disturbance, or the "greening" of a desert after rain? Blanket prescriptions cannot be provided for all scenarios, and consistent procedures should be defined and documented before an inventory is commenced.

4.2 PLACEMENT OF SAMPLES

H.C. Dawkins (pers. comm.) developed a key to assist the design of forest inventories (Table 4), but it also serves to guide biodiversity assessments. This binary key is based on a few decisions at each of up to four steps and indicates one suitable sampling strategy and some possible caveats. Note that although cheap, subjective sampling (or "expert assessment") is subject to a number of caveats: it is more likely to confirm than reject preconceptions; and the subjective element makes it unsuited to spatial and temporal comparisons, since differences may be due to personal as well as other factors.

Systematic sampling may be the best option if no prior data exist and interpolation is required (e.g. to draw species distribution maps; see Table 4). However, prior data are usually available, and can be used as a basis for stratifying and improving the sample in other ways.

If the origin of specimens is known with some certainty and data on selected environmental variables are available, spatial modelling packages such as DOMAIN (Carpenter et al. 1993) may be used to make inferences about the likely theoretical distribution of a species. These in turn may be used as a basis for sampling. It seems likely that predictions from packages such as DOMAIN may err on the optimistic side, so it maybe useful to test predictions with reliable survey data indicating absence of target species. Such evidence may be drawn from existing survey results, or supplementary surveys may be conducted near the margins of predictions.

If both presence and absence data are available, logistic regression may be used to predict from various environmental variables, the likelihood of finding the species in a given location (e.g. Austin et al. 1984; Nicholls 1989). Again in this case, supplementary sampling is useful for iteratively testing and improving the model. The estimated variance can be mapped, providing a sound basis for supplementary sampling (i.e. take more samples in areas where predictions have high variance).

While this approach is useful for limited numbers of species, it seems unlikely that logistic equations could be constructed for each of the thousands of species found in tropical regions. In such cases, it seems appropriate to focus on selected species of greatest concern (e.g. rare, endangered, keystone, etc.), or to aggregate species into groups for modelling. Groups may be formed by aggregating on environmental relationships evident in statistical models (cf. Vanclay's 1991 work on growth patterns), or by classifying on the inter-species contagion observed in species distributions.

Table 4 does not canvass the options of multi-stage and clustered sampling. These strategies may offer efficient ways to reduce travel costs associated with sampling (see e.g. Schreuder et al. 1993), and can be used in conjunction with all the sampling strategies mentioned in the key.

TABLE 4. Key to some alternative sampling designs showing selected criteria and some possible consequences (Vanclay 1994).

Criteria	Inventory alternatives & possible consequences	
Step 1		
Nature of estimate	Critical	Unimportant/personal
Forest characteristics	Unknown/diverse	Familiar or uniform
Representative selection	Unreliable	Reliable
Time and resources	Sufficient	Very limited
	Objective ⇒ Go to 2	Subjective sampling
Bias	Absent	Unavoidable
Precision	Can be estimated	Unknown
Step 2		
Periodicity	Possible/unknown	Unlikely or known
Interpolation	Not required	Necessary
Estimate of precision	Required	Unimportant
	Random ⇒ Go to 3	Systematic sampling
Sampling error	Correct estimate	Probably inflated
Periodic bias	Unlikely	Possible
Step 3		
Pattern in population	Clear or likely	Absent or unlikely
Sampling intensity	Relatively low	High
	Stratified random ⇒ 4	Unrestricted random
Inherent risks	Misjudge pattern	Sample clustering
Step 4		
Pattern in population	Obscure/unknown	Visible or well known
	Geometrical blocks	Statistical blocking
Calculations	Simple	Possibly complicated

N.B., Stratification need not be geographic, but may be based on size (diameter or height classes, under-/over-storey), seasons (winter vs. summer, day/night), etc. Similarly, systematic sampling need not be geographic, but may be according to variance.

4.3 NATURE OF SAMPLES

We have already seen from Figure 3 that the size, shape and placement of sample units may affect estimates. Thus, it is not difficult to find conflicting advice regarding samples, for example:

- use plots of a standard size (e.g. 1-ha or 50-ha plots);
- use nested plots, according to organism size;
- use evidence of an asymptote to determine sample size;
- use many small rather than few large plots.

Any of these strategies may or may not suit a particular need in a given location.

In theory, when the dimensions of target organisms varies greatly, the best sample strategy may be to use point samples of zero area (e.g. by dropping a pin or projecting a laser, and recording all organisms intersected). If repeated many times, this procedure should give a good sample of presence and relative abundance, for all organisms, irrespective of size and shape (the biomass-weighted relative abundance might be overestimated for flat organisms like water-lilies, but energy-weighted abundance should be OK). Unfortunately, such point samples are rather impractical, not least because few organisms are registered at each sample, so that many samples are needed. For an efficient analogue, step with Alice "Through the Looking Glass" (Carroll 1872; i.e. into a land of make-believe), and inflate each organism to say, 100 times it usual size, allowing them to overlap without interference. Then each point sample would register 100 times as many individuals. This is not the same as replacing the pin with a frame (cf. fixed-area plot), since inflation maintains the probability of registration proportional to size, whereas a frame is likely to sample more small organisms. In reality, we cannot inflate organisms, but we can inflate the probabilities to efficient levels, as is customarily done in sampling with probability proportional to size (PPS), a technique which is widely practised in forestry, with many variants (e.g. angle gauge sampling, see e.g. Schreuder et al. 1993).

With forest trees, the use of angle gauge sampling to sample with probability proportional to stem cross-sectional area is effective and efficient, but it is not clear how to sample other organisms exhibiting a greater range of size and shape. For organisms that are not "tree-like", biomass may be estimated and sampled with probability proportional to prediction (PPP; Schreuder et al. 1993). Continuing this analogy, if a formula enables objective estimates of biomass, model-based sampling may be effective.

Sometimes area-based sampling is desired for comparisons with other work. Notional areas sampled by PPS can be easily computed, or samples may be based on nested plots. However, even minor variations in the design of nested plots may result in substantial differences in apparent richness (e.g. Stohlgren et al. 1995), so careful consideration and some preliminary trials are recommended. Further complications arise with highly mobile organisms (e.g. birds), for which point and area-based samples are inefficient, so that other dimensions may be needed (e.g. per hour, per thousand individuals).

One dilemma with sampling is how to deal with rare organisms of particular interest, especially when these are encountered through serendipity rather than in a formal sample. Subjective sampling outside the standard sampling frame may provide a better (higher) estimate of richness, but a worse (biased) estimate of abundance. For example, suppose that a rare plant is encountered while travelling between plots comprising a systematic sample. Ignoring the observation preserves the sampling scheme and may offer the "best" estimate of abundance, but will bias the estimate of richness. Conversely, establishing an additional plot deliberately located to include the target individual should provide a better estimate of richness, but may bias estimates of abundance (the recorded frequency of the rare plant may be too high). However, the "true" abundance of the rare plant is likely to be between zero and 1/n, where n is the number of plots (assuming random distributions of plants, species and plots), so assigning such subjectively-located plots a weight of 0.5 (relative to objective samples) may be a reasonable approach. If the assessors have an intimate knowledge of the locality, they may be able to offer more realistic weights for these subjective plots, offering better precision, but introducing the risk of some bias. Such subjectivity is likely to influence the Shannon index more than the Simpson index.

The converse situation may also apply. Through chance, a sample may encounter any given species with a frequency that is considered representative of the population (although this true frequency may never be known). There are two satisfactory ways to deal with such samples that are considered unrepresentative: one is to conduct more (objective) sampling until the combined sample looks more representative; the other is to assign subjective weights to the existing samples to bring them into line with expert opinion. The latter approach is more controversial, and is discussed at some length in the literature (cf. Bayesian estimation; see e.g. Maritz 1970; Ek and Issos 1978; Hawkes et al. 1983).

4.4 NUMBER OF SAMPLES

A recurring dilemma in inventory design is the determination of sample size: frequently clients cannot afford sufficient samples to attain the precision they initially request. The usual compromise is to sample until funds are exhausted, but this may lead to a sub-optimal outcome (e.g. Hamilton 1979), so careful consideration of the options is warranted. For surveys of biodiversity, the question is particularly significant, as it concerns bias as well as precision. In a synthesis of diversity indices, Patil and Taillie (1982) defined one family as

$$s(m) = \sum_{i=1}^{s}(1-\pi_i)\left\{1-(1-\pi_i)^m\right\} \quad , \qquad 1 \le m < \infty \tag{4}$$

which includes the Simpson index (m=1) and species richness (less 1, as $\mu \to \infty$), and in which s(m) is the expected number of species in a hypothetical random sample of size m. One immediate implication of this is that s(m) will only begin to approximate s (minus 1) as m, the sample size, becomes large. Thus it seems inevitable that random sampling will always underestimate species richness.

This is consistent with the notion that species exhibit a log-normal distribution, the left-most portion (i.e. the less common species) of which will be obscured (or veiled) to an extent determined by the sample size, so that only the largest samples will reveal the full number of species and their relative abundance (Preston 1948; Magurran 1988). If the distribution can be assumed, and the veil line or truncation point can be recognized, then true species richness can be estimated (Slocomb et al. 1977), but the identity of these additional species remains unknown. Although the use the log-normal distribution in this way has been challenged (e.g. Hughes 1986), the shape of this distribution is consistent with species-area curves for many tropical forests.

In this context, it seems difficult to reconcile the common advice to "construct a diversity curve for what is considered likely to be the most diverse site and plan the sampling regime accordingly ... the point at which the curve flattens indicates the minimum viable sample size" (Magurran 1988), especially since evidence of an asymptote cannot always be detected in tropical forests, despite sample sizes of up to 50 hectares (e.g. Hubbell and Foster 1983; Lieberman et al. 1996).

In my personal inventory experience, I favour an initial objective sample, supplemented if necessary by additional samples to ensure that the combined sample matches expectations of experts familiar with the locality (or until the experts change their opinion; Vanclay 1994). To minimise the subjective element (i.e. whether the initial sample is "representative"), it is critical that no "unrepresentative" samples are discarded, and that the aggregate is made more "representative" only by taking additional objective samples. This strategy, modified to include chance encounters of rare species, should also be effective for biodiversity assessments.

4.5 WHERE AND WHEN TO MEASURE

The distribution of species has temporal as well as spatial dimensions, and species will only be detected if the observer is using the right method, at the right place, at the right time. Luck may play some part, but careful preparation and planning are important. Sampling on more than one occasion may be necessary, since single samples may seriously underestimate total species richness (e.g. Oliver and Beattie 1996). Frequently, it is efficient to deliberately sample extremes and gradients (multidimensional, including spatial, temporal and environmental - see e.g. Gillison and Brewer 1985; Beetson et al. 1992; Vanclay 1994).

5. Selected Case Studies

I conclude with a brief discussion of three case studies to illustrate some of the issues canvassed. These examples are presented for discussion, not necessarily for emulation! Collectively, they are intended to reinforce some of the discussion above, including the issues relating to clear identification of objectives, sufficient stratification, appropriate methodology and modest extrapolation.

5.1 ERWIN'S ESTIMATE OF GLOBAL ARTHROPOD RICHNESS

In a brief but much-quoted paper, Erwin (1982) extrapolated from a collection of some 955 species of Coleoptera on 19 trees of Luehea seemannii in Panama, and speculated that there may be 30 million insect species in the tropical forest (Table 5). In doing so, he posed several testable hypotheses and invited researchers to challenge his figures with better data. Key assumptions relate to the proportion of insect species that are host-specific, and the number of hosts. From an inventory perspective, some of the contentious issues are:

- Do trees such as L. seemannii form an appropriate sample, or is there excessive scope for bias?
- Are 19 trees a sufficient sample? (Unfortunately, tree-to-tree variability was not stated).
- Could the estimate be improved by stratifying by structure (e.g. upper-lower canopy, live-dead tissue, dry-moist substrate), taxonomic groups, or by time (e.g. day-night, wet-dry season, summer-winter, early-late successional forest)?

TABLE 5. Estimating global arthropod richness (Erwin 1982).

Source	Estimate	
Coleoptera on Luehea seemannii	955	
Weevils	206	
Total insects	c. 1,200	
L. seemannii host-specific (13.5%)	163	
Transients (non-host-specific)	1,038	
Assumptions	**per hectare**	**Global (million)**
Number of tree species	70	0.050
Total host-specific insects	11,410	8.150
Transient insects	1,038	0.001
Total canopy insects	12,448	8.151
Total canopy arthropods (x 2.5)	31,120	c. 20.000
Total arthropods (incl. forest floor, x 1.33)	41,398	c. 30.000

- Is it realistic to extrapolate from one species in one locality to thousands of other tree species globally? Even given the constraint of 19 samples in one locality, would it not have been more enlightening to sample e.g. 3-4 trees each of five different types (e.g. a palm, emergent, spreading canopy tree, compact canopy tree, understorey tree)?

I leave these questions for the reader to ponder, both in the context of Erwin's estimate, and in their own work. Many researchers have responded to Erwin's challenge, so there is a rich literature addressing this topic (see e.g. Stork 1993).

5.2 HAWTHORNE'S BOTANICAL SURVEY OF GHANA

Hawthorne and Juam Musah (1993) studied the conservation status of forests in Ghana, and presented specific recommendations for improved management supported by an innovative decision support system, FROGGIE, the Forest Reserves of Ghana Graphical Information Exhibitor. The aim of their botanical survey was to assess "what vascular plant lives where?", but this was framed within the broader objectives to resolve "is there much good forest in Ghana?; should it be protected? and if so, how?". Although Ghana is a relatively small country (200,000 km^2), a thorough inventory of the plant diversity of the 4 million hectares in 200 forest reserves is a considerable task. One of the difficulties in devising such a survey is to find a balance between cost-saving shortcuts and scientific rigor. Hawthorne and Juam Musah (1993) adopted several innovative initiatives, but their survey still consumed more than 7,000 man days over 30 months. They conducted both a subjective appraisal of overall reserve status, and a more formal botanical survey. The reserve appraisal involved visiting each forest reserve, and scoring it as a whole on a 6-point scale to summarise evidence of disturbance and degradation (1: few signs of disturbance, intact canopy, primary or late secondary forest; 6: mostly deforested), based on observation, local advice, inventory (in conjunction with a national forest inventory) and remote sensing.

The botanical survey involved some 700 samples (1-20 samples in each reserve) and a total of 30,000 voucher specimens to confirm some 80,000 identifications. The samples comprised some formal 1/16 ha plots (five 25×5 m strips, usually side-by-side viz. 25×25 m, but sometimes end-to-end) and many casual samples based on a walk through relatively homogeneous parts of the reserve (e.g. ridge-top, steep slope, overgrown skid trail, swampy area). These samples usually continued until some 40 canopy trees were included. The latter casual samples allowed more flexibility in selecting vouchers (e.g. fertile or mature rather than juvenile specimens). A comparison of the two sample methods, and with an earlier formal sample (Hall and Swaine 1981), indicated reasonable agreement, but results were most consistent when at least 80 species were sampled (Hawthorne and Juam Musah 1993).

For each reserve, a so-called "genetic heat index" was computed from formal, casual and existing samples:

$$I = \frac{100 \sum n_i w_i}{\sum n_i}$$

where w_i is a weight and n_i is the number of species in the ith of 5 categories (Table 6). This represents the mean weight for the species observed on each reserve. In practice, the index ranges from 0 to 406, and seems to be relatively independent of total number of

TABLE 6. Categories used by Hawthorne and Juam Musah (1993) to rank species.

Category (star-rating)	Weight	No of species	Range (degree-squares)	Description
Black	27	52	2	Rare internationally, uncommon in Ghana
Gold	9	208	8	Fairly rare internationally and/or locally
Blue	3	414	24	Rare in Ghana
Red	1	73	40	Common but exploited
Green	0	1044	70	No concerns
Total		1791		

species (Figure 4). Many reserves have multiple samples yielding rather consistent indices despite unequal samples, different times and different methodologies.

Notice that this index does not take into account relative abundance by number or biomass, but merely notes that e.g. "black stars" are present among other species. These indices, and an analogous economic index (with different weights) were incorporated into FROGGIE, a GIS able to display these data in a variety of easily-digested forms. Hawthorne and Juam Musah succeeded in this work because

- the vascular flora of Ghana was well documented (e.g. Hall and Swaine 1981);
- they did not attempt to quantify biodiversity, but chose the easier and more robust approach of highlighting where endangered species were relatively abundant;
- they managed to avoid a tangential debate about the composition of categories, and focused discussion on the implications of their work;
- they were fortunate that in many cases the "genetic heat index" and the economic index were inversely related;

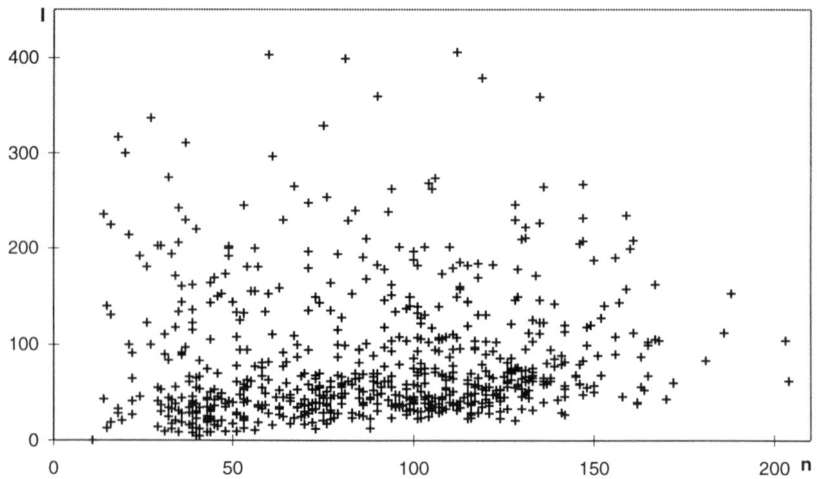

Figure 4. "Genetic heat index" (I) is relatively independent of the number of species (n), with a correlation of 0.07. Two points based on check lists are omitted: n=517, I=76 and n=320, I=265 (data from Hawthorne and Juam Musah 1993).

- they presented their results skilfully in reports, concise recommendations, decision support systems and clever icons (the black star is a national symbol).

It remains to be seen if their recommendations are implemented and effective. It is also interesting to speculate how a study based on logistic regression with existing data e.g. for the 52 black star species would have compared in terms of cost and outcome.

5.3 GILLISON'S SURVEY OF BIODIVERSITY IN SUMATRA

As part of a larger project to develop and demonstrate tools for natural resource assessment, CIFOR and several collaborators (Gillison et al. 1996) are surveying biodiversity in the vicinity of Kerinci Seblat National Park in Sumatra. Several aspects of this work are noteworthy:

- not a complete inventory, but a transect spanning the range of habitats;
- gradsect based sampling (Gillison and Brewer 1985);
- sampling both species and structure (via plant functional attributes or PFAs; Gillison and Carpenter 1994);
- involving many different experts to get "the big picture";
- coupling ground-based survey with remote sensing.

Other, more controversial issues include the decision to use comparatively small plots (40×5 m), and not to enumerate individuals (i.e. trees are not measured, and the plots merely provide a frame for sampling one individual per species). Nonetheless, results suggest a strong correlation between the PFA summary and species richness (Gillison et al. 1996).

6. Synthesis

The literature lacks concrete advice on "bio-inventory", in part, because there are no "off-the-shelf" solutions for perfect assessments. The "best" approach depends heavily on your needs and your situation. The literature also reflects a preoccupation with computing indices rather than gathering reliable data: with decorating an edifice without first laying the foundations! Thus I offer a plea for fewer numerical gymnastics and more thought about what, when and how to measure basic data. Care is required to ensure that "bio-inventory" procedures are

- complete and comprehensive,
- aligned with objectives
- reliable and repeatable, and with
- clear statements of assumptions and scope.

How can one tell a "good" estimate of biodiversity from an inferior one? Look for, and expect to find:

- well documented statement of purpose, design, method and assumptions;
- use of existing data supplemented by new surveys;
- incorporation of both presence and absence data;
- stratification where appropriate;

- sampling systems appropriate for target organisms;
- consultation between biological and biometric specialists.

7. Acknowledgements

Thanks are due to Michael Köhl for urging me to prepare this contribution; to Jon Lovett for helpful advice, and to Tim Boyle and Jeff Sayer for comments on the draft manuscript.

References

Austin, M.P., Cunningham, R.B. and Fleming, P.M. 1984. New approaches to direct gradient analysis using environmental scalars and statistical curve-fitting procedures. Vegetatio. 55: 11-27.

Beattie, A.J. 1995. Commercial exploration of biodiversity. Ecological Economics Conference Papers, Coffs Harbour, 19-23 Nov 1995, ANZSEE & CARE, Armidale, Australia, pp. 251-256.

Beetson, T., Nester, M. and Vanclay, J.K. 1992. Enhancing a permanent sample plot system in natural forests, The Statistician 41, 525-538.

Box, G.E.P. and Cox, D.R. 1964. An analysis of transformations, Royal Stat. Soc. J. B 26, 211-252.

Boyle, T.J.B. and Sayer, J.A. 1995. Measurement, monitoring and conservation of biodiversity in managed tropical forest, Commonwealth Forestry Review 74, 20-25.

Brown, T.C., Richards, M.T., Daniel, T.C., King, D.A. and Vining, J. 1990. Scenic beauty and recreation value: assessing the relationship, in J. Vining (ed) Social science and natural resource recreation management, Westview, Boulder CO, pp. 281-299.

Brush, S.B. 1995. In situ conservation of landraces in centers of crop diversity, Crop Science 35, 346-354.

Burgman, M.A., Ferson, S. and Akçakaya, H.R. 1993. Risk Assessment in Conservation Biology, Chapman and Hall, London, pp.314.

Burley, J and Gauld, I. 1995. Measuring and monitoring biodiversity: a commentary, in T.J.B. Boyle and B. Boontawee (eds) Measuring and Monitoring Biodiversity in Tropical and Temperate Forests, CIFOR, Bogor, pp. 19-46.

Carpenter, G., Gillison, A.N. and Winter, J. 1993. DOMAIN: a flexible modelling procedure for mapping potential distributions of plants and animals, Biodiversity and Conservation 2, 667-680.

Carroll, L. (C.L. Dodgson) 1872. Through the Looking Glass and what Alice found there, http://www.literature.org/Works/Lewis-Carroll/through-the-looking-glass/index.html (30 Sep 1996).

Chippendale, G.M. 1981. The natural distribution of Eucalyptus in Australia,. Australian NPWS, Special Publication No. 6, 192 pp.

Ek, A.R. and Issos, J.N. 1978. Bayesian theory and multi-resource inventory, in H.G. Lund, V.J. LaBau, P.F. Ffolliott and D. Robinson (coords) Integrated inventories of renewable natural resources: proceedings of the workshop (Tucson 8-12 January 1978), USDA Gen. Tech. Rep. RM-55, pp. 291-298.

Erwin, T.L. 1982. Tropical forests: Their richness in Coleoptera and other Arthropod species, The Coleopterists Bulletin 36, 74-75.

Faith, D.P., and Walker, P.A. 1996. Environmental diversity: an the best-possible use of surrogate data for assessing the relative biodiversity of sets of areas, Biodiversity and Conservation 5, 399-415.

Fisher, R.A., Corbet, A.S. and Williams, C.B. 1943. The relation between the number of species and the number of individuals in a random sample of an animal population, J. Anim. Ecol. 12, 42-58.

Gaston, K.J., Blackburn, T.M. and Loder, N. 1995. Which species are described first?: the case of north American butterflies, Biodiversity and Conservation 4: 119-127.

Gillison, A.N. and Brewer, K.W.R. 1985. The use of gradient directed transects or gradsects in natural resource surveys, J. Environ. Manage. 20: 103-127.

Gillison, A.N. and Carpenter, G. 1994. A generic plant functional attribute set and grammar for vegetation description and analysis, CIFOR Working Paper No. 3, 12 pp.

Gillison, A.N., Liswanti, N and Arief Rachman, I. 1996. Rapid ecological assessment in Kerinci Seblat National Park Buffer Zone: Preliminary report on plant ecology and overview of biodiversity assessment, Manuscript, CIFOR, 22 pp.

Grubb, P.J. 1977. The maintenance of species-richness in plant communities: the importance of regeneration niche, Biol. Reviews 52: 107-145.
Hall, J.B. and Swaine, M.D. 1981. Distribution and Ecology of Vascular Plants in a Tropical Rainforest: Forest Vegetation of Ghana, Geobotany 1, Junk, The Hague, 383 pp.
Hamilton, D.A. 1979. Setting precision for resource inventories: the manager and the mensurationist, J. For. 77, 667-670.
Harper, J.L. 1977. Population Biology of Plants, Academic, London, 892 pp.
Hawkes, C.L., Chalk, D.E., Hoekstra, T.W. and Flather, C.H. 1983. Prediction of wildlife and fish resources for national assessments and appraisals, USDA Gen. Tech. Rep. RM-100, 21 pp.
Hawthorne, W. and Juam Musah, A. 1993. Forest Protection in Ghana, ODA, Kumasi, 72 pp + appendices.
Hill, M.O. 1973. Diversity and evenness: a unifying notation and its consequences, Ecology 54, 427-431.
Hubbell, S.P., and Foster. R.B. 1983. Diversity of canopy trees in a neotropical forest and implications for conservation, in S.L. Sutton, T.C. Whitmore and A.C. Chadwick (eds) Tropical Rain Forest: Ecology and Management, Blackwell Scientific, Oxford, pp. 25-41.
Hughes, R.G. 1986. Theories and models of species abundance, Amer. Nat. 128, 879-899.
Hurlbert, S.H. 1971. The nonconcept of species diversity: a critique and alternative parameters, Ecology 52, 577-586.
Iltis, H.H. 1988. Serendipity in the exploration of biodiversity: What good are weedy tomatoes? in: E.O. Wilson (ed) Biodiversity, National Academy Press, Washington DC, pp. 98-105.
Jaccard, P. 1902. Lois de Distribution Florale dans la Zone Alpine, Soc. Vaud. Sci. Natl. Bull. 38, 69.
Lawton, J.H. 1976. The structure of arthropod communities on bracken, Bot. J. Linn. Soc. 73: 187-216.
Lieberman, D., Lieberman, M., Peralta, R. and Hartshorn, G.S. 1996. Tropical forest structure and composition on a large-scale altitudinal gradient in Costa Rica, Journal of Ecology 84: 137-152.
MacArthur, R.H. 1955. Fluctuations of animal populations, and a measure of community stability, Ecology 36, 533-536.
MacArthur, R.H. and MacArthur, J.W. 1961. On bird species diversity, Ecology 42, 594-598.
Magnussen, S. and Boyle, T.J.B. 1995. Estimating sample size for inference about the Shannon-Weaver and the Simpson indices of species diversity, Forest Ecology and Management 78, 71-84.
Magurran, A.E. 1988. Ecological Diversity and its Measurement, Princeton University Press, NJ, 179 pp.
Margules, C. and Usher, M.B. 1981. Criteria used in assessing wildlife conservation potential: a review, Biological Conservation 21, 79-109.
Maritz, J.S. 1970. Empirical Bayes Methods, Methuen, London.
May, R.M. 1975. Patterns of species abundance and diversity, in M.L. Cody and J.M. Diamond (eds. Ecology and Evolution of Communities, Harvard University Press, Cambridge MA, pp. 81-120.
Moeur, M. 1993. Characterizing spatial patterns of trees using stem-mapped data, Forest Science 39, 756-775.
Morista, M. 1959. Measuring of interspecific association and similarity between communities, Memoirs Faculty of Science Kyushu University, Series E, Bio. 3, 65-80.
Namkoong, G. 1995. What do measures of biodiversity tell us? in: T.J.B. Boyle and B. Boontawee (eds) Measuring and Monitoring Biodiversity in Tropical and Temperate Forests, CIFOR, Bogor, pp. 87-95.
Nicholls, A.O. 1989. How to make biological surveys go further with generalized linear models, Biological Conservation 50, 51-75.
Odum, E.P. 1968. Energy flows in ecosystems: a historical review. Amer. Zool. 8: 11-18.
Oliver, I. and Beattie, A.J. 1996. Designing a cost-effective invertebrate survey: a test of methods for rapid assessment of biodiversity, Ecological Applications 6, 594-607.
Palmer, M.W. and White, P.S. 1994. Scale dependence and the species-area relationship, American Naturalist 144, 717-740.
Patil, G.P. and Taillie, C. 1979. A study of diversity profiles and orderings for a bird community in the vicinity of Colstrip, Montana, in G.P. Patil and M. Rosenzweig (eds) Contemporary Quantitative Ecology and Related Ecometrics, International Cooperative Publishing House, Fairland MD, pp. 23-48.
Pearson, S.M. 1993. The spatial extent and relative influence of landscape level factors on wintering bird populations, Landscape Ecol. 8, 3-18.
Peet, R.K. 1974. The measurement of species diversity, Ann. Rev. Ecol. Syst. 5, 285-307.
Peet, R.K. 1975. Relative diversity indices, Ecology 56, 496-465.
Pielou, E.C. 1975. Ecological Diversity, Wiley, NY.
Prendergast, J.R., Quinn, R.M., Lawton, J.H., Eversham, B.C. and Gibbons, D.W. 1993. Rare species, the coincidence of diversity hotspots and conservation strategies, Nature 365, 355-337.
Primack, R.B. 1993. Essentials of Conservation Biology, Sinauer, Massachusetts, 564 pp.

Ripley, B.D. 1977. Modelling spatial patterns (with discussion), J. Roy. Stat. Soc. B 39: 172-212.
Scheiner, S.M. 1992. Measuring pattern diversity, Ecology 73: 1860-1867.
Schreuder, H.T., Gregoire, T.G. and Wood, G.B. 1993. Sampling Methods for Multiresource Forest Inventory, Wiley, NY, 446 pp.
Scott, J.M., F. Davis, B. Csuti, R. Noss, B. Butterfield, C. Groves, H. Anderson, S. Caicco, D. D'Erchia, T.C. Edwards, J. Ulliman and R.G. Wright 1993. Gap analysis: A geographic approach to protection of biological diversity, Wildlife Monographs 123: 1-41.
Solomon, D.L. 1979. A comparative approach to species diversity, in J.F. Grassle, G.P. Patil, W. Smith and C. Taillie (eds) Ecological Diversity in Theory and Practice, International Cooperative Publishing House, Fairland MD, pp. 29-35.
Southwood, T.R.E. 1978. Ecological Methods, Chapman and Hall, London.
Southwood, T.R.E. and Kennedy, C.E.J. 1983. Trees as islands, Oikos 41, 359-371.
Southwood, T.R.E., Brown, V.K. and Reader, P.M. 1979. The relationship of plant and insect diversities in succession, Biol. J. Linn. Soc. 12, 327-348.
Spellerberg, I.F. 1991. Monitoring Ecological Change, Cambridge University Press, UK, 334 pp.
Stork, N.E. 1993. How many species are there? Biodiversity and Conservation 2, 215-232.
Stork, N.E. 1995. Measuring and inventorying arthropod diversity in temperate and tropical forests, in T.J.B. Boyle and B. Boontawee (eds) Measuring and Monitoring Biodiversity in Tropical and Temperate Forests, CIFOR, Bogor, pp. 257-270.
Taylor, L.R. 1978. Bates, Williams, Hutchinson - a variety of diversities, in I.A. Mound and N. Waloff (eds) Diversity of Insect Fanuas, Blackwell, Oxford, pp. 1-18.
United Nations Environment Programme 1992. Convention on Biological Diversity, UNEP Na. 92-7807, Nairobi.
Valencia, R., Balslev, H. and Paz-y-Mino, C.G. 1994. High tree alpha diversity in Amazonian Ecuador, Biodiversity and Conservation 3, 21-28.
Vanclay, J.K. 1991. Aggregating tree species to develop diameter increment equations for tropical rainforests, For. Ecol. Manage. 42: 143-168.
Vanclay, J.K. 1993. Environmentally sound timber harvesting: logging guidelines, conservation reserves and rehabilitation studies, in H. Lieth and M. Lohmann (eds) Restoration of Tropical Forest Ecosystems, Tasks for Vegetation Science 30, Kluwer Academic, Dordrecht, pp. 185-192.
Vanclay, J.K. 1994. Modelling Forest Growth and Yield: Applications to Mixed Tropical Forests, CABI, Wallingford, UK, 312 pp.
Whittaker, R.H. 1960. Vegetation of the Siskiyou Mountains, Oregon and California, Ecol. Monogr. 30, 279-338.
Whittaker, R.H. 1972. Evolution and measurement of species diversity, Taxon 21, 213-251.
Whittaker, R.H. 1977. Evolution of species diversity in land communities, in M.K. Hecht, W.C. Steere and B. Wallace (eds) Evolutionary Biology, Vol. 10, Plenum, NY, pp. 1-67.
Williams, J.T. 1988. Identifying and protecting the origins of our food plants, in E.O. Wilson (ed) Biodiversity, National Academy Press, Washington DC, pp. 240-247.
Williams, P.H. and Humphries, C.J. 1996. Comparing character diversity among biotas, in K.J. Gaston (ed) Biodiversity: a biology of numbers and difference, Blackwell, Oxford, pp. 54-76.
Wilson, E.O. 1988. Biodiversity, National Academy Press, Washington DC, 521 pp.
Wolda, H. 1981. Similarity indices, sample size and diversity, Oecologia 50, 296-302.

MONITORING OF FOREST BIODIVERSITY FROM FOREST RESOURCE INVENTORY DATA

ULF SÖDERBERG AND JONAS FRIDMAN
Department of Forest Resource Management and Geomatics
Swedish University of Agricultural Sciences
Umeå, Sweden

Abstract

Forest biodiversity has become a new concern for society. It has introduced new regulations and new ways of thinking in areas of forest use and planning. In Sweden the Forestry Act of 1993 states that timber production and maintenance of biodiversity are objectives of equal dignity. This results in a need for monitoring the maintenance of biodiversity in order to know if the policy intentions are followed. The problem of maintenance of biodiversity can be looked at from different scales, such as species, ecosystem, landscape, regional and national.

On the national and regional level it is possible to use forest resource inventory data to get an overview of the forest biodiversity changes. The method used was to analyse the relative changes of variables related to biodiversity over time. The changes are compared to a reference period (1983-1987) and can be calculated as either an annual change or as a moving average of three years data. The variables studied were proportion of the broad-leaved species in middle aged forests, amount of dead or windthrown trees, amount of big trees, amount of old unmanaged forest, proportion of wet forests, clear-cut area and habitat diversity. These variables are considered important substrata to produce and maintain a high forest biodiversity.

The results show that it is possible to follow the changes of the substrata (prerequisites) which indicate a high forest biodiversity on a broader scale. By introducing new variables more directly related to biodiversity and combining different information sources more effective methods can be developed.

1. Introduction

A growing concern for the environment and the use of forests during the last decades was confirmed by the agreements on the environmental program Global Biodiversity Strategy signed in 1992 in Rio de Janeiro. In Sweden this was followed by a new Forestry Act in 1993 environment (SOU 1992:76) which sets up two equally ranked objectives, one of forest production and the other for the environment (Lämås and Fries 1995). The objective of is not operational but is described in general terms as in the Global Biodiversity Strategy Program.

P. Bachmann et al. (eds.), Assessment of Biodiversity for Improved Forest Planning, 233–239.
European Forest Institute Proceedings no. 18. © 1998 *Kluwer Academic Publishers. Printed in the Netherlands.*

In order to fulfil the intentions of the ratification and the law and understand the changes of biodiversity several measures are required. There is no single solution to the problem. One reason for this is the fuzzy definition of biodiversity (Kangas and Kuusipalo 1993). In order to be able to handle the question of maintenance of biodiversity and how it can be measured one simply have to rely on more or less assumptions about what is good for maintaining biodiversity. However, a lot of research is going on to fill the gaps of knowledge.

Maintenance of biodiversity can be looked at from different scales such as genes, species, ecosystem, landscape, regional, national and global (Noss 1990). Today much research is done within these fields. Management or monitoring activities are done (and needed) on each of these scales and give complementary information to the decision makers. The methods of monitoring at the different scales have to be different to be cost-effective.

With the possibility to use adequate information already collected, cost-effective methods for assessing biodiversity on the regional and national level can be established. That is, if adequate information exists, and/or to what extent it is useful for that purpose. It has been found that the determining factor for about 75 per cent of the threatened species in boreal forests is if there is a sufficient amount of old forest, big (thick) trees, coarse woody debris and broad-leaved species (Berg et al. 1994). A sufficient amount of broad-leaved species has been quantified to be that more than 20 per cent of volume are broad-leaved trees (Gustafsson 1993).

If we consider these characteristics as important substrata to produce and maintain a high forest diversity, and are able to quantify the amount and changes over time of the substrata, we can develop a method to monitor the changes of biodiversity.

2. Methods

In this study, data from the Swedish National Forest Inventory (NFI) were used to quantify the changes of suitable substrata for biodiversity. The design of the NFI is a systematic cluster sampling with partial replacement of plots (Ranneby et al. 1987). The sampling units are clustered into tracts (with a side length of 300-1800 m for temporary tracts and 300-1200 m for permanent tracts). The tracts are systematically distributed over the total land base. The information recorded comprises extensive site, stand, and tree data (Söderberg 1996) on plots of forest land and less intensive assessments on other land use classes. Annually about 13 000 plots are assessed.

The variables of interest for describing different strata of importance for forest biodiversity were subjectively selected after discussion with different experts in forestry as well as in ecology. In some cases several variables could be used to express the same (similar) conditions. Data from 1983 to 1992 were used in this study. Because of this, variables that have a closer relationship to forest biodiversity and that have been introduced after 1992, could not be used.

The following variables were considered to best describe different aspects of biodiversity and chosen for the study:

- proportion of the area of middle-aged forests with more than 20% volume of broad leaved species to the total forest area
  Not only the amount of broad-leaved species, but the area of mixed forests with a proportion of at least 20 % of broad-leaved species is considered to be an important habitat containing a high biodiversity.

- number of big trees (diameter at breast height >30 cm) of Scots pine (Pinus sylvestris) and birch (Betula ssp.) in young and middle-aged forests
  There is a high probability that these trees will remain in the stand in the future and thereby contribute to better conditions for biodiversity.

- area of old unmanaged forests
  Old forests are considered to have a higher degree of biodiversity than other forests. The more specific definition of this group is that the forest is older than the normal age for final cut and no treatments have been performed during the previous 25 years. Existing nature reserves are excluded.

- amount of dead or windthrown trees (diameter at breast height ≥ 10 cm) in clear cut areas and older forests
  The amount of coarse woody debris is considered to be one of the best indicators of biodiversity. Dead wood is a substrate for many species of mosses, lichens, fungi and insects. In this study only wood that can be used for fuelwood was assessed. That means that wood with a higher degree of decomposition is not considered.

- annual cutting area
  A final cut implies a drastic change in the living conditions for flora and fauna. Changes of the area of final cut is therefore significant for forest biodiversity.

- area of wet forests
  Wet forests are of great importance for the species composition in the landscape. Generally, there is a greater number of species in wet forest than in dry or mesic forests. The area of wet forests is mainly affected by drainage and final cutting.

- habitat diversity
  Habitats were defined from combinations of land use class, stage of forest, vegetation type, bottom layer and forest type. In total 30 habitats were defined. The plots from the NFI were classified into these habitats using not only plots from productive forest land but also from four land use classes related to productive forest land.

Habitat diversity was calculated according to the following method. The country was divided into squares with a size of 25 by 25 km. For these squares the number of existing habitats and the adherent proportions were calculated using the NFI-plots within the square.

A modified Shannon-Wiener index (Krebs 1989; Shannon 1948) was calculated as follows;

$$B = N^{\alpha} \bullet SW \tag{1}$$

$$SW = -\sum_{i=1}^{N} (p_i \bullet \ln(p_i)) \tag{2}$$

where p_i = proportion of habitat i
$\alpha = 1$
N = number of habitats
B = habitat diversity

For the variables above, the estimates were made for four different regions (Figure 1) as well as for the whole country. The regions used are administrative units, at the same time they represent different ecological and other conditions. Data from 1983-1987 were used as a reference period except for habitat diversity for which a three year period was used. The reason was to get the reference period to be of the same representation as the other periods, as a five year period increase the probability of more habitats. Changes were then studied using a moving three year average starting with 1986 and ending at 1992.

The arithmetic mean of B for the squares within each region was used in the calculation of habitat diversity.

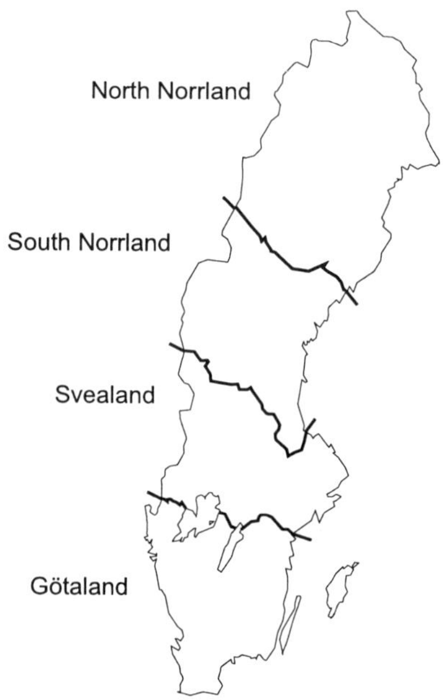

Figure 1. The regions used in the study.

3. Results

The variables studied show various patterns for the entire country (Figure 2) as well as for the different regions. The area of wet forest is fairly small for the reference period compared to the other time periods studied. It is questionable whether this is reflecting reality or is caused by some change in assessment for some variable involved. The other studied characteristics show more stability. Trends indicate a reduction of old forests, dead trees and area cut. However, a reduction of the cut area should be positive from a biodiversity perspective.

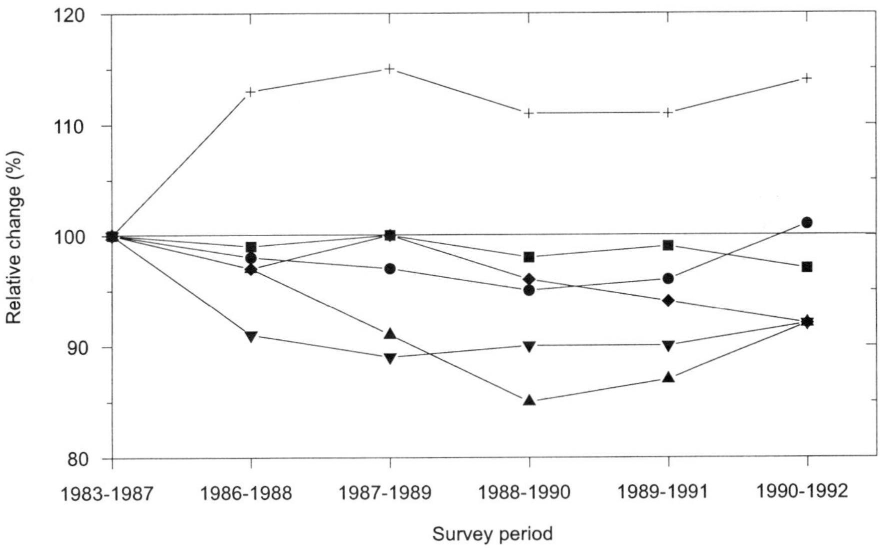

Figure 2. The relative change in the studied variables Broadleaves (■), Big trees (●), Old forest (▲), Dead trees (▼), Area cut (◆) and Wet forest (+) from the reference period 1983-1987 for the whole country.

There are some tendencies of somewhat different trends in the four regions for a specific variable. An example is given in Figure 3 where the area of broad-leaved species shows deviating trends for Northern and Southern Norrland.

Habitat diversity show an increasing trend for all regions (Figure 4) and by that also for the whole country. One explanation to this could be changes in silvicltural regimes together with the habitat definitions which increase the number of habitats when young forests are passing 7m mean height

4. Discussion

Forest biodiversity must be viewed from different scales. A complete picture according to the general definition is impossible to produce. Therefore a lot of tools taking different

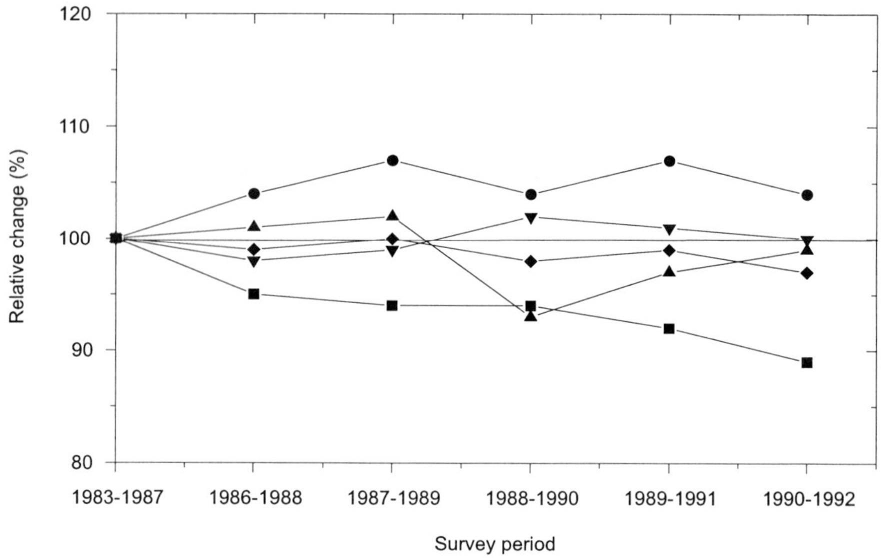

Figure 3. The relative change in proportion of area of middle-aged forests with more than 20% volume of broad-leaved species in North Norrland (■), South Norrland (●), Svealand (▲), Götaland (▼) and for The whole country (◆).

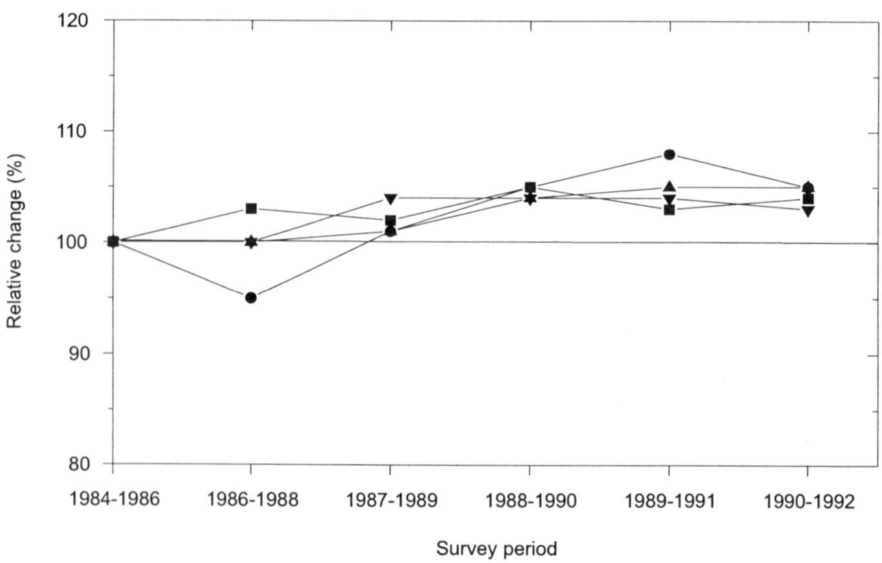

Figure 4. The relative change of habitat diversity for North Norrland (■), South Norrland (●), Svealand (▲) and Götaland (▼).

aspects of biodiversity into account, have to be used to gather information to get some hold of the changes of forest biodiversity to support the decision-makers. For an overview, the use of forest resource inventory data can be a cost effective method to obtain information for large areas.

This study was restricted to analysing existing data, mainly collected before the big interest for forest biodiversity had emerged, and were available for a 10 year period. Data from a longer period of time would have been preferred but was not possible due to changes in design and variables. Variables more related to biodiversity have been introduced to the NFI and collected after the studied period. More variables can be taken into use in the future when more knowledge in ecological relationships are available.

The results indicates only minor changes and fluctuations of the indicators during the study period. The conclusion from this is that there has not been any significant change of the amount suitable substrata for biodiversity when expressed as the studied variables during this period.

Also the design of the inventory was given, using clustered plots. For the purpose of assessment of forest biodiversity, other more effective methods could be used. As forest biodiversity is often connected to characteristics which are not so common, methods like line transect sampling, adaptive cluster sampling, strip survey and area classification by line transect (or strip) survey could be studied, and if found appropriate, used between the plots.

Methods using remote sensing techniques have advantages on the landscape level concerning patterns, fragmentation and habitats. Especially in the last mentioned case a combination with field sampling methods would be effective.

References

Berg, Å., Ehnström, B., Gustafsson, L., Hallingbäck, T. Jonsell, M. and Weslien, J. 1994. Threatened Plant, Animal, and Fungus Species in Swedish Forests: Distribution and habitat Associations. Conservation Biology. 8(3): 718-731.

Gustafsson, L. 1993. Naturvårdshänsyn vid gallring. Skogstyrelsen. Rapport 8 1993.

Kangas, J. and Kuusipalo, J. 1993. Integrating biodiversity into forest management planning and decision-making. Forest Ecology and Management. 61(1-2): 1-15.

Krebs, C.J. 1989. Shannon-Wiener function. Ecological Methodology. University of British Colombia.

Lämås, T. and Fries, C. 1995. Emergence of a Biodiversity concept in Swedish Forest Policy. Water, air and Soil Pollution. 82: 57-66.

Noss, R.F. 1990. Indicators for Monitoring Biodiversity: A Hierarchical Approach. Conservation Biology. 4(4): 355-364.

Ranneby, B., Cruse, T., Hägglund, B., Jonasson, H. and Swärd, J. 1987. Designing a new national forest survey for Sweden. Studia Forestalia Suecica. No. 177.

Shannon, C.E. 1948. A mathematical theory of communication. Bell Syst. Tech. J. Vol. 27.

SOU. 1992: 76. Skogspolitiken inför 2000-talet. Huvudbetänkande av 1990 års Skogspolitiska kommitté. Jordbruksdepartmentet.

Söderberg, U. 1996. Country report for Sweden. In: Study on European Forest Information and Communication System (EFICS). II Country Reports. Reports of the Working Package 1.2. Vol. IV. European Forest Institute.

ASSESSMENT OF COARSE WOODY DEBRIS
A comparison of probability sampling methods

GÖRAN STÅHL AND TOMAS LÄMÅS
Swedish University of Agricultural Sciences
Department of Forest Resource Management and Geomatics,
S-901 83 Umeå, Sweden.

Abstract

In this paper, different probability sampling methods for assessing coarse woody debris (CWD) are compared. The methods are circular plot sampling, strip surveying, line intersect sampling, and transect relascope sampling. The two latter methods are suited primarily for the assessment of CWD on the ground.

While the first three methods should be well known, the transect relascope method is new. It can be considered a combination of line intersect sampling and relascope sampling. A wide angle relascope is used along survey lines and all downed logs that fill the angle of the instrument are included in the sample.

The methods were compared with respect to their cost-efficiency for estimating the total length and the total volume of CWD on the ground in forest compartments. Monte-Carlo simulation was used for the purpose. Regarding volume estimation, strip surveying was found to be an efficient method in most compartments. Concerning the estimation of the total length of logs, transect relascope sampling was found to be efficient. It should, however, be stressed that the knowledge of the performance of the transect relascope in the field is limited. It is likely that measurement errors make the method less accurate than is indicated in this theoretical study.

Assessing the total length of CWD is cheaper than assessing the total volume of CWD. In case line intersect or transect relascope sampling is used, the length can be estimated by only counting the logs included in the sample. The usefulness of the total length of logs as an indicator of habitat conditions can, however, be questioned. Still, if logs are classified into different dimension classes, a description of this kind should not be too different from using the total volume of logs.

1. Introduction

Dead decaying wood has been identified to be an important substrate for many species in the forest ecosystems (e.g. Samuelsson et al. 1994). Since forestry of today acknowledges the preservation of biodiversity as an important goal, management programmes that account for biodiversity must be developed. Consequently, there is a need to describe the habitat conditions for the species considered.

P. Bachmann et al. (eds.), Assessment of Biodiversity for Improved Forest Planning, 241–248.
European Forest Institute Proceedings no. 18. © 1998 *Kluwer Academic Publishers. Printed in the Netherlands.*

Regarding wood-dependant species, a first approach would be to quantify the amount of woody debris present. In forestry of today, this is often made subjectively in terms of "high" or "low" occurrence of woody material. Subjective information of this kind is in many cases appropriate since it is cheaply obtained, and since it allows for holistic views on what should be considered valuable pieces of land from a biodiversity preservation aspect. In some cases, however, subjective methods have considerable drawbacks. In case comparisons of conditions between different regions are to be made or in case trends are to be monitored, region- and time-dependant bias must be avoided. Generally, the solution is to use probability sampling methods in the surveys.

The study of the latter kind of methods for assessing woody debris involves some aspects that are not present in studies of traditional timber inventories. The woody material is generally rather scarce, a fact which makes the standard methods (e.g. circular plot inventory) less efficient. Also, downed material extends in the plane in a way that allows for intersect sampling techniques to be used.

The aim of this article is to compare different probability sampling methods with respect to their efficiency for estimating downed coarse woody debris in compartments. The methods are circular plot sampling, strip surveying, line intersect sampling, and transect relascope sampling. The first three methods are well known from the literature (e.g. De Vries 1986), while the last one is newly developed. Therefore, brief attention is also given to the principles of transect relascope sampling.

2. Material and Methods

The methods were compared with regard to their cost-efficiency in surveys of coarse woody debris in forest compartments, both when the parameter of interest was the total volume of woody material and when it was the total length. Only downed material was considered.

The total volume or the biomass of woody debris is often mentioned to be correlated with the habitat conditions for wood-dependant species (e.g. Harmon et al. 1986). The total length, on the other hand, is not. The motivation for studying the latter parameter all the same is that it should provide at least a rough estimate of habitat conditions, in the mean time as it is very simple to estimate. Also, if the total length is estimated separately for different dimension classes, the use of length should not be too different from the use of volume.

A sampling simulator (Ståhl and Lämås 1995) was used for the evaluations. In the simulator, the methods were applied in six different compartments. Four of these were artificial, two were real. The latter were old-growth forests from the county of Västerbotten in Sweden, in which all downed logs had been recorded for volume and length. The volumes were determined using formulas for living trees (with diameter and height as predictor variables) or by sectioning in case a tree was broken. The coordinates of the logs were determined using a differential GPS and the fall directions recorded using a compass.

In the construction of artificial compartments, the aim was to build objects of different types, permitting studies of which methods are useful in different kinds of forests. Summary information about the compartments is given in Table 1.

Although only a limited number of artificial compartments were studied, they were selected in such a way that pairwise comparisons of different characteristics should be

TABLE 1. A description of the compartments used in the simulations.

Compartment	Origin	Area (ha)	Number of logs (ha^{-1})	Average log length (m)	Log volume ($m^3\ ha^{-1}$)	Clustered
A	Real	2.8	7.6	11.3	0.81	No
B	Real	2.9	35.9	10.2	4.47	No
C	Artificial	3.0	10.0	5.0	0.06	No
D	Artificial	3.0	10.0	13.3	1.05	No
E	Artificial	3.0	30.0	13.5	3.20	No
F	Artificial	3.0	30.0	13.6	3.22	Yes

possible. The impact of log length could be (briefly) studied using compartments C and D, the impact of log density using D and E, and the impact of clustering using E and F. A comparison of real compartments with high and low density of logs could be made using compartments A and B. It should be stressed, however, that the study is very limited. A full study would involve a large number of compartments of different kinds.

In all the artificial compartments, the logs' fall directions followed a uniform distribution (0 to 360 degrees). In the real compartments, the fall directions did not deviate too far from a uniform distribution in the interval -90 to 90 degrees. This is sufficient for estimators that assume random orientation of logs (see below) to work appropriately.

Log locations in all artificial compartments (except F) were generated from a Poisson process. In compartment F, ten cluster centres were generated according to a Poisson process. The number of objects in each cluster was randomly determined and the probability of a log (butt end) to be situated more than 25 m from a cluster centre was set to zero. The length of a log was generated from a (truncated) normal distribution with 30% standard deviation about the mean. Log volumes (V, dm^3) were established using the formula: $V = 2.5 \cdot 10^{-4} \cdot l^{2.6} \cdot (1+\varepsilon)$, where l is the length (dm) of a log, and ε is a normally distributed random number with expectation zero and standard deviation 0.1. The formula was derived empirically using data from living trees.

The shape of all compartments was rectangular, the sides oriented in east-west and north-south.

As mentioned in the introduction, the methods compared were circular plot inventory, strip surveying, line intersect sampling, and transect relascope sampling. The first three methods are well known from the literature (e.g. de Vries 1986). The last method is newly developed and will therefore be briefly described. A wide angle relascope is used along survey lines. All downed logs that fill the angle of the relascope, from any point on the line, are counted. By only counting the logs, an estimate of the total length of objects (in the population) can be obtained, since the probability of inclusion of a log is proportional to its length. If the total volume of logs is the parameter of interest, measurements of log volumes and lengths must be made, at least on a sub-sample of logs. For a more detailed description, see Ståhl (1995). Figure 1 is provided as an illustration of the use of the method in the field.

Each of the methods can be varied to some extent. The circular plot method, e.g., can be varied with regard to the size of the plots, their distribution in a compartment, and the number of plots used. Similar variations can be made for the other methods. Below, it is described how the methods were specified in the study.

Figure 1. The use of the transect relascope in the field. A downed log is included in the sample when a survey line crosses its area of inclusion (indicated by dashed lines).

For the circular plot method, the plot radius used was 15 metres. This size is suggested to be efficient for this kind of inventory by Lämås (1996). The plots were laid out entirely randomly in the compartments. A downed log was classified to belong to a plot once its butt end was located on a plot. Boundary conditions were taken care of using the mirage method (Schmid 1969).

In the strip inventory, the width of strips was 15 meters. As in the circular plot case, a log was classified to belong to a strip once its butt end was on a strip. The strips were laid out in a north-south direction; the locations were entirely randomly determined.

In line intersect sampling, all logs intersected by a survey line are included in the sample. Using the method, two different types of estimators can be used (e.g. Kaiser 1983; De Vries 1986). One of them implies measurement of the angle between survey lines and logs. This estimator is unbiased both when the orientation of the survey lines is determined on beforehand and when it is random. For the second type of estimator, the above angle is not measured. Using this estimator, unbiasedness can be achieved either if the orientation of the survey lines is random or if it is predetermined and the angles between logs and lines are randomly distributed (usually they are assumed to be uniformly distributed). In the simulations, an estimator of the second type was used. The orientation of the survey lines was determined on beforehand (north-south); their locations were entirely randomly determined. All lines traversed the entire compartment under study.

A 40 degrees relascope angle was used for the transect relascope sampling. As in the line intersect case, the type of estimator used assumed log orientations to be random, since the direction of survey lines was determined on beforehand (north-south). Transects were randomly laid out, always to traverse the entire compartment under study. In the transect relascope case, survey lines were allowed to fall also outside the compartment (in a close vicinity). Due to this, no boundary bias was introduced.

For all the methods, the number of plots/strips/lines was varied in order to study the efficiency of the methods given different budgets.

When line intersect or transect relascope sampling was used for estimating the total volume, every third log was subsampled and a ratio estimator with log length as an ancillary variable was applied.

Using Monte-Carlo simulation in forest populations with known object locations and attributes, the bias and variance of different inventory methods and estimators can easily be calculated. The bias is obtained as the average deviation between the estimated values and the true value and the variance as the average squared deviation between the estimated values and the mean of the estimated values. If the number of (random) repetitions is high, the bias and precision can be very adequately determined.

Apart from bias and precision, the cost of a method is important to consider when surveys are designed. The average cost of a method can also be obtained by simulation, provided the time consumptions for the different phases of the inventory work are specified. The total time consumption for a method is obtained through addition of partial times, and the total cost through multiplication by per unit time cost.

The time specifications for the study are given in Table 2. Note that the figures are based on limited time studies only (Ståhl 1995; Lämås 1996). Other specifications could have lead to other results regarding the cost-efficiency ordering of the methods.

TABLE 2. Time consumption for log measurements

Method	Measurement* of log length (seconds)	Measurement of log volume (seconds)
Circular plot	40	60
Strip survey	40	60
Line intersect	5	60
Transect relascope	15	85

*) To estimate the total log length with line intersect or transect relascope sampling, it is sufficient to count the logs sampled.

To estimate the total length of logs, measurements on sampled logs must be made when circular plot or strip inventory is used. When line intersect or transect relascope sampling is used, it is sufficient to count the objects for estimating the total log length in the population. To estimate the volume, measurements of length and diameter were assumed to be made on the logs. The higher figure for the transect relascope method is due to the time required for classification and walking to objects located at some distance from the survey line. All measurements were assumed to be made without errors.

The time required for walking 100 m was set to 150 seconds for all the methods, and the time required for establishing one plot/strip/transect was set to 120 seconds. To obtain costs, the average total time was multiplied by 175 SEK h^{-1} (SEK is short for Swedish crowns). Only variable costs were considered. All results are based on 400 simulations of a certain design.

3. Results

In Figures 2 and 3, the cost-efficiency of the methods in terms of cost versus standard error (%) is given. The outcomes from simulations using different plot/strip/line numbers are connected to lines. The size of bias is not reported, since the bias was always very limited.

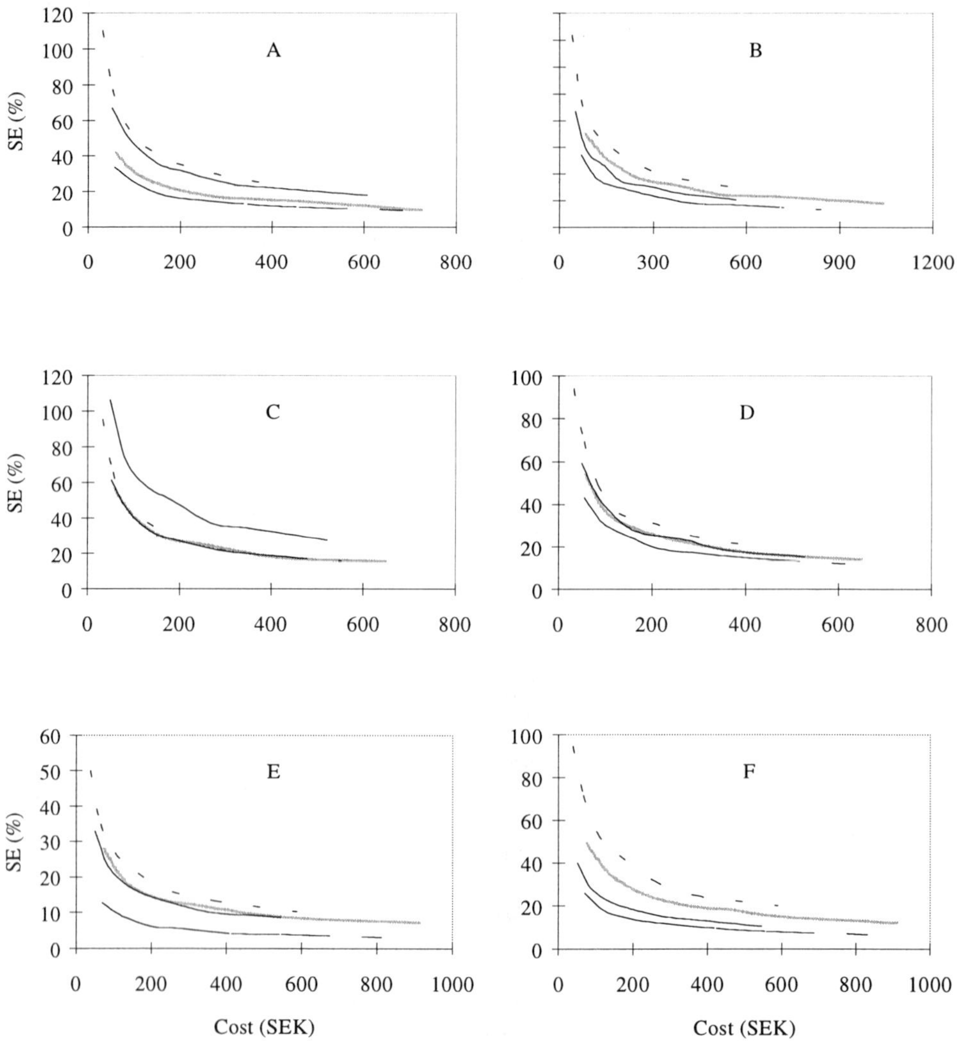

Figure 2. Cost-precision relationships for the methods in compartments A-F when the total *length* of downed logs was estimated. The results for strip surveying are given by the thick solid line, for line intersect by the thin solid line, for circular plots by the short-dashed line, and for transect relascope by the long-dashed line. The precision is given as the standard error (SE) in percent of the true parameter value in a compartment.

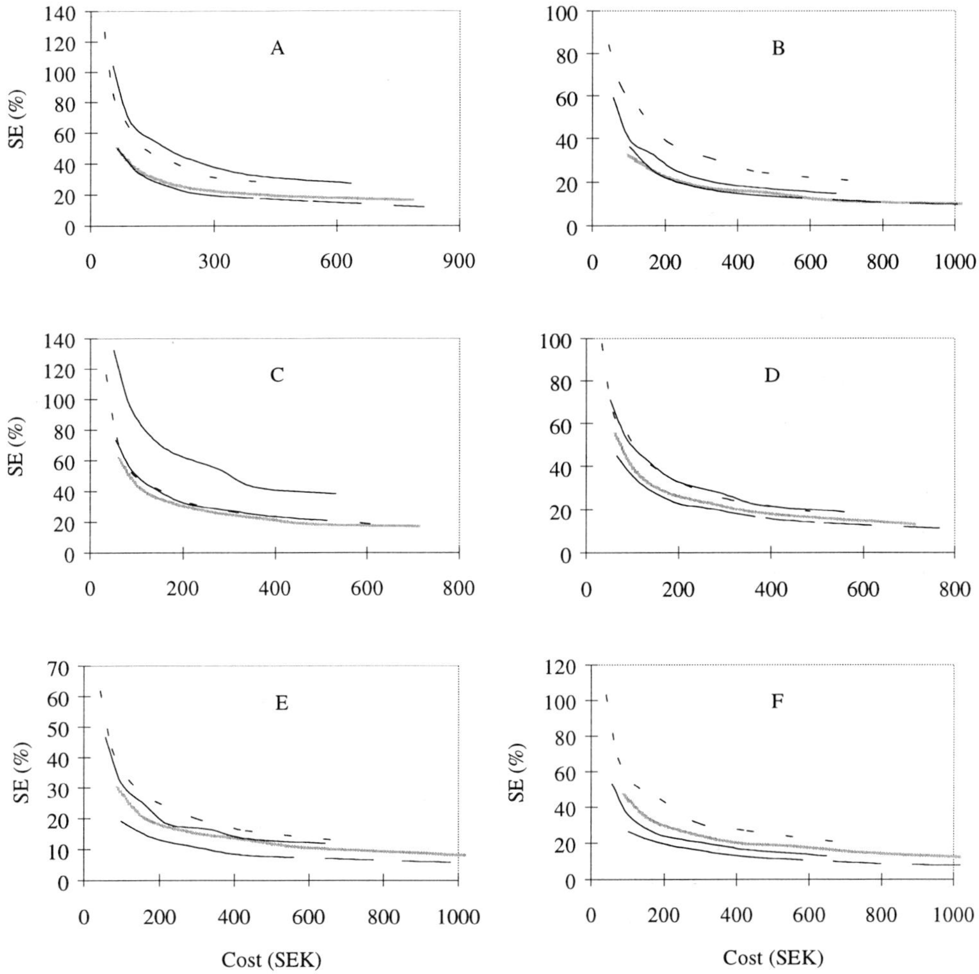

Figure 3. Cost-precision relationships for the methods in compartments A-F when the total *volume* of downed logs was estimated. The results for strip surveying are given by the thick solid line, for line intersect by the thin solid line, for circular plots by the short-dashed line, and for transect relascope by the long-dashed line. The precision is given as the standard error (SE) in percent of the true parameter value in a compartment.

4. Discussion

Strip surveying was found to be an efficient method for estimating the total volume, and also the total length, of woody debris. Although the transect relascope method was sometimes found to be more efficient, too little is known about its properties in the field for the method to be recommended.

As expected, line intersect and transect relascope sampling were less efficient when the logs were short (compartm. C). Concerning the density of logs (D vs. E), the general pattern of the results were similar although the standard errors (%) decreased when the occurrence of logs was high. The clustered forest raised the standard error levels for most methods. The increase for the line intersect method was, however, not as high as the increase for the others. The circular plot method generally performed poorly.

The same general results were obtained both when the volume and the total length of logs were studied. A certain level of precision for the volume is, however, more expensive to obtain than the same level of precision for the length.

In this kind of study, there is a problem that only one replica of each forest type is tested. Peculiarities in that particular forest may lead to erroneous general conclusions.

No substantial bias was found for any of the methods. For line intersect and transect relascope sampling, this was due to the (almost) uniform distribution of log fall directions in the compartments. For the particular estimators used and predetermined survey line orientations, this unbiasedness does not hold in general.

In practical forestry there is often a need for many different parameters describing the state of, e.g., a compartment. Since different methods are efficient for different parameters, a practise in which methods are combined should often be a good solution. E.g., surveys for timber parameters can be made on plots while habitat conditions for wood-dependant species are determined along transects between the plots (cf. Lämås and Fries 1995).

5. Acknowledgements

The kind assistance of MSc J. Wallerman, allowing us to use some data from real compartments in the study, and the constructive comments given by Prof. L. Fattorini and Dr. M. Marchetti are gratefully acknowledged.

References

De Vries, P.G. 1986. Sampling theory for forest inventory. Springer. New York.

Harmon, M.E., Franklin, J.F., Swanson, F.J. et al. 1986. Ecology of coarse woody debris in temperate ecosystems. Advances in ecological research 15: 133-302.

Kaiser, L. 1983. Unbiased estimation in line-intercept sampling. Biometrics 39: 965-976.

Lämås, T. and Fries, C. 1995. An integrated forest inventory in a managed north Swedish landscape for estimating growing stock and coarse woody debris. In: Köhl, M., Bachmann, P., Brassel, P. and Preto, G. (eds.). The Monte Veritá conference on forest survey designs. May 2-7, 1994. Monte Veritá. Switzerland. Swiss Federal Inst. for Forest, Snow, and Landscape Research. Birmensdorf.

Lämås, T. 1996. Circular plots versus strips for assessing rare forest objects. In: Lämås, T. Forest management planning for biodiversity and timber production. SLU. Department of Forest Resource Management and Geomatics. Umeå. Sweden. Report 3.

Samuelsson, J., Gustafsson, L. and Ingelög, T. 1994. Dying and dead trees, a review of their importance for biodiversity. Swedish Environmental Protection Agency. Report 4306.

Schmid, P. 1969. Stichproben am Waldrand. Mitt. Schweiz. Anstalt für Forstliches Versuchswesen. 45: 233-303.

Ståhl, G. and Lämås, T. 1995. A simulator for sampling of rare forest objects. In: Hyttinen, P., Kähkönen, A. and Pelli, P. Multiple use and environmental values in forest planning. Proceedings 4. European Forest Institute. Joensuu. Finland

Ståhl, G. 1995. The transect relascope, an instrument for the quantification of coarse woody debris. Working paper 4. SLU. Department of Forest Resource Management and Geomatics. Umeå. Sweden.

DIGITIZED AERIAL PHOTOS FOR ASSESSING FOREST BIODIVERSITY

MARKUS HOLOPAINEN
University of Helsinki
Department of Forest Resource Management, PO Box 24, SF-00014, Helsinki, Finland

GUANGXING WANG
University of Helsinki
Department of Forest Resource Management, PO Box 24, SF-00014, Helsinki, Finland

Abstract

In addition to the volume and assortment distribution of timber, new multiple-use characteristics, such as biodiversity, have emerged in recent forest inventories. One method of assessing biodiversity is to quantify the structures of forests at a landscape or biotope level. Remote-sensing techniques may provide an efficient tool for assessing and monitoring forest biotopes and their structural diversity. Due to the high spatial resolution of digitized aerial photographs they may offer an accurate and effective means of mapping, inventorying, and monitoring forests. In the present paper, we propose the use of remote sensing indicators, especially digitized aerial photo indicators, for assessing forest biotopes and their structural diversity.

1. Introduction

Growing interest exists in the field of forest inventory and monitoring for deriving information on multiple-use forest characteristics such as biodiversity; however, inventoring of biodiversity with field measurements is time consuming. On the other hand there is a tendency to minimize the amount of expensive field work in the inventories, expressed as a need for reliable and standardized information that can be produced at reasonable cost. Remote sensing has the great advantage, compared with field measurements, of providing exhaustive and objective information on large forest areas cheaply. In addition, it can be used for updating forest information at short intervals.

Biodiversity is an issue that first emerged about fifteen years ago, and can be defined as the total variability of life on Earth (Lovejoy 1980; Heywood 1995). During recent years biodiversity has become a major scientific and political issue (Wilson and Peters 1988; Solbrig 1991; Global Biodiversity Strategy 1992). The three commonly referred components of biodiversity are genetic, species, and ecosystem diversity. Other proposed

P. Bachmann et al. (eds.), Assessment of Biodiversity for Improved Forest Planning, 249–254.
European Forest Institute Proceedings no. 18. © 1998 *Kluwer Academic Publishers. Printed in the Netherlands.*

components include species concentrations and landscape diversity containing all levels of biological hierarchy, from ecosystems to species and genes, that are targeted for biodiversity inventories and conservation (Solbrig 1991). Landscapes are composed of a mosaic of patches (Urban et al. 1987), that, within the context of forest inventory can be understood as forest stands (McGarigal 1994). According to Uuttera et al. (1995), it is possible to divide forest stands with simple and easily measurable characteristics e.g. tree species, tree age, and soil fertility, into classes that effectively reflect the stand structure. In the present paper, these classes, or patches, are hereafter referred to as forest biotopes.

By tradition, biodiversity conservation has been focused on protecting individual threatened species, however, this species-by-species approach is expensive and inefficient. Due to the complex relationships between species and their biotopes, a biotope approach to the conservation of biodiversity is often preferable (Kennedy et al. 1994). Of the many means of describing vegetation diversity, the most common are extent, structure, composition, biomass or production, and condition. Each of these can be assessed on the ground, but they may be assessed more effectively from various forms of remote-sensing data (Boyle and Boontawee 1995).

Some case studies have been undertaken in which remote-sensing data coupled with ground truth data have been used to monitor vegetation cover, land use, forest loss, and other aspects of biodiversity (Heywood 1995). However, no general description or method for assessing biodiversity or detecting changes in it by remote sensing exists. The objective of the present paper is thus to describe possible remote sensing indicators, especially digitized aerial photo indicators, for estimating forest structural diversity at the biotope and landscape levels.

2. Digital image features and indicators of structural diversity

Several digital image features can be used in numerical classification and in the assessment of the diversity of a forest (Table 1. Parmes 1996). The features can be derived from the digital numbers of single pixels, segments, or groups of segments.

Traditionally, texture has been divided into statistical and structural texture (Haralick 1979). Statistical texture can be understood as the randomlike variation of tone values in the neighborhood, while structural texture quantifies the structure, form, and location of pixels or segments. If geometric, contextual, and pattern features are used,

TABLE 1. Digital image features (Parmes 1996). Features in column F(i) quantify the variability of reflectance values, or labels in an interpreted image. The features in column f(x,y,z) are based on the location of pixels and segments and their geometric structures, respectively.

Feature unit	F(i)	f(x,y,z)
Pixel	reflectance values from various sources, dates, or bands	location
Segment or artificial neighborhood	texture = statistical variability of tone values	geometry = size and shape, direction
Group of segments segments	context = labels of neighboring	pattern = geometrical structure, fragmentation,

presegmentation of the image is needed. Presegmentation divides the image area into units of homogeneous tone values. Once the image has been segmented, geometric features such as the size and shape of land cover units and their distribution can be calculated. The frequency of segment boundaries describes the texture or diversity of a landscape. In contrast, the ratio of the size of the segments to their boundary lengths is correlated with the complexity of the landscape (Parmes 1996).

When the structure becomes more clear and shows better spatial resolution, it is possible to use structural contextual features, in which the units are homogeneous stands, or labels of a group of stands. If location of these stands is known, spatial heterogeneity can be characterized by fragmentation. In practice, the natural distribution of ecosystems is heavily fragmented, mainly as a result of human intervention (Boyle and Boontawee 1995).

The fragmentation of a forest area can be characterized by using some simple metrics of landscape, e.g. biotope areas, density, size, and variability. Edges, shapes, core areas, and nearest neighbors can also be used as indicators of landscape diversity. A group of diversity indices exists, traditionally used in the assessment of species diversity, that can also be used to determine landscape diversity. The most popular of these is the Shannon diversity index (Shannon and Weaver 1949), which represents the amount of information per individual or biotope. It can be used for comparing various landscapes or the same landscape at different times. Another popular measure, the Simpson index (Simpson 1949), represents the probability that any 2 observations selected at random would be of different types.

It may be possible to assess the structural diversity of an image area without any ground truth data, however, with field measurements it is possible to acquire information on the representation of segmented biotopes and derive even more informative indicators of diversity, e.g. number of biotopes, their rarity, amount of old growth, and key biotopes. In addition to biotopes, the diversity of vegetation or fungi is dependent on the topography, slope, and aspect of the site and number of minor biotopes such as springs and streams (Kellomäki and Kolström 1995). These indicators can be derived by using other auxiliary data than remote-sensing material, such as digital maps and elevation models.

In summation (Table 2), spatial features or texture features can be used to quantify the degree of homogeneity or heterogeneity of a landscape. Haralick texture features and fragmentation indicators could be the basis of a method for quantifying and monitoring the biodiversity of a landscape, and for comparing the biodiversity of various landscapes. In addition to remote sensing material, available ground truth data and other auxiliary data can be utilized.

Most of the structural diversity indicators referred to here should be independent of data resolution, suggesting that the indicators can be applied for a 1-m (digitized aerial photo) resolution as well as for a 1000-m (NOAA satellite image) resolution. The diversity calculated from e.g. NOAA 1-km pixels does not, however, describe the diversity of forests, but is an indication of the diversity in land use patterns in general.

3. Classification of forest biotopes by digitized aerial photographs

Digitized aerial photographs have already been used to classify forests into biotopes (Holopainen 1995; Holopainen and Wang 1996). Their high spatial resolution makes

TABLE 2. Assessing biodiversity by remote sensing. Available indicators.

Indicators that are measurable directly from the images	Indicators of biodiversity when ground truth data are available	Other auxiliary data
- Haralick texture	- Number of biotopes	- Topography
- Fragmentation indicators	- Rarity of biotopes	- Nature conservation areas
• Biotope density, size, and variability	- Amount of old growth	- Key biotopes e.g. springs and streams
• Area	- Key biotopes	
• Edge		
• Shape		
• Core area		
• Nearest neighbor		
• Shannon diversity index		
• Simpson diversity index		
• Modified Simpson index		
• Biotope richness		
• Relative biotope richness		
• Shannon evenness index		
• Simpson evenness index		

even tree-, plot- and standwise inventories possible. The use of digitized aerial photos, however, is quite complicated mainly due to the low elevation of the camera and the central projection of the photographs. The digital numbers of the pixels are affected by their location within the photo, thus 2 similar sample plots in different parts of the photo may possess quite dissimilar tone and texture features. In addition, photos taken in different areas or on different dates are spectrally inconsistent. Holopainen and Lukkarinen (1994) developed regression calibration method to account for the central projection of the photos, while the accuracy of digitized aerial photos for estimating plotwise forest data was also tested. In regression calibration the original tone values are corrected to the principal point level with linear regression models. The predicted variable is the observed window mean value of the sample plot, and the predictor variables are calculated from the sun location at exposure time.

Holopainen (1995) assessed forest biotopes from digitized color infrared (CIR) aerial photos with neural networks, mean pixel values, and texture features of Haralick (1979). His results indicated that with aerial photographs some special biotopes, e.g. old conifer stands, could be distinguished quite accuratelly, indicating that certain key biotopes could be located by remote sensing. Holopainen and Wang (1996) further developed the inventoring of forest biotopes with digitized aerial photos, and showed that when using forest biotopes derived by tree species and tree age, the accuracy of the stratification was probably sufficient for calculating diversity indices with estimated biotopes. The overall correct percentage in the best biotope stratification of their study was about 90%, while for the 4 dominant biotope types, including middle-aged and mature pine and spruce stands, the correct percentages varied 80% - 95%. The most difficult task was the separation of mature mixed stands from pure spruce stands. Stratification was performed using a digitized CIR photo at a scale of 1:30 000 and with a pixel size of 1.7 m. The calibrated means of the red, green, and blue bands were applied in the stratification.

The stratification accuracy of aerial photos could be improved by using texture features. Wang et al. (1996) developed a texture analysis module for digitized aerial photos, and concluded that texture features of digitized aerial photographs were useful for revealing forest types and improving the stratification. The texture method showed promise as a means of estimating some forest characteristics, but selection of the best combination of tone or texture features was problematic.

4. A method for assessing structural diversity with digitized aerial photos

To account for the possibilities of remote sensing in present forest inventorying and monitoring methods, forest diversity should be determined on the landscape level, a process that leads to the assessment of forest structure and biotope distributions. If it is possible to determine the biotopes by remote sensing, it is also possible to use remote sensing for mapping and monitoring the diversity of nature.

The method for assessing the structural diversity of forests by remote sensing could be based on GIS in which available remote-sensing, ground truth, and other auxiliary data could be analyzed (Figure 1). The first step in the method is the verification of classification accuracy, which can be done by comparing remote-sensing estimates with ground truth. When the classification accuracy has been evaluated, available biodiversity indicators can be analyzed with the GIS. The indicators to be used should be selected according to the area and available ground truth data. If remote-sensing data are the only

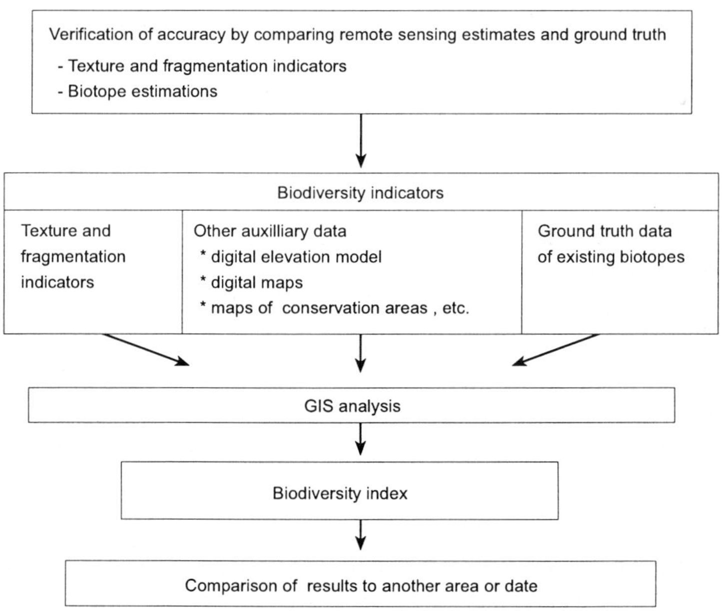

Figure 1. Suggested method for assessing forest diversity with remote sensing

data available, diversity can be estimated by texture and fragmentation features, which can be calculated for any part of the study area or for the entire image. If ground truth data (sample plot or forest compartment data) are available, the quantitative estimates (species proportions, age classes, etc.) of forest diversity can also be derived. The result of a quantitative classification is a satellite or aerial photo map of the biotopes in the area. With the aid of this map it is possible to determine several biodiversity indicators, e.g. number of biotopes, their rarity, number of old growth and deciduous forests, and key biotopes. With this system, it is possible to estimate the diversity of a given image or its portion. The results (indices, values) can be compared with any other image processed in the same manner.

References

Boyle, T.J.B. and Boontawee, B. 1995. Measuring and Monitoring Biodiversity in Tropical and Temperate Forests. Center of International Forestry Research.

Global Biodiversity Strategy 1992. Guidelines for action to save, study, and use earth's biotic wealth sustainably and equitably. World Resource Institute (WRI). The World Conservation Union (IUCN). United Nations Environment Program (UNEP).

Haralick, R. M. 1979. Statistical and structure approaches to texture. Proceedings of the IEEE. 67(5): 786-804.

Heywood, V.H. (ed.) 1995. Global Biodiversity Assessment. UNEP. Cambridge University Press.

Holopainen, M. 1995. Kaukokartoitus luonnon monimuotoisuuden inventoinnissa. Remote Sensing for Inventory and Monitory of Forest Biodiversity. Thesis for master degree. Helsinki University of Technology. 60 p. (In Finnish, summary in English).

Holopainen, M. and Lukkarinen, E. 1994. The use of digital aerial photographs in a forest inventory. Digitaalisten ilmakuvien käyttö metsävarojen inventoinnissa. University of Helsinki. Department of Forest Resource Management. Publications 4. (In Finnish, summary in English).

Holopainen, M. and Wang, G. 1996. Estimation of Forest Biotopes by Digitized Aerial Photographs. University of Helsinki. Department of Forest Resource Management. Manuscript.

Kellomäki, S. and Kolström, T. (eds.) 1995. Metsikön rakenteen vaikutus monimuotoisuuteen: Tutkimuksen linjauksia. University of Joensuu. Publications 31. (In Finnish).

Kennedy, P.J., Päivinen, R. and Roihuvuo, L. (eds.) 1994. Designing a System of Nomenclature for European Forest Mapping. European Commission. EUR 16113. Office for Official Publications of the European Community. Luxemburg. 458 p.

Lovejoy, T.E. 1980. Foreword. In: Soule, M.E. and Wilcox, B.A. Conservation Biology: An Evolutionary-ecological Perspective. V-IX. Sinauer Associates. Sunderland. Mass.

McGarigal, K. 1994. Fragstats. Spatial Pattern Analysis Program for Quantifying Landscape Structure. Forest Science Department. Oregon State University.

Parmes, E. 1996. Landscape analysis from satellite imagery. In: Bachmann, P., Kuusela, K. and Uuttera, J. (eds.). Assessment of Biodiversity for Improved Forest Management. Proceedings 6. European Forest Institute. Joensuu. Finland.

Shannon, C. and Weaver, W. 1949. The mathematical theory of communication. University of Illinois Press. Urbana.

Simpson, E.H. 1949. Measurement of diversity. Nature 163: 688.

Solbrig, O.T. (ed.) 1991. From Genes to Ecosystems: A Research Agenda for Biodiversity. IUBS/SCOPE/ UNESCO. Paris.

Urban, D.L., O'Neill, R.V. and Shugart, H.H. Jr. 1987. Landscape ecology: A hierarchical perspective can help scientist understand spatial patterns. BioScience 37: 119-127.

Uuttera, J., Maltamo, M. and Kuusela, K. 1995. An approach to a method for monitoring habitat diversity in boreal coniferous forests. Comparative study North Karelia, Finland vs. Republic of Karelia, Russian Federation. Manuscript. European Forest Institute. Joensuu. Finland.

Wang, G., Poso, S., Waite, M.-L. and Holopainen, M. 1996. A texture module for using digitized aerial photo for forest inventory. Manuscript. University of Helsinki. Department of Forest Resource Management.

Wilson, E.O. and Peters, F.M. (eds.) 1988. Biodiversity. National Academy Press. Washington DC.

A FRACTAL APPROACH FOR THE CHARACTERIZATION OF NOAA-AVHRR NDVI PROFILES OF BROAD SCALE ECOREGIONS

C. RICOTTA[1], R. D. RAMSEY[2], A. FALCONER[2] & M. MARCHETTI[3]
[1] *University of Rome "La Sapienza", Piazzale Aldo Moro 5, 00185 Roma, Italy.*
[2] *Department of Geography and Earth Resources, Utah State University, Logan, Utah 84322-5240.*
[3] *Eurimage, Via Edoardo D'Onofrio 212, 00155 Roma, Italy.*

Abstract

Regional variations in climate, vegetation and landform are important in the development of ecosystems. For this reason, it is necessary to recognize regional differences in ecologically relevant factors at the highest level of spatial scale in the ecosystem classification.

With the operational program of the National Oceanic and Atmospheric Administration (NOAA), begun in 1990, periodical time series of the Normalized Difference Vegetation Index (NDVI) from the Advanced Very High Resolution Radiometer (AVHRR) are available for large area monitoring of Earth's vegetation phenology and long-term changes resulting from human interactions or global climate alterations.

We show here that the fractal dimension (D) of average NDVI profiles associated with broad scale ecoregions can be used to summarize their phenological pattern better than the simple visual examination of NDVI profiles or previous profile-derived indexes do.

1. Introduction

During the last two decades, resource managers are increasingly involved in an effort to develop a comprehensive ecosystem classification as an aid to achieving quality land management while maintaining sustainability of biodiversity (Bailey 1983). Regional variations in climate, vegetation and landform are important in the development of ecosystems. For this reason, it is necessary to recognize regional differences in ecologically relevant factors at the highest level in the classification.

Ecoregion classification is based on perceived patterns of a combination of natural and anthropogenic characteristics, including land use, potential natural vegetation, physiography, soils and climate (Omernik 1987). This synoptic approach in ecosystem geography invites a comparison of ecoregion map to remotely sensed data. Satellite spectral information is a function of Earth's surface patterns resulting from a combination of biotic and abiotic factors and is therefore an expression of the same constituents that define ecoregion maps.

P. Bachmann et al. (eds.), Assessment of Biodiversity for Improved Forest Planning, 255–261.
European Forest Institute Proceedings no. 18. © 1998 *Kluwer Academic Publishers. Printed in the Netherlands.*

Since 1990, the EROS Data Center in Sioux Falls, South Dakota, has provided an annual set of biweekly 1-km resolution maximum Normalized Difference Vegetation Index (NDVI) composites (Holben 1986) of the conterminous United States derived from the National Oceanic and Atmospheric Administration (NOAA) satellite series carrying the Advanced Very High Resolution Radiometer (AVHRR)(Eidenshink 1992).

The NDVI is a ratio of near-infrared (AVHRR channel 2) and visible (AVHRR channel 1) reflectances normalized over the sum of channels 1 and 2. NDVI yields a measure of plant biomass. The computed NDVI ranges from -1 to 1 but were rescaled over a 0-200 byte range with a value of 100 equal to a computed NDVI of 0.

For a given region, the average NDVI index can be plotted against time to show the broad scale remotely sensed vegetation phenology pattern. This visual approach has been used extensively for several years. To summarize the phenological cycle of vegetation with a single index, Tucker (1981) proposed the calculation of the integral of the curve. However, the integral of the curve represents only the cumulative effect of NDVI. It does not provide any information on the global shape of NDVI profiles.

Fractal geometry has become increasingly popular among ecologists largely because most spatial patterns of nature are so irregular and fragmented that Euclidean geometry cannot provide tools for the analysis of their forms (Palmer 1988). The term fractal refers to a series in which the Hausdorff-Besicovitch dimension (D) exceeds the topological dimension (Mandelbrot 1982). For a linear fractal function D may vary between 1 (if the dependent variable shows full spatial dependence from the independent variable) and 2 (if there is a complete spatial independence, so that it effectively takes up the whole of a two dimensional topological space (Burrough 1981). Much graphs of ecological data, like the graph of NDVI as a function of time, are likely to fall between these extremes.

We show here that the fractal dimension (D) of average NDVI profiles associated with broad scale ecoregions can be used to summarize their phenological pattern better than the simple visual examination of NDVI profiles or previous profile-derived indexes do (e.g., Samson 1993).

2. Methods and Results

The 1991 georeferenced biweekly AVHRR-NDVI data sets for the state of Utah (Ramsey et al. 1995) were extracted from the larger conterminous United States NDVI data set and overlaid on the Utah portion of Omernik's ecoregion map in order to determine the average NDVI profiles for each ecoregion (Figure 1).

The Utah map contained major portions of three ecoregions: Wasatch-Uinta Mountains (WUM), Colorado Plateau (CP) and Northern Basin and Range (NBR). Smaller parts of the Southern Basin and Range (SBR) and the Wyoming Plateau (WP) were encountered in the southwest and northeast portions of Utah, respectively.

Figure 2 shows the average NDVI time-series plot associated with the five Utah's ecoregions in 1991. The fractal dimension of the NDVI profiles was obtained from variogram analysis. Imagine a dependent variable *Z(i)* takes a given value at any point *i* along a transect. The semivariance g(*h*) is defined as:

$$\gamma(h) = \frac{1}{2N(h)} \sum_{i=1}^{N(h)} (z(i) - z(i+h))^2 \tag{1}$$

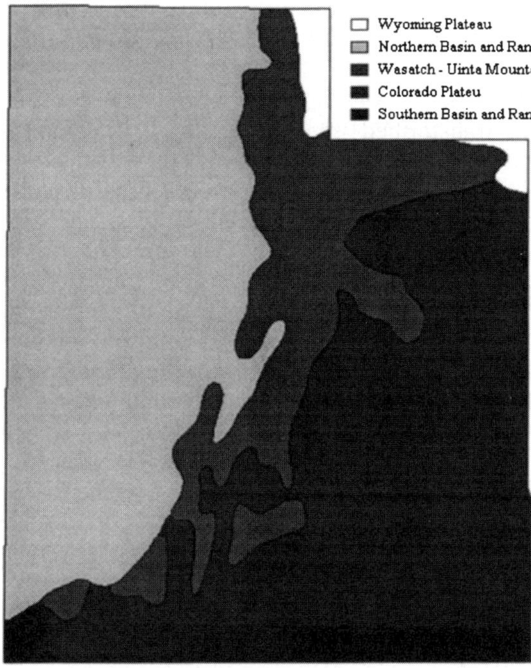

Figure 1. Distribution of ecoregions in the state of Utah as described by Omernik (1987). From Ramsey et al. (1995).

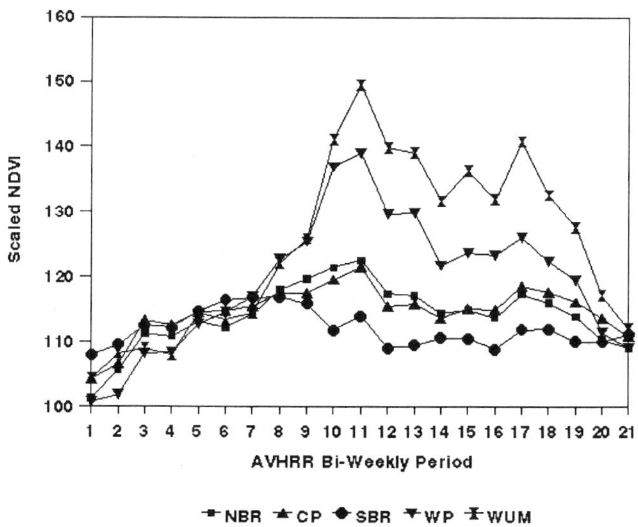

Figure 2. Time series plot of average NDVI values for five ecoregions in Utah for 1991.

where $Z(i+h)$ is the value of the dependent variable at a point separated from point i by distance h (the lag distance), and $N(h)$ is the number of pairs of points separated by distance h. The semivariogram is the plot of $g(h)$ as a function of h (Palmer 1988).

For each NDVI profile, D was calculated from the slope of the best fitting line of the double log-transformed semivariogram plot computed by the function D = (4-slope)/2 (Figure 3). When constructing a semivariogram, the length of the lag determines the number of those lags in any transect (Curran 1988). As a result, the confidence that can be ascribed to the average semivariance decreases with increasing lag. For this reason, only lags up to a third of the transect length were used in this study.

Table 1 presents the values of the integral of the curve of the five Omernik's ecoregions of Utah in 1991 along with their fractal dimension and break distance. The break distance (Klinkenberg 1992) is the maximum distance to which a least-square line could be fitted to the first linear segment in the log-log variogram plot with a correlation greater than 0.90 and represents the distance of spatial independence of the data. Due to the small number of lags used for the calculation of the least-squares fitted line, the break distance of the NDVI profiles is still coincident with the maximum lag. To highlight comparisons between the integral of the curve values of the five ecoregions, the integral value of the SBR was set to 100 and the other values were calculated proportionally.

TABLE 1. Integral under the curve, fractal dimension and break distance values of the 1991 biweekly NDVI profiles of Utah's Omerink Ecoregions.

EcoregioN	Integral under the curve	Fractal Dimension	Break Distance (Lag)
WUM	111.973	1.421	7
WP	106.996	1.363	7
CP	102.637	1.624	7
NBR	102.084	1.463	7
SBR	100.000	1.521	7

In general, the mean NDVI for each ecoregion increases through to the peak of the growing season and then falls to a minimum value as shown in Figure 2. As previously highlighted, the integral of the curve values of Utah's ecoregions, which act as indicators of the cumulative effects of NDVI, increase following the increase in phenological variation from SBR to WP. However, both, visual comparison of the NDVI profiles of Figure 2 and accurate statistical analysis (Ramsey et al. 1995), show that CP and NBR are nearly identical in the distribution of NDVI. This fact is largely due to the similarities in the vegetation components and climate for these ecoregions. However, previous observations of field ecologists (Johnson 1989; Harper 1994) have identified a shift in plant growth among those two ecoregions with a second surge of greenness during the late summer and early fall in the CP due to monsoonal type rains coming from the south.

This second growing peak in greenneess during late summer strongly influences the fractal dimension of the 1991 CP NDVI profile. Note that a late summer greenness peak is present in the 1991 WUM NDVI profile too. As a consequence, WUM phenological pattern shows a lower time dependence and thus a higher fractal dimension than the WP phenological profile which shows a lower amount of NDVI annual dynamics.

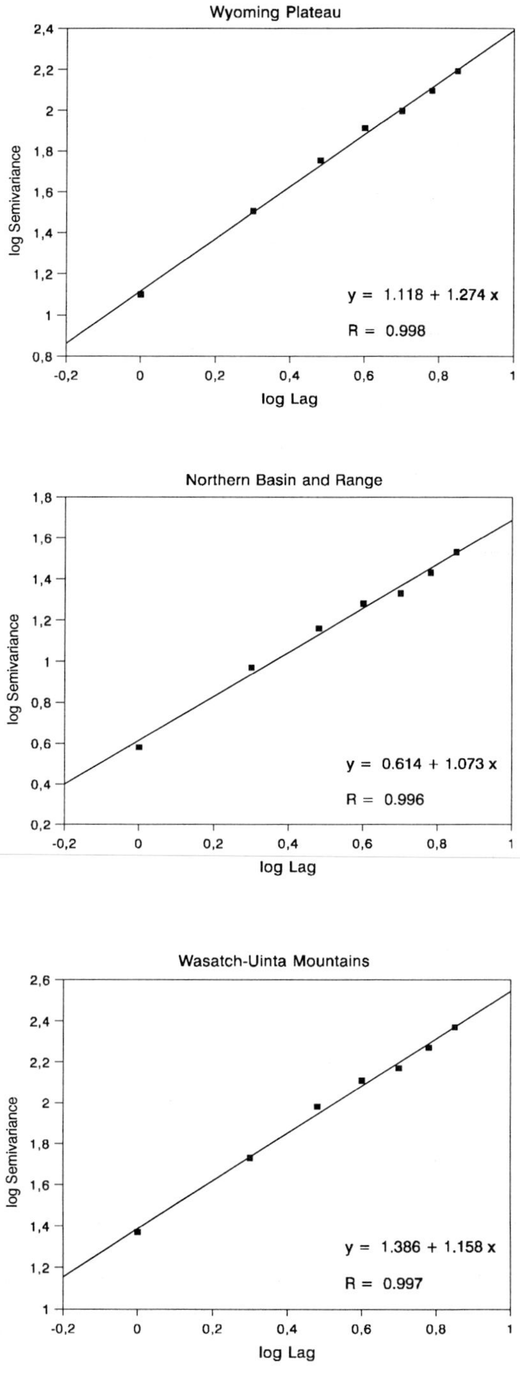

Figure 3. Double log-transformed semivariogram plots of mean NDVI values for Utah's ecoregions in 1991.

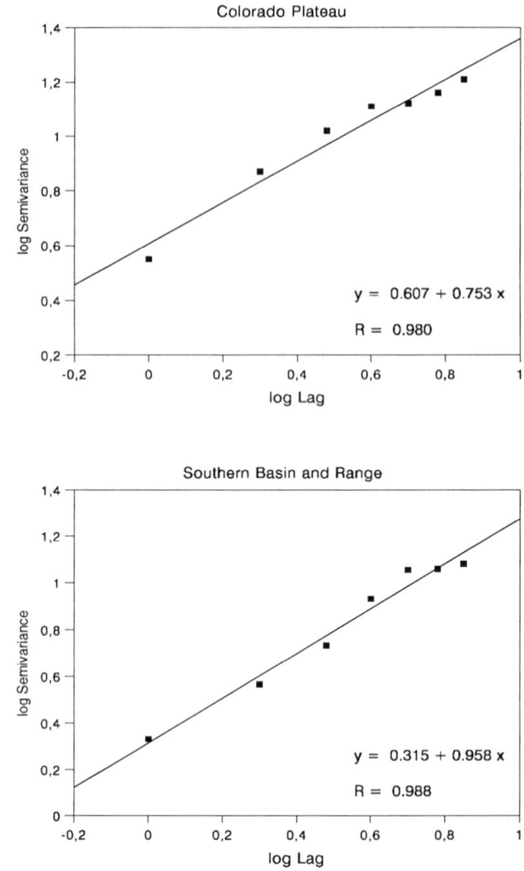

Figure 3. Double log-transformed semivariogram plots of mean NDVI values for Utah's ecoregions in 1991.

3. Conclusion

These results suggest not only that broad scale remotely sensed NDVI average profiles show fractal behavior, but that their fractal dimension can be used to summarize the general shape of the phenological cycle as an effective complement of the integral under the curve. However, these results should not be accepted uncritically. First, due to the biweekly periodicity of the AVHRR-NDVI data set of the conterminous United States (Eidenshink 1992), only 21 NDVI measures are available for the year 1991. This small number of measures strongly affects the confidence limit of semivariances and thus the number of points used for the calculation of the least-squares fitted line. Second, like simple NDVI profiles, the fractal dimension of vegetation phenological cycles calculated on annual basis could be affected by substantial differences driven by climatic variations (Townshend and Justice 1986), particularly if the analysis is restricted to small portions of a larger ecoregion. In this paper, this is the case of WP and SBR. However, despite these inconveniences, we think that fractal dimension of broad-resolution NDVI profiles offers the scientific community a new perspective for the study of vegetation behavior over time and also gives resource managers an useful tool for more efficient management of large ecosystems and their vegetated components.

References

Bailey, R.G. 1983. Delineation of ecosystem regions, Environmental Management 7, 365-373.

Burrough, P.A. 1981. Fractal dimensions of landscapes and other environmental data, Nature 294, 240-242.

Curran, P.J. 1988. The semivariogram in remote sensing: an introduction, Remote Sensing of Environment 24, 493-508.

Eidenshink, J.C. 1992. The 1990 conterminous U.S. AVHRR data set, Photogrammetric Engineering and Remote Sensing 58, 809-813.

Harper, K.T. 1994. Natural history of the Colorado Plateau and Great Basin, University Press of Colorado, Boulder.

Holben, B.N. 1986. Characteristics of maximum-value composite images from temporal AVHRR data, International Journal of Remote Sensing 7, 1417-1434.

Johnson, K.L. 1989. Rangeland resources of Utah, Cooperative Extension Service, Utah State University, Logan.

Klinkenberg, B. 1992. Fractals and morphometric measures: is there a relationship?, Geomorphology 5, 5-20.

Mandelbrot, B.B. 1982. The fractal geometry of nature, Freeman, San Francisco.

Omernik, J.M. 1987. Ecoregions of the conterminous United States, Annals of the Association of American Geographers 77, 118-125.

Palmer, M.W. 1988. Fractal geometry: a tool for describing spatial patterns of plant communities, Vegetatio 75, 91-102.

Ramsey, R.D., Falconer, A. and Jensen, J.R. 1995. The relationship between NOAA-AVHRR NDVI and Ecoregions in Utah, Remote Sensing of Environment 53, 188-198.

Samson, S.A. 1993. Two indices to characterize temporal patterns in the spectral response of vegetation, Photogrammetric Engineering and Remote Sensing 59, 511-517.

Townshend, J.R.G. and Justice, C.O. 1986. Analysis of the dynamics of African vegetation using the normalized difference vegetation index, International Journal of Remote Sensing 7, 1435-1445.

Tucker, C.J. 1981. Remote sensing of total dry-matter accumulation in winter wheat, Remote Sensing of Environment 11, 171-189.

BIODIVERSITY ASSESSMENT FOR CONSERVATION PLANNING IN UGANDA'S FORESTS

P. C. HOWARD, P. VISKANIC AND F. W. KIGENYI
Forest Department
P.O.Box 1752, Kampala, Uganda

Abstract

The Uganda Forest Department recently completed a major national inventory of forest biodiversity, aimed at providing the information necessary to design a representative protected area system for the country. The inventory covered 5 National Parks and a further 60 Forest Reserves, and involved the collection of data on five indicator taxa of plants and animals. The project involved approximately 100 man-years of work, during which 17,600 plant site records were made, 100,000 trap-nights of small mammal work undertaken, 57,000 large moths, 21,000 butterflies and 14,000 birds trapped.

The analysis of data generated by the inventory has involved the development of a scoring system, by which the biodiversity and socio-economic values of different sites were compared, and nature conservation priorities established.

More than 95% of species belonging to the five indicator taxa are represented in the present 10 National Parks combined with 11 selected Forest Reserves, and a more complete network of 43 sites would encompass more than 98% of species. The proposed network of forest Nature Reserves is presently being put in place.

1. Background and Justification

The Uganda Forest Department manages approximately 1.2 million hectares of land (7% of Uganda) as a permanent forest estate, distributed across the country in more than 700 forest reserves. These reserves encompass a wide range of vegetation types: more than two thirds of the associations recognised by Langdale-Brown et al. (1964) are represented in the major reserves, including forest, woodland, and more open communities. The reserves are managed to satisfy a variety of purposes, including the protection of fragile environments (mountain catchments), environmental services and biodiversity; the sustained production of timber and other forest products; and meeting the subsistence needs of local communities. Operationally, multiple-use management objectives are satisfied by designating particular reserves to specified uses, or defining different management zones within each reserve.

In the late 1980s, a decision was taken to manage 50% of the natural forest land primarily for environmental protection, with 20% maintained as Nature Reserves and 30% as low-impact use 'Buffer Zones'. The remaining 50% of the estate is to be managed primarily for the sustained supply of quality tropical hardwoods and other forest

P. Bachmann et al. (eds.), Assessment of Biodiversity for Improved Forest Planning, 263–271.
European Forest Institute Proceedings no. 18. © 1998 *Kluwer Academic Publishers. Printed in the Netherlands.*

products. The Forest Department seeks to achieve these broad allocations by applying the principles embodied in the Man and Biosphere concept of reserve design in which management zones are defined around a totally protected core (Nature Reserve), with zones of increasingly intensive use towards the periphery of each reserve (Batisse 1985).

This paper describes how decisions have been made over the selection of sites for Nature Reserve establishment. There is a rapidly expanding literature on the subject of biological reserve selection, and the related problem of priority-setting in conservation programme planning and resource allocation. Well known examples at a global scale include WORLDMAP (Vane-Wright et al. 1991), and the ICBP priority-setting programme based on endemic bird areas (ICBP 1992); while at the national and sub-national level examples include work in Australia (e.g. Margules et al. 1988; Pressey et al. 1994), South Africa (Rebelo 1994) and India (Daniels et al. 1991) amongst many others. Useful recent reviews have been provided by Pressey et al. 1993; Johnson 1995 and McNeely 1996. The procedures developed for Nature Reserve selection in Uganda's forests build on this experience, adapting it to local conditions.

When the decision to establish new forest Nature Reserves was taken there was very little information available on the suitability of different areas for particular uses, and information on biodiversity values was limited to very few sites (Howard 1991). A major programme of biological inventory work was therefore undertaken between 1990 and 1995, focusing on five 'indicator groups' of plants and animals sampled from all reserves exceeding 50 km^2, and some smaller ones (where these included vegetation types not otherwise represented). The inventory programme involved about 100 man-years of work, during which 17,600 plant site records were made; 100,000 trap-nights of small mammal work undertaken; 57,000 large moths, 21,000 butterflies and 14,000 birds trapped. The results of this work now provide a very strong basis from which to identify 'biodiversity hotspots', and ensure that the selection of areas for designation as Nature Reserves is technically sound.

This paper aims to describe how this new biodiversity information is being used to ensure that decisions over the allocation of forest land to nature conservation are made as objectively as possible. A scoring system is developed as a means of evaluating each site in terms of its potential for various alternative uses, and ensuring that decisions over land allocations are made accordingly. The aim is to adjust the 50-30-20 percent (national average) management zone allocations according to the specific characteristics of each site.

In respect of land to be designated as Nature Reserves, the selection procedure incorporates the following principles:

A) Nature Reserves should be designated to protect:
- 'biodiversity hotspots' – areas which support an unusually large number of species; and/or concentrations of rare species;
- sites that encompass a large number of habitat/vegetation types; and/or unusual vegetation associations;
- sites that are natural, undisturbed and likely to remain so.

B) Nature Reserves should contribute to an efficient national protected areas system, which protects the full range of biodiversity represented in Uganda without unnecessary (and costly) duplication.

C) Nature Reserves should be selected in the most cost-efficient way, avoiding areas that are well-suited to commercial forestry or community-use wherever this is possible. However where species and/or habitats occur only in such areas, they should be designated for protection.

2. Methods

The procedure followed in the selection of areas to be designated as Nature Reserves involved three phases, starting with field data collection, followed by analysis of national priorities and selection of a suitable network of sites, and finally the designation of management zones within individual forests. The last of these phases has only just commenced.

2.1 DATA COLLECTION

2.1.1 Selection of sites for evaluation

The forest estate as a whole covers 1.2 million hectares, comprising 713 separate Forest Reserves. From this total, biological inventory work was carried out in all the larger forests (54 reserves exceeding 5,000 ha), plus 10 smaller reserves selected to represent vegetation types that do not occur in the larger reserves. Altogether these 64 forest reserves account for 75% of the total area of the forest estate. A map of the sites that were surveyed is provided in Figure 1.

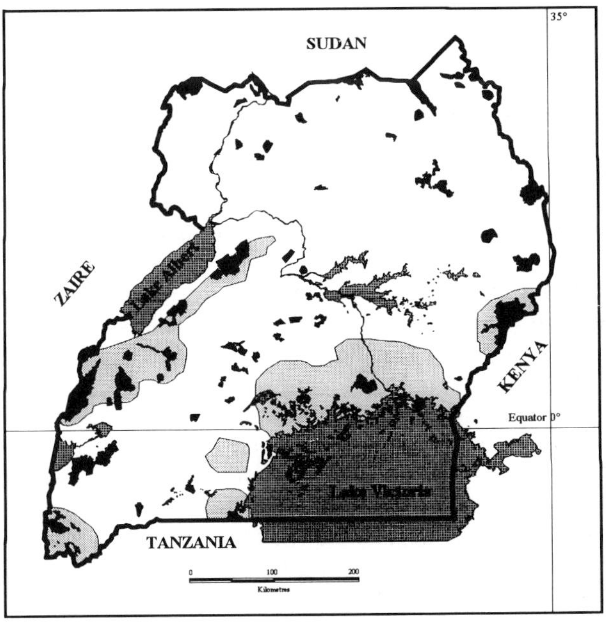

Figure 1. Map of Uganda showing the location of forests included in the biodiversity inventory programme, and areas of the country that naturally support closed canopy forest (stippled).

2.1.2 Selection of indicator taxa
Ideally conservation planning should be based on a thorough and complete knowledge of all groups of flora and fauna, but this is obviously impossible, so indicator taxa have to be used. Recognising that priorities established for particular groups of organisms are often quite different (e.g. Daniels et al. 1991; Howard 1991; Pomeroy 1993; Prendergast et al. 1993a) we selected five groups to represent the widest possible taxonomic and ecological diversity, consistent with the need for rapid assessment, and reliable identification. Woody plants, birds, small mammals, butterflies and large moths were used to define an overall 'importance score for biological conservation' for each forest.

2.1.3 Fieldwork
Having decided on the forests to be surveyed, and the indicator taxa to be sampled, the field programme was initiated by training 25 technical grade Forest Rangers for three months in biological inventory techniques. These men subsequently formed four field teams, each accompanied and supervised in the field by an expatriate biologist. The four field teams worked simultaneously in different forests, establishing base camps at strategic locations so as to provide maximum possible coverage of each reserve. In principle the field sampling regime was designed to cover the full range of altitude, vegetation types and aspect represented in each reserve (full details are provided in Howard and Davenport 1996).

2.2 DATA ANALYSIS

The analysis of data was carried out as an iterative process, aimed at selecting a network of sites for designation as forest Nature Reserves in which all species and habitats are represented in the most cost-effective way. The various steps in this process are summarised in Figure 2 and described in more detail below:

STEP 1. Rank sites according to their relative biological value. An overall biological importance score was derived for each forest, by combining scores for each of the five indicator taxa. These scores take into account both species richness (standardised for differences in sampling intensity using rarefaction; Krebs 1989; Prendergast et al. 1993b), and the rarity of the 'average' species (considered at continental and national scales). Further details are provided in Howard (in prep.).

STEP 2. Combine biological and socio-economic considerations to rank sites according to their suitability for Nature Reserve designation. Nature Reserve suitability scores were derived to help identify sites of high conservation value where Nature Reserves could be designated without conflicts over land use. Initially, the biological importance score (STEP 1) was combined with a score for compatible non-consumptive uses (namely tourism and recreation potential, watershed value and importance for education and research) to derive an overall conservation value for each site. This was then 'traded off' against the socio-economic pressures impinging on the forest, which reduce the feasibility of long-term site protection. These pressures are scored on the basis of an evaluation of commercial forestry prospects, and the needs of local communities around the periphery of each forest, to derive a score for alternative use potential (details of which are provided in Howard, in prep.). Thus a final Nature Reserve Suitability score is derived for each forest as: Nature Reserve Suitability =

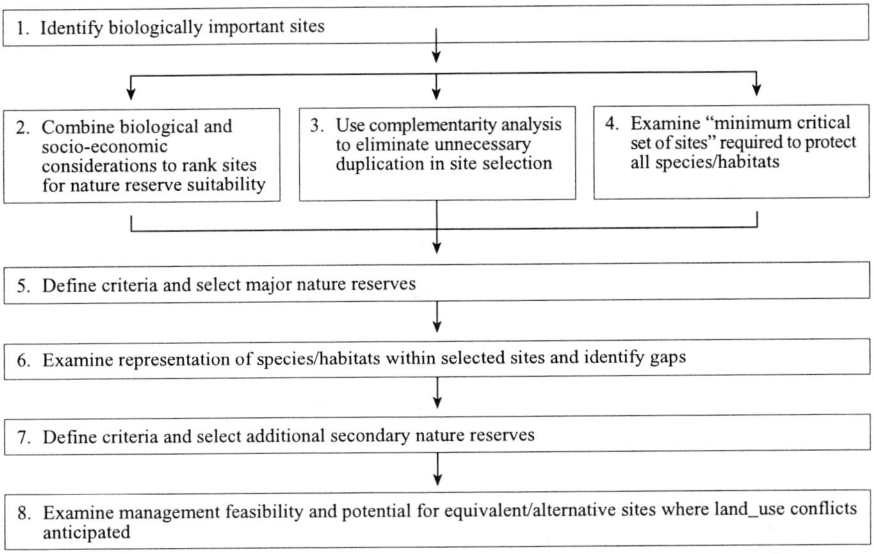

Figure 2. Diagram indicating the methodology used to select the sites for the establishment of Nature Reserves (see text for further explanation)

Conservation Value – Alternative Use Potential. Figure 3 shows the relative values of the scores derived in this way for the 64 forests that were evaluated.

STEP 3. Rank sites according to the complement of species contributed to a representative national network of protected areas. A complementarity table (Johnson 1995) was derived for each of our five indicator taxa to rank forests according to the number of additional species they contribute to a network of sites, after taking into account species already represented in forests higher up the table. Such a table always starts with the most species-rich site, followed by the site which adds the greatest number of 'new' species, not known from the richest site. Initially the 64 forests were arranged in this way, based entirely on the representation of species at each site. Subsequent analyses, however, were designed to take into account the existence of National Parks (already under protective management), and the socio-economic considerations incorporated into the Nature Reserve Suitability scores (STEP 2). Thus, an alternative complementarity table was derived by combining five subsets of sites, arranged sequentially and starting with the National Parks subset at the top of the table. Next were added all steep mountain catchment forests (since these will inevitably be protected), followed by subsets of sites characterised by high, medium and low Nature Reserve Suitability scores.

STEP 4. Define the minimum critical set of sites necessary to protect all species. In doing this, species represented in National Parks were taken into consideration, so that only forests which make a unique contribution to the protected area system as a whole were included in the minimum critical set. Furthermore, species that are (or may be) of

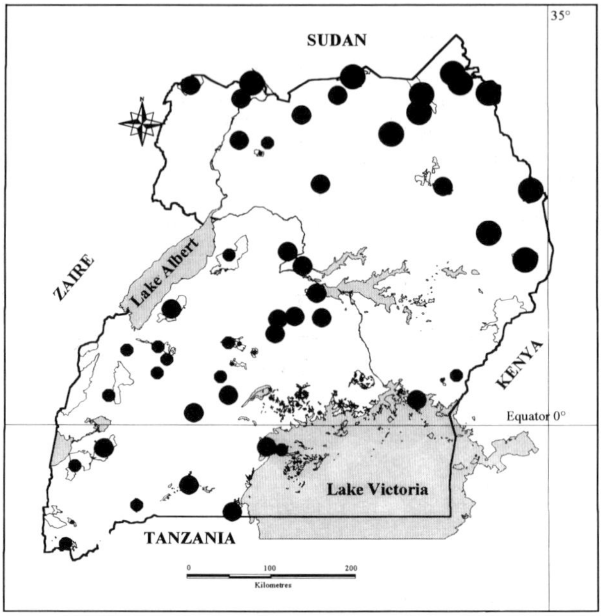

Figure 3. Relative values for Nature Reserve Suitability of the forests surveyed. Symbol size is scaled according tao the value of derived scores (see text for explanation). National Parks are not considered for the calculation of the suitability scores as they provide already the highest degree of protection.

national, regional or global conservation concern were distinguished. Thus a minimum critical sites matrix was derived, listing each site and the number of species of each taxon that are unique to it (distinguished as 'conservation important' and 'less concern').

STEP 5. Define criteria and select key sites for the designation of major Nature Reserves. Key sites were taken to be those which either:

- support a large proportion of unique species, found in no other Ugandan forest; more than 1% of species within any one taxon was used as the qualifying criterion; or
- contribute more than 1% of species of any taxon in the complementarity analysis after taking species represented in National Parks and forests satisfying the 1% uniqueness criterion into account. Within these key sites, the aim is to designate significantly more than the national average (as a proportion of area) as Nature Reserves.

STEP 6. Define criteria and select secondary sites for the designation of additional Nature Reserves. Secondary sites were then added on the basis of three further criteria:

- adding at least 0.5% to the complement of species within any taxon;
- contributing at least one species considered to be of conservation concern on account of its limited range;

- contributing a vegetation type (based on Langdale-Brown et al. 1964) not otherwise represented (or seriously under-represented) in the protected area system.

STEP 7. Examine potential for alternative/equivalent sites. Cluster analysis (TWINSPAN) was used to compare forests on the basis of species lists for each taxon, and examine the potential for designating alternative Nature Reserves to those selected during STEP 5 and STEP 6. This was particularly valuable where a major Nature Reserve was indicated that appeared incompatible with alternative demands on the land, or where the condition of the forest did not warrant Nature Reserve status. The results of cluster analysis were used to check that each major group of forests was represented in the proposed national system.

2.3 DESIGNATION OF NATURE RESERVES AND OTHER MANAGEMENT ZONES WITHIN INDIVIDUAL FORESTS

Now that the relative importance of different forests has been analysed and a suitable national network of sites selected for Nature Reserve establishment, appropriate management zones must be selected within each forest, mapped at 1:50,000, demarcated on the ground and managed appropriately. Where feasible a substantial undisturbed core area of each forest, covering the widest possible range of altitude, and variety of vegetation types is required as a Nature Reserve. An important consideration will be the location of natural features (such as streams, ridges) that can serve as clear internal boundaries between Nature Reserves and adjacent management zones. Where Nature Reserves are to be established for the benefit of particular species, the preferred habitat and breeding ranges of those species (where known) will be taken into account. Special consideration will be given to the needs of local people, and areas within 2 km of a forest boundary avoided, with a preference for areas at least 5 km from forest-adjacent communities.

3. Results

Nationally a network comprising 14 key sites and 20 secondary sites has been identified for designation of forest Nature Reserves to complement the existing National Parks. Based on available information, at least 77% of species are represented in Uganda's National Parks, and more than 95% are included with the addition of the 11 forests listed in Table 1. The full network of protected areas (National Parks and forest Nature Reserves) would protect more than 98% of species belonging to the five indicator taxa investigated under this programme.

The Biodiversity Inventory Programme has produced a vast amount of information on Uganda's forests which has been presented as a series of 33 biodiversity reports (Howard and Davenport 1996). Each report provides a summary of the data collected at one major forest, or a group of similar adjacent ones. The subsequent data analysis and national protected area systems planning exercise (outlined above) has only recently been completed, and the results are to be documented more fully in a Nature Conservation Masterplan, currently in preparation.

TABLE 1. Summary complementarity table listing forests which individually contribute at least 0.5% to the total No. of species represented in the national protected area network. Note that species represented in the county's National Parks are considered to be protected and incorporated collectively in the top line of the table. Forests added below are not sorted for Nature Reserve Suitability (see text STEP 3 for explanation)

Reserve	% of species added	Cumulative % species represented
National Parks (10)	77.0	77.0
Budongo	6.1	83.1
Otzi	3.3	86.4
Moroto	2.6	89.0
Mount Kei	1.1	90.1
Sesse Islands	1.1	91.2
Labwor Hills	0..9	92.1
Sango Bay	0.7	92.3
Kasyoha-Kitomi	0.7	93.6
Nyangea-Napore	0.6	94.2
Kalinzu-Maramagambo	0.5	94.7
Timu	0.5	95.1

4. Acknowledgments

This programme was supported by the EC-funded 'Natural Forest Management and Conservation Project' (No. 6100.37.42.015) and the project of 'Institutional Support for the Protection of East African Biodiversity' (UNO/RAF/006/GEF), financed through UNDP/FAO by the Global Environment Facility. A great many people contributed to this work and it is impossible to mention them all by name. Particular thanks are however extended to colleagues Michael Baltzer, Tim Davenport, Chris Dickinson, David Duli, David Hafashimana, Isaac Kapalaga, Tony Katende, Robert Kityo, Roger Matthews, Edward Mupada, and Robert Nabanyumya, all of whom contributed greatly to the management of the field programme.

References

Batisse, M. 1985. Action Plan for Biosphere Reserves. Environmental Conservation. 12(1): 17-27

Daniels, R.J., Hegde, M., Joshi, N.V. and Gadgil, M. 1991. Assigning conservation value: A case study from India. Conservation Biology. 5: 464-475.

Howard, P.C. 1991. Nature Conservation in Uganda's Tropical Forest Reserves. IUCN. Gland. Switzerland and Cambridge.

Howard, P.C. in prep. Nature Conservation Masterplan. Uganda Forest Department. Kampala. Uganda.

Howard, P.C. and Davenport, T.R.B. (eds.) 1996. Forest Biodiversity Reports. Series of 33 reports. Uganda Forest Department. Kampala. Uganda.

ICBP 1992. Putting biodiversity on the map: Priority areas for global conservation. International Council for Bird Preservation. Cambridge. UK

Johnson, N.C. 1995. Biodiversity in the Balance: Approaches to Setting Geographic Conservation Priorities. Biodiversity Support Program. Washington DC. USA.

Krebs, C.J. 1989. Ecological Methodology. Harper and Row. Publishers. New York. USA

Langdale-Brown, I., Osmaston, H.A. and Wilson, J.G. 1964. The vegetation of Uganda and its bearing on land-use. Uganda Government Printer. Entebbe. Uganda.

Margules, C.R., Nicholls, A.O. and Pressey, R.L. 1988. Selecting networks of reserves to maximise biological diversity. Biological Conservation 43: 63-76.

McNeely, J.A. 1996. Assessing methods for setting conservation priorities. Paper presented to OECD Conference on Biodiversity. March 1996. Cairns. Australia.

Pomeroy, D. 1993. Centers of high biodiversity in Africa. Conservation Biology 7(4): 901-907

Prendergast, J.R., Quinn, R.M., Lawton, J.H., Eversham, B.C. and Gibbons, D.W. 1993a. Rare species, the coincidence of diversity hotspots and conservation strategies. Nature 365: 335-337.

Prendergast, J.R., Wood, S.N., Lawton, J.H. and Eversham, B.C. 1993b. Correcting for variation in sampling effort in analyses of diversity hotspots. Biodiversity Letters 1: 39-53.

Pressey, R.L., Humphries, C.J., Margules, C.R., Vane-Wright, R.I. and Williams, P.H. 1993. Beyond opportunism: key principles for systematic reserve selection. Tree 8(4): 124-128.

Pressey, R.L., Bedward, M. and Keith, D.A. 1994. New procedures for reserve selection in New South Wales: Maximising the chances of achieving a representative network. In: Forey, P.L., Humphries, C.J. and Vane-Wright, R.I. (eds.). Systematics and Conservation Evaluation. Clarendon Press. Oxford. Pp 351-374

Rebelo, A.G. 1994. Using Proteaceae to design a nature reserve network and determine conservation priorities for the Cape Floristic Region. In: Forey, P.L., Humphries, C.J. and Vane-Wright, R.I. (eds.). Systematics and Conservation Evaluation. Clarendon Press. Oxford. Pp 375-396.

Vane-Wright, R.I., Humphries, C.J. and Williams, P.H. (1991). What to protect? – systematics and the agony of choice. Biological Conservation 55: 235-254.

ON THE ACCURACY OF LINE TRANSECT SAMPLING OF RARE FOREST OBJECTS

T. LÄMÅS AND G. STÅHL
Swedish University of Agricultural Sciences
Dept. of Forest Resource Management and Geomatics
S-901 83 Umeå, Sweden

Abstract

For timber production purposes, plot-based inventory methods, such as circular plot sampling, are widely used. For rare objects, however, other methods are likely to be more cost-efficient and, consequently, studies of alternative methods are needed. In line transect sampling, the surveyor walks through the forest along inventory lines, and when an object of interest is noticed, the perpendicular distance between the object and the line is measured. After the inventory, a probability-of-detection function is estimated. Thereby, unbiased estimates are obtained, although objects in the neighbourhood of the line are not always noticed. The method has, for example, been used for inventory of bird populations and should also be a useful method for acquiring data on rare forest objects of interest for nature conservation (e.g. individual plant and animal species, snags and logs).

The shape of the (true) probability-of-detection function depends on, for example, sighting conditions and number of objects. The accuracy of the method depends, among other things, on how well this function can be estimated. If only a few objects are found in the inventory, the function is likely to be poorly estimated.

In this study, some characteristics of the line transect sampling method were investigated by Monte-Carlo simulation. Effects of the number of objects found on the estimation of the probability-of-detection function and, consequently, the accuracy of the inventory method were investigated. Furthermore, the influence of the spatial distribution of the objects (Poisson or clustered distribution) was studied.

1. Introduction

Apart from valuable and sustainable timber yield, the objectives for forestry have broadened to include viable forest ecosystems and to maintain biodiversity. This has brought about a need for broader information about the forest state. For planning purposes, unbiased data of known accuracy, not only of timber stock but also of plants and animals, are required. Among probability sampling methods, circular sample plot surveys have been frequently used in forestry. As live trees are abundant in forests, this method is appropriate for surveys of growing stock. Many objects of interest in multi-purpose forestry and nature conservation, such as snags, fallen trees, and individual plant

P. Bachmann et al. (eds.), Assessment of Biodiversity for Improved Forest Planning, 273–281.
European Forest Institute Proceedings no. 18. © 1998 *Kluwer Academic Publishers. Printed in the Netherlands.*

and animal species are, however, rare. For surveys of such objects, circular plot surveys are likely to be less appropriate. Normally, only a small proportion of the area is inventoried, and, when walking between the plots, no information is collected whether objects of interest are noticed or not.

For rare forest objects, line transect sampling provides an alternative probability sampling method (e.g. Burnham et al. 1980). Using this method, the surveyor walks through the forest along lines, and when an object of interest is noticed, the perpendicular distance between the object and the line is measured (Fig. 1). After the inventory, a probability-of-detection function is estimated. Thereby, unbiased estimates are obtained, although objects in the neighbourhood of the line are not always noticed.

This paper presents parts of ongoing work aiming at finding cost efficient inventory methods for surveying (probability sampling) rare forest objects. The accuracy of line transect sampling, for estimations of number of objects, is presented. The studies are performed by means of Monte-Carlo sampling simulations which is a simple and straightforward alternative to analytical methods.

Practical implications are not discussed. Moreover, neither the cost-efficiency of this method, nor its relation to other probability sampling methods, were considered.

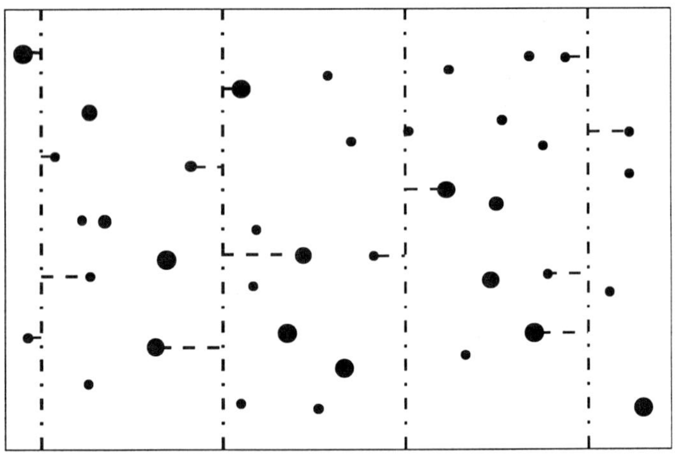

Figure 1. A general view of a line transect sample. Legend: —• —• — inventory line, - - - perpendicular distance between the inventory line and the object.

2. Material and methods

2.1 THE SAMPLING SIMULATOR

A sampling simulator constructed by Ståhl and Lämås (1995) was used. The simulator is developed for evaluating the cost-efficiency of different inventory methods. By means of Monte-Carlo simulation, the performance of different inventory methods for surveying rare objects can be studied. The simulator consists of three parts. In the first part, different forests are simulated. In the second part, Monte-Carlo simulations of different inventory strategies (inventory method and design of the inventory such as plot radius and number of plots) are performed in the simulated forests. It is also possible to include real forests where the location of objects has been determined (using e.g. global positioning system (GPS) technique). The last part is a graphical unit which permits forests and forest inventories to be visualised on the computer screen.

In the simulator, for a given forest, an inventory strategy is repeatedly simulated and the parameter of interest is estimated. The variance of the parameter using the specific inventory strategy is then estimated as the variance of estimated values in the simulations. This is usually called Monte-Carlo simulation.

Apart from line transect sampling, six other inventory methods have been implemented in the simulator. These methods are circular plot inventory, distance methods, strip surveys, line intersect sampling, adaptive cluster sampling, and transect relascope sampling. A description of the latter method is given by Ståhl (1995).

2.2 LINE TRANSECT SAMPLING

Among area based probability sampling methods, a variety of designs of sample plots exists. The strips can be viewed as long rectangular plots. In forestry, strip survey was for example used in the first Swedish National Forest Inventory in the 1920's (Matérn 1984).

In some surveys, it was obvious that not all objects (e.g. animals) within the strip were recorded. In order to compensate for this loss, different approaches to make estimations based on the 'effective strip width' appeared. Estimations of effective strip width were, in turn, based on measurements of the perpendicular distance between the object and the strip centre line.

In 1949, Hayne presented an estimator based on the *sighting distance*, i.e. the distance from the point of observation on an inventory line to the object (Hayne 1949) (Fig. 2).

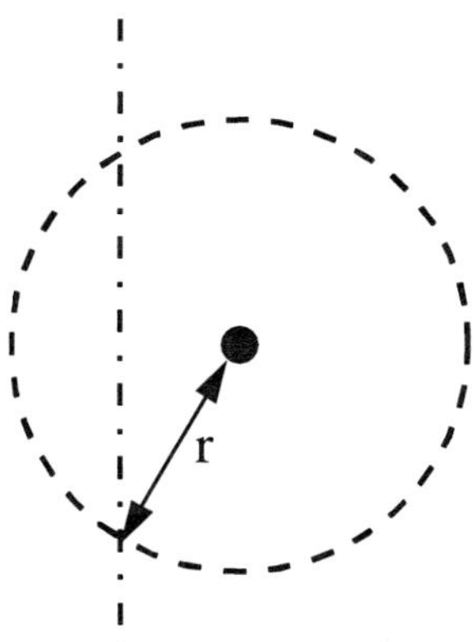

Figure 2. The principle of the Hayne estimator. The sighting distance, i.e. the distance between the object and the observer at the moment of detection, is denoted *r*. Legend: — • — • — inventory line, - - - area of inclusion.

According to this model, each object will be recorded once an observer traverses the circular area of a radius equal to the sighting distance (r) around the object. In case m inventory lines are located within an area of width L (perpendicular to the inventory lines), an unbiased estimator of some population attribute total, $\hat{Y}$, is then:

$$\hat{Y} = \frac{L}{2m} \sum_{i=1}^{m} \sum_{j=1}^{k_i} \frac{y_{ij}}{r_{ij}} \tag{1}$$

where k_i is the number of recorded objects on line i, and y_{ij} is the attribute of the j:th object on the i:th line. In case the true area is known, the estimator can be enhanced by using ratio estimation with area as an ancillary variable.

A possible drawback of this approach is that, in case an object is noticed at a very short distance, i.e. r_{ij} is small, this object will have a very large influence on the estimate. Moreover, it is not evident that the 'area of inclusion' really is a circle (Burnham et al. 1980).

A general formulation of line transect sampling based on the probability of detecting an object was presented by Burnham and Anderson in the 1970's (Burnham and Anderson 1976; Burnham et al. 1980). According to this approach, the probability for an object to be noticed is a function of the perpendicular distance between the inventory line and the object, x_i. Population attribute totals can then be estimated as follows:

$$\hat{Y} = \frac{L}{2m \int_0^w f(x)dx} \sum_{i=1}^{m} \sum_{j=1}^{k_i} y_{ij} \tag{2}$$

where w is the maximum distance for detection and $f(x)$ is the probability-of-detection (PoD) function. As the Hayne estimator above, this latter estimator can also be improved by ratio estimation in case the true area is known.

It is likely that objects on the line are detected with a probability equal to one. The probability of detection for other objects, i.e. the shape of the PoD function, depends on a number of factors. It depends on, among other things, the skill and experience of the surveyor, attributes of the objects, and the environment in which the objects are located, which affects the sighting conditions (cf. Fig. 4 and 7).

After the field survey, the parameters of the true PoD function have to be estimated. If, in a survey, only a small number of objects are found, the parameters are likely to be poorly estimated, and, consequently, the accuracy of the inventory method will be low. In this study, the effect of the number of objects found on the estimation of the PoD function and the accuracy of the inventory method were investigated. Also, the influence of the spatial distribution of the objects (distribution according to a Poisson process or clustered distribution) was investigated (cf. e.g. Burnham et al. 1980).

In the sampling simulator, the parameters in the PoD function were estimated in an iterative manner based on the result of the simulated inventory.

2.3 SIMULATION OF FORESTS AND INVENTORIES

By simulation, forests comprising 5 and 25 objects per hectare were generated. For each density, forests were generated where the objects were located according to a Poisson process (conditioned on a fixed number of objects) and located according to a clustered distribution. Moreover, in order to study the effects of different numbers of objects detected, quadratic forests of 4, 9, 16 and 25 ha were simulated. A single replicate of each forest was used (Fig. 3).

In the clustered forests, the number of cluster centra per hectare was set to $N / 5$ where N was the number of objects per hectare. The maximum radius of clusters was 30 m. First, cluster centra were established according to a Poisson process (conditioned on a fixed number of cluster centra). Then, locations of potential cluster members were chosen entirely at random. The probability of a potential object to be permanented was a decreasing linear function of the Euclidean distance from the nearest cluster center to the object. For details, see Ståhl and Lämås (1995).

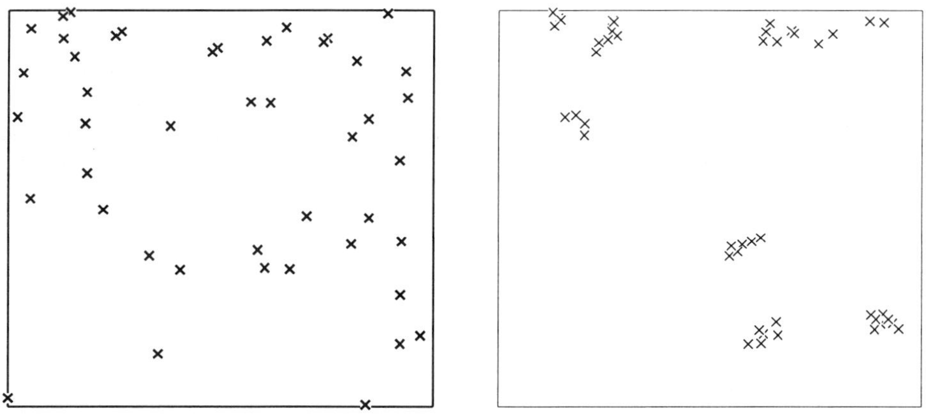

Figure 3. Examples of the 'forests' used in the study. The examples shown are areas of 9 ha with 5 objects ha^{-1}. The objects were located according to a Poisson process (left) or a clustered spatial distribution (right).

Inventory lines (oriented in a 'north-south' direction) corresponding to 200 m lines per hectare were used. As only one replica of a forest was used for each simulated inventory, systematic sampling tended to give uneven results. Therefore, simple random sampling was used in the study (cf. Lämås 1996). The number of inventory lines used corresponds to a distance between the lines of 50 m in a systematic design. In order to avoid boundary bias the inventory lines were allowed to be located also outside the area (up to the maximum detection distance). In this study, the following PoD function was used:

$$f(x_i) = \left[1 - \left(\frac{x_i}{w}\right)^p\right]^{\frac{1}{p}} \tag{3}$$

where x_i is the perpendicular distance from a survey line to the i:th object. The shape of the curve is determined by the parameters p and w. The parameter w, the maximum detection distance, was set to 20 m. Two values for the parameter p, $p = 0.7$ and $p = 1.5$, were used (Fig. 4).

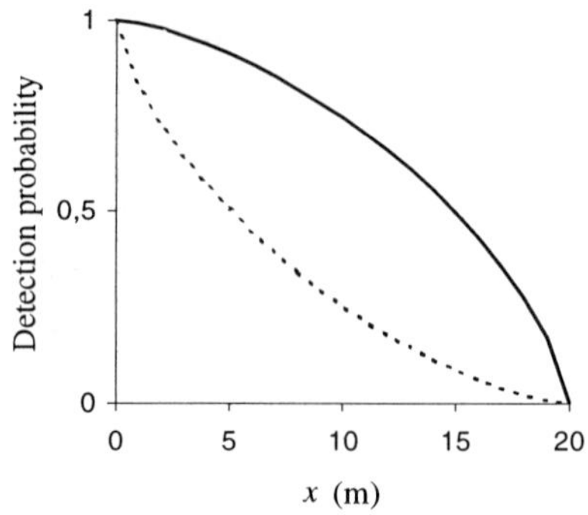

Figure 4. The probability-of-detection function used in the study. Maximum detection distance, perpendicular to the inventory line (w) was set to 20 and two values, 0.7 and 1.5, for parameter p were used. Legend: - - - - $p = 0.7$, —— $p = 1.5$.

In the simulated surveys, the probability of detecting an object was simulated according the true PoD function, i.e. $p = 0.7$ or $p = 1.5$. After each simulated survey, the parameter p was estimated based on the perpendicular distances between the inventory line and the observed objects. This was done in an iterative manner using the method of maximum likelihood. The estimated PoD function was then used in the estimator of total number of objects. In order to evaluate the influence of the (erroneous) estimation of the PoD function, the number of objects was also estimated using the true function in the estimator.

The expected number of objects included in the sample (E(n)) varies depending on the density of objects, the size of the inventoried area, and the parameters p and w in the PoD function (Table 1).

TABLE 1. The expected number of objects included in the sample for the different alternatives ($w = 20$)

Density		Inventoried area (ha)			
(obj. ha^{-1})	p	4	9	16	25
5	0.7	5	11	20	32
5	1.5	11	25	44	68
25	0.7	25	57	102	159
25	1.5	55	123	219	342

3. Results

The standard error (SE) of the estimated number of objects decreased with the increased number of objects detected. With some exceptions, clustered spatial distribution of objects generated higher SEs than objects randomly distributed (distribution according to a Poisson process). In the case of $p = 0.7$, the inventoried area 4 ha, and a density of objects of 5 ha^{-1}, the SEs were very large (Fig. 5).

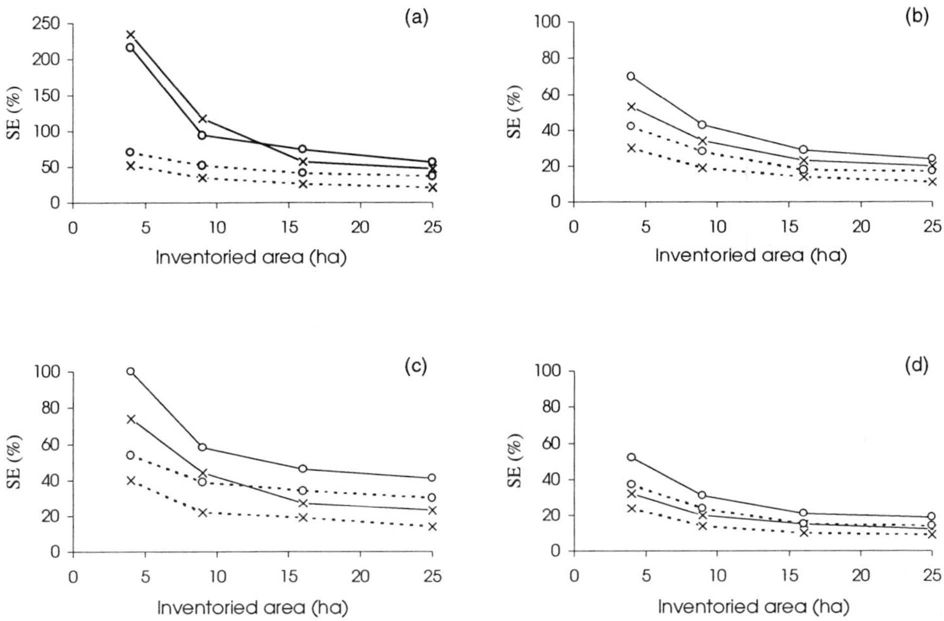

Figure 5. Standard errors (SE, in percent of true values) of the estimated number of objects in the case of p = 0.7 (a and b) and p = 1.5 (c and d). The density was 5 (a and c) and 25 (b and d) objects ha^{-1}, respectively. The objects were located according to a Poisson process (×) or located in clusters (O).The parameter in the probability-of-detection function was either estimated based on the sample (—) or the true function was used (- - -). Note the different scale on the y-axis in (a).

In the case of p = 0.7, the highest density of objects (25 ha^{-1}), and the largest inventoried area (25 ha), E(n) was equal to 159. In this case, the SEs were 20 and 24% for random distribution of objects and clustered spatial distribution, respectively. Given the same forests and p = 1.5, E(n) was equal to 342, and the corresponding SEs were 12 and 19%, respectively. Approximately, the SEs were proportional to $1/\sqrt{E(n)}$ (Fig. 6).

4. Discussion

In this study, the accuracy of line transect sampling was studied by means of Monte-Carlo simulation. In practical use for surveying rare forest objects, numerous aspects of the method have to be considered. Consequently, field studies, and especially time studies, have to be made in order to allow comparisons of the cost-efficiency of alternative methods. As time consumption can be included in the simulator, also the cost-efficiency can be studied

A single PoD function was used in the study. It is likely that, in many cases, the true probability function has a shape different from the one used. Moreover, Burnham et al. (1980) argues that two criteria of a PoD function is that $f(0)$ = 1 and $f'(0)$ = 0. Only the first of these two criteria is met by the function used. However, in the case of p = 1.5, the

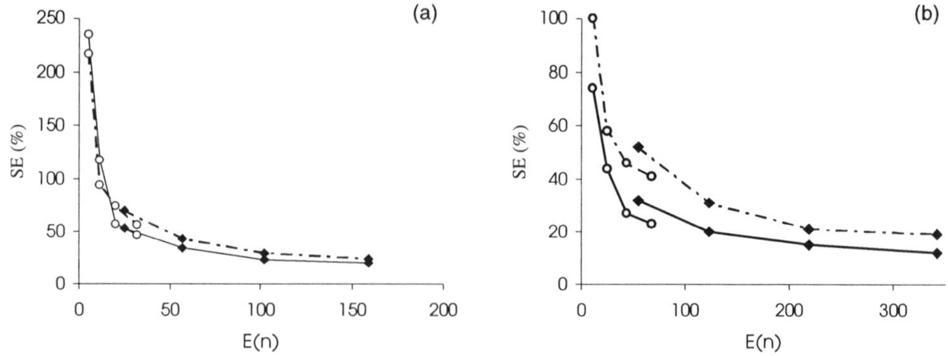

Figure 6. Standard errors (SE, in percent of true values) of the estimated number of objects depending on expected number of objects detected (E(n)) in the case of $p = 0.7$ (a) and $p = 1.5$ (b). The density was 5 (O) and 25 (◆) objects ha^{-1}, respectively. The objects were located according to a Poisson process (—) or located in clusters (– • –). Note the different scales on the y-axes.

second criteria is not violated to any large extent. In some cases, it ought to be appropriate to truncate the PoD function. In practice, this corresponds to the establishment of a maximum distance for which objects are recorded, although some objects are detected on longer distances (cf. Fig. 7).

Of the parameters in the PoD function, only parameter p was estimated in the simulations. The maximum distance of detection (w) was assumed to be known. In a real situation, this is certainly not the case.

To conclude, line transect sampling is a potentially useful method for surveying rare forest objects. The method, frequently used in some countries, is rarely used in Sweden. Therefore, it is of interest to study its applicability in Scandinavian boreal forests for surveying objects of interest in the setting for Swedish forestry of today.

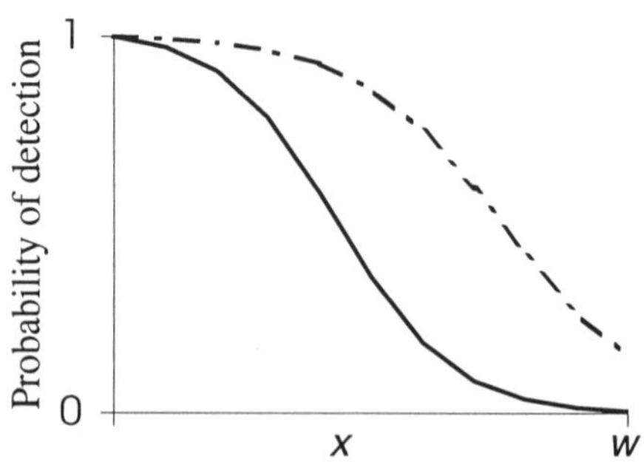

Figure 7. Examples of alternative probability-of-detection functions (reverse logistic functions) of which one is truncated. The perpendicular distance between an object and the inventory line is denoted x and w is the maximum distance of detection.

4.1 FURTHER RESEARCH

The accuracy of line transect sampling depends on how well the PoD function can be estimated. If the inventory lines in a survey transect different forest types with different sighting conditions, different PoD functions for the different forest types are likely. Although estimates can be obtained using an average function, the accuracy of the method ought to be improved if the forest types could be separated and separate functions estimated. Inventory lines and locations of detected objects can easily be georeferenced using GPS technique. In complex situations, classification of forest type and sighting conditions can then be aided by remotely sensed data. In forthcoming studies, the potential for improving the accuracy of line transect sampling by means of such auxiliary data will be investigated.

References

Burnham, K.P. and Anderson, D.R. 1976. Mathematical models for nonparametric inferences from line transect data. Biometrics 32(2): 325-336.

Burnham, K.P., Anderson, D.R. and Laake, J.L. 1980. Estimation of density from line transect sampling of biological populations. Wildl. Monographs 72.

Hayne, D.W. 1949. An examination of the strip census method for estimating animal populations. J. Wildl. Manage. 13(2): 145-157.

Lämås, T. 1996. Circular plots versus strips for assessing rare forest objects. In: Forest management planning for biodiversity and timber production. Swedish University of Agricultural Sciences. Department of Forest Resource Management and Geomatics. Report 3. Dissertation.

Matérn, B. 1984. Four lectures on forest biometry. Swedish University of Agricultural Sciences. Section of Forest Biometry. Report 23.

Ståhl, G. 1995. The transect relascope. An instrument for the quantification of coarse woody debris. Swedish University of Agricultural Sciences. Department of Forest Resource Management and Geomatics. Working Paper 4.

Ståhl, G. and Lämås, T. 1995. A simulator for sampling of rare forest objects. In: Hyttinen, P., Kähkönen, A. and Pelli, P. (eds.). Multiple use and environmental values in forest planning. Proceedings of the International Summer Course. June 5-10, 1995. Tohmajärvi. Finland. EFI Proceedings No. 4. European Forest Institute. Joensuu. Finland. Pp. 185-196.

MONITORING EUROPEAN FOREST BIODIVERSITY AT REGIONAL SCALES USING SATELLITE REMOTE SENSING

NIALL MCCORMICK AND STEN FOLVING

Space Applications Institute,
Environmental Mapping and Modelling Unit,
Commission of the European Communities,
Joint Research Centre, I-21020 Ispra (VA), Italy

Abstract

As part of the Space Applications Institute's FIRS (Forest Information from Remote Sensing) Project, a satellite-based methodology for assessing and monitoring forest biodiversity over wide areas, has been developed. The methodology, which is suitable for use with both optical and microwave satellite imagery, is based on the estimation of the three main components of forest biodiversity – i.e. composition, structure and development. Central to the methodology is a powerful technique for automatically delineating the natural spatial units of the landscape (including forest stands), by image segmentation. The actual forest biodiversity is estimated using the following three major, complementary indices: (a) Fractal Dimension; (b) Dominance Index; (c) Contagion Index. The methodology provides a cost-effective, objective system for obtaining regular, up to date information on the major environmental indicators which determine forest biodiversity, in a form which facilitates comparisons between different geographical areas and between different time periods.

1. Introduction

In view of the importance of forests in Europe, and the increasing pressures, both natural and anthropogenic, on their existence, the European Commission has recognised the need for new, more efficient environmental monitoring systems – particularly those using satellite remote sensing – which can provide accurate, up-to-date information on forests over large geographic areas and at regular time intervals. Such information is essential for the formulation of policies aimed at promoting the sustainable management of forests, and the conservation of their biological diversity or biodiversity.

An operational methodology for assessing and monitoring European forest biodiversity, using satellite remote sensing, has been developed as part of the FIRS (Forest Information from Remote Sensing) Project, at the JRC's Space Applications Institute. The methodology was developed primarily for implementation in the CEEC, where there are severe environmental problems, under the Commission's PHARE Regional Environment Programme.

P. Bachmann et al. (eds.), Assessment of Biodiversity for Improved Forest Planning, 283–289.
European Forest Institute Proceedings no. 18. © 1998 *Kluwer Academic Publishers. Printed in the Netherlands.*

In this paper, the main components of forest biodiversity are defined, the indicator variables of forest biodiversity which can be measured from satellite imagery are summarised, and the satellite-based methodology for assessing and monitoring forest biodiversity is outlined.

2. Components of Forest Biodiversity

For the purposes of this paper, forest biodiversity is considered as having three main components: composition; structure; development. The compositional component of forest biodiversity refers to the identity, distribution and relative proportions of the landscape elements (i.e. forest types) present in a forested area. The structural component of forest biodiversity refers to the spatial pattern (numbers, sizes, shapes, dispersion, etc.) of the forest blocks or patches present in a forested area. The development or temporal component of forest biodiversity refers to the changes (e.g. forest clearance, forest growth) which take place in the composition and structure of a forested area over time.

The indicator variables of forest compositional, structural and temporal biodiversity, which can measured from satellite image data, are summarised in Table 1. The key GIS-image processing operations for estimating these indicator variables, are also shown in Table 1.

TABLE l: Satellite-based indicators of forest biodiversity

Component of forest biodiversity	Corresponding indicator variables measurable from satellite imagery	Key GIS - image processing operation
Composition	ldentity, distribution and relative proportions of forest patches	Segment-based image classification
Structure	Spatial pattern (density, size, shape, dispersion, etc.) of forest patches	Spatial analysis of image classes
Development	Temporal changes (e.g. clearance, growth, phenology) in forest composition and structure	Change detection

3. Description of Methodology

An overview of the satellite-based methodology for assessing and monitoring forest structural diversity is presented in Figure 1. The methodology is illustrated for a hypothetical forested area of interest (e.g. a national forest park), and sample values have been assigned to the indicator variables. The primary inputs into the system are satellite images (e.g. Landsat TM, SPOT, RESURS, ERS) of the area of interest. The forest types which are shown in Figure 1 are classes which can be quite easily distinguished from satellite image data. Other forest classes can also be used. The main stages of the methodology are summarized as follows:

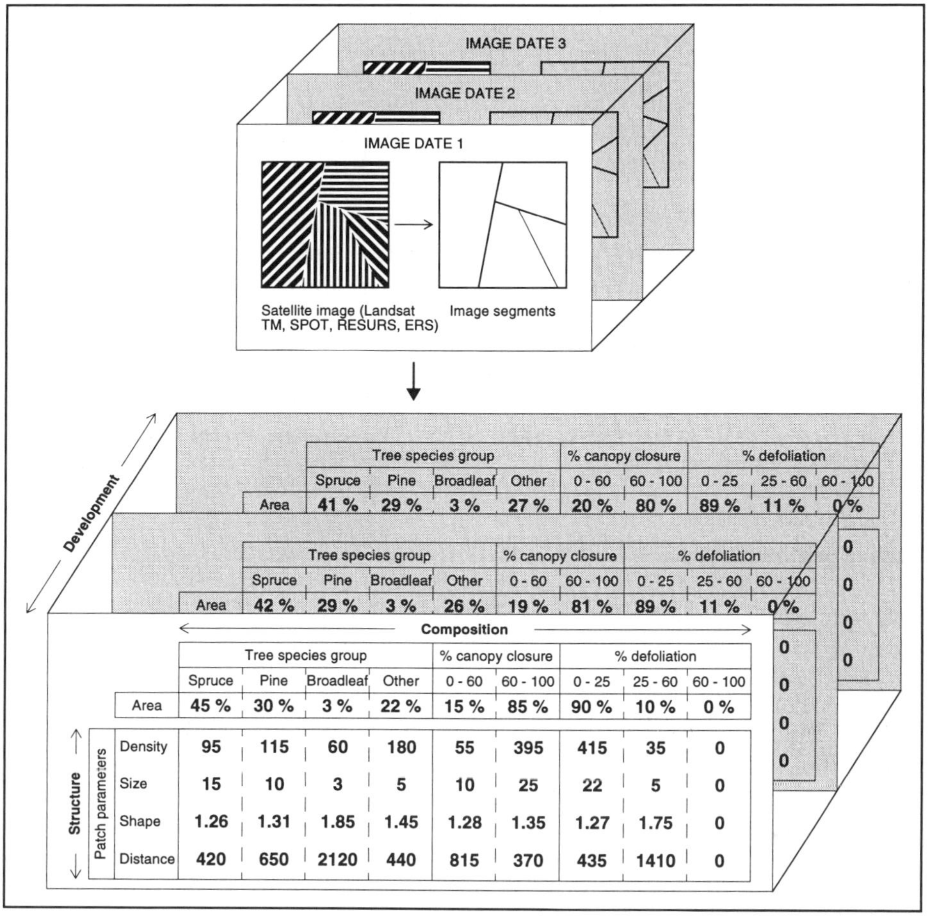

Figure 1. Overview of methodology for monitoring forest biodiversity

1. The forested area of interest is identified, and any existing administrative forest maps, aerial photographs, and ancillary geographic data are compiled and entered into a GIS. These will be used for calibrating the data processing and for testing the results.

2. As an image pre-processing stage, geometric and radiometric corrections are applied to the satellite imagery, in order to remove the effects of perspective displacements and varying illumination and atmospheric conditions (McCormick et al. 1995).

3. The natural spatial units of the landscape (including forest stands) are automatically delineated by edge-preserving smoothing and segmentation of the satellite imagery. This stage is described in greater detail in Section 4.

4. The forested area of interest is classified into its constituent cover types, using a segment-based method of image classification (McCormick 1996).

5. Forest structural diversity is assessed by computing the following three major, complementary indices of landscape structure: Fractal Dimension; Dominance Index; Contagion Index. These three indices are described in greater detail in Section 5.

6. The following basic spatial parameters are also computed for each forest class: area of cover type; patch density, patch size, patch shape, inter-patch distance. These are used to assist in the interpretation of variations in the three major landscape structural indices.

As part of the methodology, a computer software package for the pre-processing, smoothing, segmentation, and segment-based classification of satellite image data, has been developed. The software is called SILVICS (Satellite Image Land Vegetation Integrated Classification System).

4. Image Segmentation

Image segmentation refers to the automatic partitioning of a satellite image into disjoint, homogeneous regions. The objective of image segmentation is to extract spatial units (e.g. forest stands) which are defined by naturally occurring boundaries, rather than by the artificial constraints imposed by the satellite's spatial resolution.

The method for image segmentation which is used in the present methodology is based on the method of Narendra and Goldberg (1980). In order to optimise the segmentation results, the spatial information content of the imagery is first enhanced using a modified version of the edge-preserving smoothing method of Nagao and Matsuyama (1979). The image is then processed using standard directional edge operators, producing a typical edge image in which high grey-level values indicate points of discontinuity (i.e. edges). The actual segmentation is based on the principle of linking each pixel in the edge image to the neighbouring pixel with the minimum edge value. The linking of all pixels in the edge image produces data structures called "directed trees". Pixels belonging to the same directed tree form a segment.

Figure 2 illustrates the automatic delineation of forest stands using the segmentation method. On the left, conventional (i.e. ground-based) forest stand boundaries are overlaid on a 2.9x2.9 kilometres (841 hectares) Landsat TM image of central Ireland. On the right, the boundaries derived by image segmentation are overlaid on a smoothed version of the image. As can be seen, the segmentation produces a more accurate and detailed forest stand delineation, while including the major boundaries of conventional forest stand maps. Note that in the Landsat TM image in Figure 2, forests are various shades of dark-green and black, peatlands are purple and pink, and agricultural fields are light-green and yellow.

The major advantage of image segmentation is that it incorporates the valuable spatial information content of the satellite imagery into the subsequent image classification, thereby allowing the computation of the major indices of forest structural diversity.

Figure 2. Forest stand boundaries derived by conventional means (left) and by automatic segmentation of Landsat TM imagery (right)

5. Indices of Forest Structural Diversity

In landscape ecology, structure is defined as the distribution of energy, materials and species in relation to the sizes, shapes, numbers, kinds and configurations of landscape elements or ecosystems (Forman and Godron 1986). Forest structural diversity, therefore, refers to the spatial pattern (cover type, density, shape, shape, dispersion, etc.) of the blocks or "patches" present in a forested area.

The structural diversity of a forested landscape has profound effects on the ecology of the area. For example: certain wildlife habitats are frequently associated with particular tree species; the relative abundances of edge- and interior-dwelling flora and fauna are strongly affected by the degree of forest fragmentation; the interaction between forest patches is largely dependent on their connectedness.

A summary of the major indices which are commonly used to measure landscape structural diversity, is presented in Table 2. The meaning of the various indices, and the literature references of major studies in which the indices have been applied, are also indicated in Table 2.

As can be seen from Table 2, the three indices which have been selected for the present study are focused on three complementary, ecologically important aspects of forest structure. The Fractal Dimension describes the **complexity of shapes** within the landscape. The Dominance Index measures the extent to which one or a few land uses dominates the landscape, and so reflects the **diversity of cover types** in the area of interest. The Contagion Index quantifies the degree to which cover types are aggregated or clumped, and so reflects the **degree of fragmentation** of the area of interest.

TABLE 2: Commonly used indices of landscape structural diversity

Landscape structural parameter	Environmental characteristic measured	References of major application Studies
Areas of cover types; patch density; patch size; patch shape; inter-patch distance	Basic structural properties of the landscape	Benson and MacKenzie (1995); Bowen and Burgess (1981); Burgess and Sharpe (1981); Luque et ai. (1994); Ritters et ai (1995); Sader (1995); Turner (1990a); Turner (1990b)
Fractal Dimension	Complexity of shapes within the landscape	Benson and MacKenzie (1 995); Hunsaker et al. (1994); LaG ro (1991); Leduc et al. (1994); 0'Neill et al. (1988); Ritters et al. (1995); Turner (1990a); Turner (1 990b)
Dominance Index	Diversity of cover types within the landscape	Hunsaker et al. (1994); Luque et al. (1994); 0'Neili et al. (1988); Turner (1990a); Turner (1990b)
Contagion Index	Aggregation of cover types within the landscape (i.e. fragmentation)	Hunsaker et al. (1994); Luque et al. (1994); 0'Neili et al. (1988); Ritters et al. (1995); Turner (1990a); Turner (1990b)
Shannon Index	Diversity of cover types within the landscape	Luque et al. (1994); Mayaux and Lambin (1995); Ritters et al. (1995); Turner (1990b)
Matheron Index	Fragmentation of the landscape	European Commission (1995); Mayaux and Lambin (1995)
Juxtaposition	Numbers and relative proportions of edge types types within the landscape	Hunsaker et ai. (1994); Mead et ai. (1991); Turner (1990a); Turner (1990b)

6. Conclusions

A simple, operational system for assessing and monitoring European forest biodiversity using satellite remote sensing, has been described. The information obtained from such a system, when combined with existing data on environmental stress (e.g. air pollution levels, grazing intensities) can be used to identify areas at high risk of deterioration in forest biodiversity. These areas can then be subject to more detailed studies – for example, at species or genetic level. The monitoring system can also be used to establish relationships between specific indicators of forest biodiversity and actual forest processes, thus facilitating the selection of optimal indicators of forest biodiversity for further monitoring studies.

References

Baudry, J. and Merriam, H.G. 1988. Connectivity and connectedness: functional versus structural patterns in landscapes. In: Schreiber, K.F. (ed.). Connectivity in Landscape Ecology. Proceedings of the Second International Seminar of the International Association of Landscape Ecology. Pp. 23-28.

Benson, B.J. and MacKenzie, M.D. 1995. Effects of sensor spatial resolution on landscape structural parameters. Landscape Ecology. 10(2): 113-120.

Bowen, G.W. and Burgess, R.L 1981. A quantitative analysis of forest island pattern in selected Ohio landscapes. Oak Ridge National Laboratory. Oakridge. Tennessee. Publication ORNL / TM - 7759. 107 p.

Burgess, R.L. and Sharpe, D.M. (eds.) 1981. Forest Island Dynamics in Man-Dominated Landscapes. Springer-Verlag. New York. 311 p.

European Commission 1995. Study of Forest Non-Forest Interface Typology of Fragmentation of Tropical Forest. Catalogue. TREES Research Report 2. Report EUR 16291 EN. 77 p.

Forman, R.T.T. and Godron, M. 1986. Landscape Ecology. John Wiley and Sons. New York. 619 p.

Hunsaker, C.T., O'Neill, R.V., Jackson, B.L., Timmins, S.P., Levine, D.A. and Norton, D.J. 1994. Sampling to characterize landscape patterns. Landscape Ecology. 9(3): 207-226.

Janssens, P. and Gulinck, H. 1988. Connectivity, proximity and contiguity in the landscape interpretation of remote sensing data. In: Schreiber, K.F. (ed.). Connectivity in Landscape Ecology. Proceedings of the Second International Seminar of the International Association of Landscape Ecology. Pp. 43-47.

LaGro, J. 1991. Assessing patch shape in landscape mosaics. Photogrammetric Engineering and Remote Sensing. 57(3): 285-293.

Leduc, A., Prairie, Y.T. and Bergeron, Y. 1994. Fractal dimension estimates of a fragmented landscape: sources of variability. Landscape Ecology. 9(4): 279-286.

Luque, S.S., Lathrop, R.G. and Bognar, J.A. 1994. Temporal and spatial changes in an area of the New Jersey Pine Barrens landscape. Landscape Ecology. 9(4): 287-300.

Matheron, G. 1970. La théorie des variables régionalisées et ses applications. Les cahiers du Centre de Morphologie Mathématique de Fontainebleau. Fasc. 5.

Mayaux, P. and Lambin, E.F. 1995. Estimation of tropical forest area from coarse spatial resolution data: a two-step correction function for proportional errors due to spatial aggregation. Remote Sensing of the Environment. 53: 1-15.

McCormick, N. 1996. Automatic forest stand mapping using segment-based classification of Landsat TM satellite imagery. In: Proceedings of the International Workshop on Application of Remote Sensing in European Forest Monitoring. Vienna. Austria. 14-16 October 1996.

McCormick, N., Kennedy, P. and Folving, S. 1995. An integrated methodology for mapping European forest ecosystems using satellite remote sensing. In: EARSEL Journal on Advances in Remote Sensing. 4(3): 87-92.

Mead, R.A., Sharik, T.L., Prisley, S.P. and Heinen, J.T. 1981. A computerized spatial analysis system for assessing wildlife habitat from vegetation maps. Canadian Journal of Remote Sensing. 7(1): 34-40.

Nagao, M. and Matsuyama, T. 1979. Edge-preserving smoothing. Computer Graphics and Image processing. 9: 394-407.

Narendra, P.M. and Goldberg, M. 1980. Image segmentation with directed trees. IEEE Transactions on Pattern Analysis and Machine Intelligence. PAMI-2(2): 185-191.

O'Neill, R.V., Krummel, J.R., Gardner, R.H., Sugihara, G., Jackson, B., De Angelis, D.L., Milne, B.T., Turner, M.G., Zygmunt, B., Christensen, S.W., Dale, V.H. and Graham, R.L. 1988. Indices of landscape pattern. Landscape Ecocolgy. 1(3): 153-162.

Ritters, K.H., O'Neill, R.V., Hunsaker, C.T., Wickham, J.D., Yankee, D.H., Timmins, S.P., Jones, K.B. and Jackson, B.L. 1995. A factor analysis of landscape patterns and structure metrics. Landscape Ecology. 10(1): 23-29.

Sader, S.A. 1995. Spatial characteristics of forest clearing and vegetation regrowth as detected by Landsat Thematic Mapper imagery. Photogrammetric Engineering and Remote Sensing. 61(9): 1145-1151.

Shannon, C.E. and Weaver, W. 1962. The Mathematical Theory of Communication. University of Illinois Press. Urbana.

Turner, M.G. 1990a. Landscape changes in nine rural counties in Georgia. Photogrammetric Engineering and Remote Sensing. 56(3): 379-386.

Turner, M.G. 1990b. Spatial and temporal analysis of landscape patterns. Landscape Ecology. 4(1): 21-30.

FOREST CLEARINGS, MARGINS AND MIXED AREA SURVEY FOR DIVERSITY ASSESSMENT IN THE LIGURIAN INVENTORY (NORTHERN ITALY)

M. MARCHETTI[1], F. CAMPAIOLA[2], G. LOZUPONE[2] AND V. TOSI[3].
[1] *Eurimage, Via Edoardo D'Onofrio 212, 00155 Rome, Italy*
[2] *Italeco, consultants, Via Carlo Pesenti 109, 00156 Rome, Italy*
[3]*ISAFA, P.zza Nicolini, Trento, Italy*

Abstract

The growing demand for a more careful use of natural resources implies a multifunctional approach to the forest resource inventory. Therefore, different methodological ways have been developed to the study of diversity and sustainable management of forest landscapes. Many new attributes are studied in forest researches to survey ecological, structural, wildlife and scenic parameters adapted to the description of habitats and to the mensuration of forest biodiversity . In Italy, a recent pilot project has been performed to inventory forest areas at regional level with a multipurpose approach. In this context the chosen attributes concerning habitat and landscape are clearings and margins, with their definition, size, shape, lenght, etc.; the choice will permit to correlate field measurement with spatial arrangements of patches using GIS. Furthermore, the specific landscape of Mediterranean areas in Italy, typically mosaic-based, led also to the survey of mixed agroforestry areas with appropriate measures. The present paper deals with the first results of data processing from the field measurements and the comments of a historical multitemporal analysis, to investigate changes in the rural landscape.

1. Introduction

Forest inventories have been developed from surveys of the wooded surfaces to get forest resource information in terms of surface area, mass, growth, population breakdown and the characteristics of the different formations. Reading and processing the collected data made it possuble to create general instructions for the use and management of the forest heritage, thereby making the inventory an instrument for the control and planning of forest resources.

In the light of these objectives, various forest inventories were made in Italy in the course of the 1980s, including the first National Forest Inventory (I.F.N.I.). Its methodology was developed by the Research Institute for Forest and Rangelands Management (I.S.A.F.A.), and, as in the regional forest inventories, it varies mostly in function of the territorial dimension of the inventory investigation, without showing substantial differences in the level of the objectives.

P. Bachmann et al. (eds.), Assessment of Biodiversity for Improved Forest Planning, 291–300.
European Forest Institute Proceedings no. 18. © 1998 *Kluwer Academic Publishers. Printed in the Netherlands.*

The growing demand for a more careful use of the environment and its natural resources throughout the world as well as in Italy (Bianchi and Tosi 1991) created a need to extend and unify the survey methods in order to achieve a multi-functional vision of the forest system. Inventories were to include a greater level of detail; without neglecting timber estimates, it was necessary to gather information about the entire biomass, with a more explicitly ecological viewpoint in the evaluation of animal habitats and biodiversity.

With these presuppositions, in 1993, a multi-resource forest inventory (I.F.M.R.) was projected and completed for the region of Liguria, the first example in Italy of a multi-objective approach to the management of forest resources in which the wood system is represented not only for coenosis in which the tree component is prevalent, but through the collection of data in natural and semi-natural systems that are in some way connected: shrubland, pastures, meadows and mixed and fragmented areas (Marchetti et al. 1993; Lalle and Marchetti 1994). This inventory was carried out as part of the project "Forest Map of Italy at scale 1:50,000 and Forest GIS", for the Ministry of Agriculture with the aim of evaluating the elements that interact with the forest system in its widest definition. This was done, because the increase and diversification in the demands concerning forests calls for their multiple use. It is not possible to do such planning without a multi-functional analysis and a thorough collection of data. The primary objective of the project was to create a map of the actual vegetation and forest coverage which will provide a basis for an economic and ecological evaluation of the forests.

To create the map covering all the land of the Liguria region, a vast amount of information was used from various data sources: Landsat images, Color Infrared Photos (CIR) at a scale of 1:33,000, Digital Terrain Models (DTM) at 30 meters, and databases from the multi-resource inventory.

The map is the product of processing carried out by an "expert system" in which the information deriving from the construction of density classes was later implemented. The result is organised according to a legend which breaks down the nomenclature in different hierarchical levels: the dynamic series of vegetation, the structural types described in the inventory with an innovative description, and the typological aspects in their physiognomic forms (Del Favero 1992; Marchetti 1994). The main features of the constructed nomenclature is thus the search for integration among the floristic-ecological, dynamic-evolutive and technical-forestal aspects (Mondino 1992; Marchetti 1994).

The core of the Forest Information System is the multi-resource inventory (IFMR). It integrates vegetational and structural types (Marchetti et al. 1993; Bianchi and Tosi 1991), with the goal of providing data in terms of vegetation, silvicultural-ecology, production and habitat capacities. The sample design called for the distribution of permanent sample plots according to a regular geometrical scheme (Lund et al. 1992) localised to the UTM grid. Two overlapping grids were created with a trikilometric cadence for detailed studies that include pedologic and dendroecologic surveys and with a kilometric cadence for the investigation of the main attributes. The trikilometric scale is the one used in the First National Forest Inventory (MAF-ISAFA 1988). This allows the use of the data for both historical comparison and for comparison of the differing approaches of various regions.

To begin with, the typology of all the vegetation in the territory was defined, using the floristic, phytosociologic and structural surveys and their technical analysis through multi-varied analysis techniques and the use of ancillary data, which was intentionally collected (environmental evidence relating to the soil, climate and geomorphology).

Successively the typologies defined were submitted to verification of discriminability on the available spectral base of space remote sensing, namely Landsat TM data.

The heterogenous data types in the IFMR include both precise quantitative information (sample area survey forms) and spatial data (digital images) as well as qualitative information, including that representing the experience of forest experts. This motivated the creation of a model to simulate the logical deductive processes used in the interpretation of a forest area.

2. Sampling designs and measurements in the permanent plots

The sample design of the IFMR in Liguria took into consideration the creation of 3,064 permanent plots located in a grid that coincides with the UTM ED-50. The 3 km scale is the one used in the National Forest Inventory (IFNI) in 1985, for which data were collected ten years later for 416 sample plots,. The investigated characteristics are described in the Tables 1 and 2.

The same plot configuration has been adopted for all the strata of the inventory. This simplified the work in the survey phase and made the data homogenous in the elaboration phase. The single elements were sampled inside the sample plot (600 m^2), with an intensity proportional to the respective variability; the number of samples studied are listed according to the different components of the survey:

Station Description - 3064; Stand Description - 3060; Floristic Census - 2873; Deadwood Survey - 2668; Height and Diameter Surveys - 2681; Shrub Pattern - 2651; Phytosociologic Survey - 958; Mixed Area - 464; Soil Survey - 416.

TABLE 1. General attributes

Description of the station			
Elevation	Slope	Aspect	Locaation
Soil type	Humus	Unevenness of soil morphology	Stoniness
Rock outcrops	Soil erosion	Accessibility	Infrastructures
Facilities	Function of the forest area	Ownership	Land use restrictions
Clearings	Margins	Superficial water	Wildlife
Stand description			
Type of stand	Stratified stand structure	Tree species composition	Stand origin
Silvicultural system and stand development	Tending of stand	Age of stand	Stand degradation
Wood production			
Diameter at breast height	Tree height	Bole height	Crown breadth
Stem quality			
Vegetation			
Floristic composition	Regeneration	Shrub pattern	
Deadwood			
Dead fuel survey	Litter	Fuel models	

TABLE 2. Detailed attributes

Phytosociological survey		
Soil survey		
Soil description	Soil Depth	Texture of soil and presence of Concretion stones
Cutans	Speckling	Presence of roots
Linear transect of mixed areas		

Usually in the forest context the nomenclature system was of a traditional kind, not suitable for the definition of non-canonic coenosis from a silvicultural aspect. Instead, in the botanical field the trend was that of producing nomenclature systems linked at least to physionomic type only on a floristic basis and without structural data, or with a particular syntaxonomic phytosociological approach. Now, due to the emergence of multi-functionality and intersectoriality of the natural and semi-natural vegetation, nomenclature systems will need to be organised in integrated forms across different subjects to meet the requirements of the various instruments and methodologies and, of these, particularly the need to use the typologies on a large scale. Current experiences also give particular attention to the hierarchic structure and collaboration among Botanists, Forest Engineers and Remote Sensing and Mapping specialists.

The hierarchic structure of the adapted Nomenclature allows flexibility of movement towards higher hierarchic levels for the classes that are difficult to define with the present technologies. Following multipurpose aims, the land use studied classes are forests and woodlands, bushlands, rangelands and mixed areas. This last class is used for the Italian territorial characteristics to assess the alternation of natural and semi-natural environments within different ecosystems (grasslands, woods) and with different utilisation for agriculture. Definitions of assessed land cover classes are (the minimum size of a forest area is 5,000 m^2):

1) Trees: tree coverage > 20% in polygons with surface >= 5,000 m^2
2) Bush: bush coverage > 40% with tree coverage < 20% in polygons with surface >= 5,000 m^2
3) Grass: permanent meadows, pastures, fallow with grass coverage > 20 tree coverage < 20% and bush coverage < 40% in polygons with surface >= 5,000 m^2
4) Mixed areas: a mosaic of classes 1/2/3 in small areas (< 5,000 m^2) in polygons with surface not less than 10,000m^2
5) Barren land: areas with grass coverage < 20%, tree coverage < 20% and bush coverage < 40% in polygons with surface > 5,000 m^2
6) Included areas: surfaces with an homogenous internal land use but not homogenous as regards the surrounding land use with an area < 5,000 m^2 and >500 m^2 for classes 1/2/3 and 40,000 m^2 for barren or arable land.

3. The survey of margins and clearings

The stational description consists of the measurement and registration of all elements which, while fully part of the bioecological studies, were not the object of specific sampling methods. For example, the census included the presence of fauna by means of the study of tracks, shelters, nests and residue that characterise the various species and the recording of the clearings, margins and bodies of water. In the area surrounding the sample unit extending to 5000 m^2 (that is: within a radius of about 40m), the presence in the landscape of the elements and microhabitats (clearings, margins and surface waters) which diversify the environment were surveyed, and were evaluated not according to the inventorial typology but rather by their ecological features.

By clearings are meant areas primarily covered by grasses surrounded by forest formations (trees and shrubs); these are classified according to the surface (Table 3 and 4). Margins are microhabitats in which the typical species of the neighbouring habitat cohabit with the ecotonal species that prefer the contact zones between two environments (Table 3). For the high environmental value and significance in increasing the biodiversity of these microhabitats, the contact components, expressed by the level of land use and the form of contact among elements, were surveyed indicating the type of limits in the passage between the two elements and the evaluation of the length of the margin. In addition, various types of present bodies of water were described (Classes: Temporary superficial, Permanent superficial , Source, Rill, Stream, Pond, Peatland).

Clearings were present in 581 sample units (Table 4), which represent 19% of the total. The majority of these clearings were small with surface areas less than 1,000 m^2, originating in accidental events like fallen trees or artificial events like openings for coal production sites. Of interest is the scarcity of large clearings, with surface area greater than 5,000 m^2 which denotes a certain continuity and homogeneity in the articulation of the land use. The reduced presence of this type of clearing has a negative impact on the biological diversity of the forest stands, in that 5,000 m^2 represents the minimum level for the use of the area by medium and large sized wild fauna.

TABLE 3. Nomenclature and definition of clearing and margin

Attribute	Definition	Measurement scale or classes and threshold values
Microhabitat - clearing: extension	Evaluation of the clearing extension	9 classes, from S < 1000 m^2; to 7 classes, each 2500 m^2 to maximum extension of 20000 m^2
Microhabitat - margins: extension	Evaluation of the margin length	m, recorded in 1-m classes, not more 99 m
Microhabitat - margins: type	Synthetic assessment of the closest different land use	7 classes: a) forest b) shrub land c) grass land d) arable land e) water bodies f) rocks g) clearing
Microhabitat - margins: limits	Description of the shape of margin	3 classes: a) clear-cut b) gradual c) not evident
Microhabitat -: superficial water:	Assessed the presence	7 classes

TABLE 4. Clearings

Classes	N° of Plots	%
$S < 1{,}000\ m^2$	357	11.67
$1{,}000\ m^2 < S < 2{,}500\ m^2$	113	3.69
$2{,}500\ m^2 < S < 5{,}000\ m^2$	58	1.90
$5{,}000\ m^2 < S < 7{,}500\ m^2$	23	0.75
$7{,}500\ m^2 < S < 10{,}000\ m^2$	10	0.33
$10{,}000 < S < 12{,}500\ m^2$	6	0.20
$12{,}500 < S < 15{,}000\ m^2$	4	0.13
$15{,}000 < S < 17{,}500\ m^2$	1	0.03
$17{,}500 < S < 20{,}000\ m^2$	9	0.29
Total	581	18.99

The achieved results are hereafter discussed. The margins are present in more than 35% of the sample units with a clear prevalence of one of the two components of the wood types. The woods in effect represent some 70% of the inventoried types and make up a sort of base matrix of the system in which other elements are inserted. Thus, even if from the analyses of data surveyed in the clearings an elevated homogeneity in the structure appears, through the analysis of the margins a diametrically opposed picture emerges. In effect, the clearings are a particular microhabitat in which the prevailing vegetal component is made up of grass elements. The margins, on the other hand, involve all of the multiple physionomical aspects present in natural and semi-natural areas, thus representing a more reliable index of the real articulation of the forest stands in the landscape. The level of fragmentation of the forest area is thus high, constituting a serious element for reflection on the efficiency of the system in relation to the various explicit functions (positive or negative? Catastrophic, according to Schirone et al. (1992), or excellent, according to green interests, for the fauna?).

With respect to the type of limits, a prevalence of clear limits appears which prefigures its artificiality and thus indicates low intrinsic stability. The more frequent contact with this typology of limits is that between tree and grass formations. It presents a maximum relative to 135 sample units, corresponding to the classes of extension greater than 50 m^2. The limit of the gradual type was sampled in around 10% of the sample areas with the maximum frequency for the contact between tree and shrub formations. This type of margin is rather interesting in the way it corresponds to the model of the frontal advancement of the forest, in which the expansion results through the formation of mantles along the perimeter of the forest. Two other models of advancement are by dispersion, in which the terrain is colonised by tree species or pre-forest species in a random way, and that of nucleozation, in which some shrub formations become sites of refuge for other species of trees forming nucleous which with time tend to reunite and augment the structural complexity: These are characteristics of margins with diffused limits. These limits are most frequent in correspondence with the contact between tree and shrub formations with 87 sample areas. By summarising gradual and diffused limits for contacts between the natural formations (tree, shrub and grasses), it is possible to quantify the evolutive phenomena present in Liguria. Three hundred and sixty-six sample units are concerned which is equal to 12% of the total in terms of surface, corresponding to 54,600 ha (total regional surface area investigated 454,700 ha). The data are very significant, considering that Liguria is the region in Italy with the highest forest index with a value of 69.1% (IFNI 1985; MAF 1988).

4. The survey in mixed areas

The inclusion of this particular inventory stratum – which is moreover very important in the analysis of possible future scenarios – led to the creation of a surveying method capable not only of describing the different elements of which it consisted, but also of supplying a quantitative indication on the relationships that elapsed.

The survey consists of two linear transects from the central point of the sample plot to the limits of the mixed area and therefore no longer than 200 m. The two transects are oriented orthogonally northward and eastward. For each intersected element, the length relative to the portion of the transect on which it was surveyed and the typology (Table 5.) with the indication of the present species were recorded. The elements of the mixed areas were described according to the inventory typology for the tree and grass formations with a surface area less than 5,000 m^2.

This category shows the extreme fragmentation of the surface areas in relation to the land use. More than 10% of the entire land presents these features, a high figure in that in reality these areas are located in restricted ambits, in particular where the intervention of man in the past has been more intense. The progressive abandonment of cultivations in the most disadvantaged areas has enabled the slow but inexorable affirmation of the natural and semi-natural areas. The consistency of this phenomenon poses management problems, especially in the delicate intermediate phase between the widespread continuous human intervention in the protection of soil from erosion and the establishment of effective natural forest stands.

On the total of 468 plots (15.2%), the prevalence of mixed agro-forestry areas (287 - 9.3%) is a clear cut confirmation of the importance of the agricultural element in the make up of this category. Instead, the make up of the natural or semi-natural areas (93 - 3%) with those that present urban ambits (88 - 2.9) is equal. Their behaviour is certainly different. While in fact it is possible to predict that areas mixed with urban ambits which are very fragmented, slowly advance towards a natural typology thanks to the continuous presence of the disturbance factor of human interaction, for the areas that are already formed by natural or semi-natural elements, the progression towards arboreal formations seems assured and rapid.

TABLE 5. Nomenclature of mixed areas, with land use components

Types of mixed areas	Description
Urban ambits	Residential or industrial, infrastructures, leisure and green areas
Surface water	see bodies of water
Crop areas	Cereals, vineyards, oliveyards
Minimum areal elements with prevalence of trees	Lone trees, nucleous, groups, small woods
Minimum linear elements with tree prevalence	Woody shelters, fences, hedges, windbreakers, etc...
Minimum linear elements with shrub prevalence	Shruby and bushy linear formations

5. Historical multitemporal analysis

In Italy, and in particular in regions like Liguria, the different land uses, the vegetation features and their distribution are above all a result of the intensive use and alteration of natural resources by man over time. This analysis is based on the quantification of the changes that have been verified over a consistent interval of time, in two different temporal periods with a relatively long interval (around 40 years) which is very significant for the variation of the relationships between society and the use of productive resources in the light of today's economic and social situation, and permits predictions on the evolution of these lands. In this sense, the investigation is configured also as an indispensable surveying and preparatory tool for territorial planning and environmental management. The goal pursued in the current work was the collection of information on the modifications in land use, quantifying the changes observed through, in particular, analyses of the forest coenosis and identifying the processes that took place and that are currently taking place. This allows, among other things, the calculation of the probability of the transition by each land use to the others present. The historical analysis also permits taking into account the processes of fragmentation and the recomposition of the forest stands that until several decades ago were less frequent. They are now rapidly increasing and for these it is important to study the arrangement of the phenomenon in the cadre of the system's tendency towards equilibrium, in itself dynamic.

The comparison was carried out between the present situation, characterised by field campaigns during the IFMR, and the situation in the year 1954, as expressed by the available aerial photography coverage of the Military Geographic Institute (flight GAI 1954, average scale 1:35,000/40,000). The investigation took place using the method of counting photo sample points in correspondence to the intersections of the UTM ED 50 kilometric grid upon which the IFMR surveys were created (Marchetti and Gusmeroli 1994). The first phase of the work consisted in verifying the situation in the year 1954, going to characterise on the photos the inventorial point and establishing the land use category. Because of the quality of the photographs, it was often difficult to recognise some of the structural features and the density of the trees.

Throughout the region, the results emphasise the evolutive tendency for each category of the land use. The most interesting piece of information, on a regional level, concerns the enlarging of forest surface area: from 1954 to today, it has increased by 15.24% while in general, the other categories have diminished in various proportions; it is opportune to signal to date the notable reduction in the category of mixed areas (-10.36%).

For a better understanding of the dynamic relationships among the different formations and a correct use of the collected information, it is important to verify what the origins of the variation are around each class in percentile terms. By analysing in greater detail the net fluctuations (that is to say the difference for each category of land use, between the points that have come from that category and those which instead have been passed to another during the period of the investigation), between the categories of land use most involved in the forest dynamic, that is those which most easily and rapidly have the possibility to evolve into forests, at a regional level, we can conclude: the wood formations are increasing steadily (84% of the formations remained unchanged since 1954), and the increased supply of the new forests is given by the shrub types (9.05% of the current Tree Formations) and from the Mixed Areas (9.77%) followed by Pastures (5.02%) and in lesser quantities by the Barren lands (1.09%), together with the Uncultivated areas (0.93%). As concerns the Shrub Formations, a slight instability is evident (only 19.27% of the formations remained unchanged since 1954) and the Pastures primarily contribute to their reconstruction (13.59% of the current Shrub Formations), followed by Uncultivated lands (3.56%), the Barren lands (3.56%) and in very limited measures, the Mixed Areas (0.65%). The pastures are characterised by a low average stability (34.80%) a and originate primarily in the Mixed Areas (11.56%) and to a lesser degree in the Uncultivated areas (2.01%).

In conclusion, it is opportune to emphasise the importance of the Mixed Areas in the dynamic of the formations studied in detail (Tree, Shrub and Meadow formations). These surfaces in fact are more or less large in territory and may represent the clearest example of mutation and of vegetal succession, since subjected to intense and active recolonization processes on the work of the natural formations. The changes which took place in the relationship between human components and activities in Liguria also produced notable mutations in the forest Landscape. The analysis undertaken supplies very interesting information, even if it could be important to have intermediate temporal data preceding the period of the recent and probably irreversible agricultural crisis. In fact, during the last ten years, the population active in agriculture has decreased considerably. The main reason for this phenomenon is the decrease in the economic basis of various activities (terrace cultivation, chestnut fruit, pastures, the use of woods for coal and firewood), both because these demanded an intensive use of labour and because the demand for certain products decreased dramatically.

The land use and the structure of the landscape thus were strongly controlled by agriculture and traditional forestry; the current phenomenon of separating "agriculture" and "nature" with respect to their previously strict integration, promises on the one hand a mid-term amplification of forest and natural areas and an improvement in the most valuable components under the natural profile, and on the other hand, a progressive banalization of the landscape ecology (that is of the configuration on the territorial unit level), and also an increase in conditions predisposing critical events (for example, fires).

From a management point of view, with the studies related to the dynamic of the succession and the structure of the populations, it will also be possible to deduce "norms" for management which permit, with the help of the tendency to equilibrium of the successive states, to maintain a good level of "biodiversity" and an improved diversification of the detailed structure of the landscape, also in this way maintaining the cultural characteristics of these environments.

References

Bianchi, M. and Tosi, V. 1991. L'inventario forestale nel rilievo globale degli ambienti naturali e seminaturali. Proceedings UNIF on Forest Management. November 1991.

Bianchi, M. and Tosi, V. 1991. Il rilevo degli ambienti naturali e seminaturali negli inventari forestali a vasta scala. Copertura Forestale e Territorio. Milano. Pp. 155-195.

Delfavero, R. 1992. Un'esperienza di studio di tipologia forestale. Annali Accademia Italiana di Scienze Forestali. 41: 65-84.

Lalle, A. and Marchetti, M. 1994. Integrazione di tecniche di telerilevamento nella realizzazione della Carta Forestale d'Italia. Proceedings A.I.T. on Il telerilevamento per lo studio e la pianificazione forestal. Brixen (BZ). 1994.

Lund, H.G., Päivinen, R. and Thammincha, S. 1992. A primer on permanent plots for monitoring natural resources. Proceedings IUFRO S 4.02.05. Pattaya. Thailand

MAF-ISAFA 1988. Inventario forestale nazionale 1985. Sintesi metodologica e risultati – Rappresentazioni cartografiche. Trento. 464 + 14 p.

Marchetti, M., Raggi, L. and Garavoglia, S. 1993. Metodi di rilievo per un inventario multiobiettivo dei boschi e di altre aree naturali e seminaturali. Proceedings UNIF: Ricerca ed esperienze nella pianificazione multifunzionale del bosco. 23-24 Novembre 1993. Centro Ricerche ENEA Brasimone (BO). Pp. 136-155

Marchetti, M. 1994. Geo-referenced Forest Information for Italy. In: Kennedy, P.J., Päivinen, R. and Roihuvuo, L. Designing a System of Nomenclature for European Forest Mapping. Proceedings of the International Workshop. 13-15 June 1994. Joensuu. Finland. Office for Official Publications of the European Community. Luxemburg. 458 p.

Marchetti M. and Gusmeroli, E. 1994. Evoluzione del paesaggio forestale dell'Appennino laziale: analisi storica e modificazione in corso. Studi sul territorio. Annali di Botanica dell'Università degli Studi La Sapienza. Roma 1994.

Mondino, G.P. 1992. La vegetazione forestale del Piemonte. Materiali per una tipologia forestale regionale. Annali Accademia Italiana di Scienze Forestali. 41: 85-137

Piovesan, G. and Marchetti, M. 1993. A physionomic-structural analysis for the study of forest stand dynamics. Annali di Botanica. LI: 245-249.

Schirone, B., Piovesan, G., Spada, F. and Fantucci, R. 1992. La frammentazione boschiva nell'Appennino: basi teoriche e criteri d'intervento. Tecniche di rinaturalizzazione e di ingegneria naturalistica. Lignano Sabbiadoro (UD). 21-23 Maggio 1992. Ed. Patron. Bologna.

Schirone, B., Piovesan, G., Borelli, S., Marchetti, M., Lozupone, G., and Daquino, C. 1993. La stima del grado di frammentazione di una cenosi forestale. Linea Ecologica. 2: 14-16.

Vos, W.E. and Rossi, R. 1992. Un paesaggio dell'Appennino che si sta trasformando: la valle del Solano, nel Casentino. Linea Ecologica. 6: 39-44.

TOWARDS A LOGIC FOR ASSESSING BIODIVERSITY IN BOREAL FOREST

PER K. ANGELSTAM
Swedish University of Agricultural Sciences, Forest faculty, Department of Conservation Biology, Grimsö Wildlife Research Station, S-730 91 Riddarhyttan, Sweden

Abstract

To maintain forest biodiversity, and if necessary to restore it, quantitative goals should be specified for all properties of a given landscape or region. Here an idea is presented on how biodiversity can be assessed in the boreal forest, which has a short history of transformation compared with other European forest types, and how quantitative nature conservation goals can be formulated in different geographical scales. Step 1 is to develop a book-keeping system for biodiversity. The starting point is the identification of the different disturbance regimes found in a natural boreal forest landscape. The links between site type and disturbance regime are critical because they shape the composition and structure of the forests, as well as many important processes, to which forest species have adapted. Step 2 is the development of practical methods to measure biodiversity in a landscape. The first task is to define the different ecological properties that can be found in a naturally dynamic boreal forest landscape as well as to identify species that are unique to each property. If it is possible to translate the species' habitat requirements into criteria based on their environmental demands, these criteria can be used as opposed to the more costly inventories of certain indicator species in the field. It is important that any system and methods are validated in areas with original biodiversity. In step 3, when sufficient knowledge is available on how the amount of each property affects the viability of its indicator specie's populations, strategic and operative goals can be formulated for each property representing the composition and structure, in different geographical scales, of naturally dynamic boreal landscapes. To implement this logic in practical management will alleviate the introduction of proactive management as well as monitoring of whether or not quantifiable criteria and indicator species change in the desired direction. The practical implementation is beyond the sphere of research but must be considered when developing methods to assess forest biodiversity.

1. Introduction

Anthropogenic alteration, fragmentation and loss of natural habitats are the causes of the decline in biological diversity on Earth (Dudley et al. 1996; Stanners and Bourdeau 1995; Heywood 1995; DeGraaf and Miller 1996). This is not only a fundamental nature conservation problem, but may ultimately affect also the long-term sustainable use of

P. Bachmann et al. (eds.), Assessment of Biodiversity for Improved Forest Planning, 301–313.
European Forest Institute Proceedings no. 18. © 1998 *Kluwer Academic Publishers. Printed in the Netherlands.*

renewable resources required by a growing world population (Apps et al. 1995; Dudley et al. 1996; Korpilahti 1996; Larsen 1996).

Assessment and management of biodiversity requires that its different components are defined as clearly as possible, and that the factors underlying the loss can be identified. Harper and Hawksworth (1995) stressed that the full definition of biodiversity must be kept in mind, and proposed the adjectives genetic, organismal and ecological to describe the three components of biodiversity. Moreover, a number of important processes which affect forest composition and structure (e.g. fire, flooding, browsing, fungal and insect infestation) have to be considered (Noss 1990; Angelstam 1992 and 1997).

To avoid the local and regional extinction of species, and unbalanced processes (e.g. too little fire, too much browsing on certain tree species), in landscapes which have long land use history requires active planning and management (Angelstam and Pettersson 1996 and 1997). It is therefore important that the environment can be assessed, and the information expressed in a way that it can be used in the regular forest management planning, such as the composition and structure of the trees and the stands that make up the forest landscape (Noss 1990, Angelstam 1997).

In this paper I outline a logic for designing a book-keeping system for the ecological and the organismal biodiversity as well as some important processes which are relevant in boreal forest in northern Europe. This forest ecosystem has a short history of anthropogenic transformation compared with central Europe where the original forest biodiversity has been severely reduced over large areas (Angelstam 1996a; Stanners and Bourdeau 1995; Mikusinski and Angelstam in press). The probability of successful maintenance of the naturally occurring species in boreal forest is still high, provided that the long-term negative habitat trends in very old forest, dead wood and old deciduous trees is broken (Kouki 1994). After a brief review of the present biodiversity problems in boreal forest, I define some requirements for the development of such a book-keeping system. I then describe the natural European boreal forest disturbance dynamics which form the main selection pressures affecting plant, animal and fungal populations. Finally, I define the main boreal forest properties, how they can be assessed and how the critical amount of these properties can be determined using studies of indicator species.

2. Biodiversity problems in boreal forest

Extinction is a process rather than an event. Today this process is initiated by rapid human-induced environmental change (as through habitat loss). Later, random factors associated with low population size, skewed sex ratios, low genetic variability and environmentally extreme conditions begin to have critical impacts on population viability and distribution. Different factors then interact in a mechanism of feedback loops resulting in a downward population spiral to extinction.

Even if the boreal forest is still relatively unexploited compared with many others of the world's ecosystems, there are problems in some parts of this biome. It is therefore reasonable that each region must take its own responsibility to maintain the local and regional biodiversity. Compared with the pristine conditions, changes in land use and forest management in the western part of the boreal forest belt have altered the composition and structure of both stands and landscapes (Kouki 1994; Angelstam et al. 1995; Angelstam and Majewski 1996; Angelstam 1996a; Mikusinski and Angelstam in press). Even, as in Norway, Sweden and Finland, the forest cover in the landscape, and wood volume in stands, is similar to, or even sometimes higher than under natural conditions, the distribution and amount of different properties and the function of important processes have been severely altered (Gamlin 1988; Esseen et al. 1992; Kouki 1994; Hansson 1997). Many of the properties which are lacking in the managed forest are typical for recently burned wood or ground, older successional stages after forest fire, and for no-fire refugia with gap-phase dynamic, a long continuity in forest cover and large amounts of dead wood, all of which are found in naturally dynamic boreal forests. The strong reduction of such properties have made forest plant and animal populations decline (Järvinen et al. 1977; Helle and Järvinen 1986; Rassi and Väisänen 1987; Virkkala 1987; Berg et al. 1994; Esseen et al. 1992; Hansson 1997).

The changes in composition and structure between the naturally dynamic and the managed forest occur at all geographical scales, ranging from the tree and stand, to the landscape and the region. Forest fragments may have become restricted in size and surrounded by a modified version of the original, thus being often of little use to the forest specialist species concerned because the fragments are too far apart for successful movement of individuals between fragments. Moreover, within forest fragments, the age class or tree species distributions may have been modified by management so that forest stands have become too simplistic, and hence unsuitable even to habitat specialists with small area requirements. Hence, a diversity of properties (e.g. very old trees, dead wood in different stages of decay, and old forest stands with a long continuity of forest cover) have to be maintained to provide for the biological diversity (Hunter 1990). Preservation and restoration of biodiversity therefore requires that several geographical scales are considered, such as the point- (tree level), alpha- (stand level) and gamma-diversity scales (landscape level) (Whittaker 1977; MacArthur 1965; Wiens 1976).

3. Natural disturbance as a basis for biodiversity book-keeping

The large number of species in any ecosystem may make it prohibitively difficult to create a biodiversity assessment system. However, the task appears feasible if one assumes that the potential properties and the related species composition in the boreal ecosystem are related to the different vegetation types and their natural disturbance regimes (Noss 1990; Angelstam 1996b; Fries et al. in press). The main argument is that species have evolved to meet the conditions, i.e. selection pressures, defined by the composition and structure as well as processes of forests with particular disturbance regimes (Bergeron et al. 1993; Bradshaw et al 1994; Angelstam 1996a).

The first step in the restoration of a degraded landscape is to maintain all remaining habitat elements of the natural landscape, i.e. the natural ecological biodiversity. This requires the maintenance, or restoration, of the full range of available site types and disturbance regimes, tree and other key-stone species and serial stage representation found in the naturally dynamic forest. It also requires a diversity of stand sizes, juxtapositions and configurations as well as the pertinent processes affecting the forest ecosystem maintenance (Hunter 1990; Angelstam 1992 and 1997).

3.1 FIRE AS A DISTURBANCE IN THE BOREAL FOREST

In the following I will present a logic for the book-keeping of ecological diversity in boreal forest. The diversity of biotopes in any naturally dynamic landscape is linked to differences in geology, soil types, topography, access to nutrients and water on the one hand, and to the disturbance regimes on the other. The dominating natural disturbance factor of the boreal landscape in Europe is fire (e.g. Goldammer and Furyaev 1996) even if flooding, wind, insect outbreaks and herbivory are also important (Angelstam 1996a).

Different fire frequencies and intensifies occur in a locally statistically predictable way, mainly according to the type of field layer vegetation, soil moisture and topography (Pyne 1984; Zackrisson 1977; Furyaev 1996; Engelmark 1984 and 1987), and form characteristic fire regimes (Pyne 1984; Goldammer and Furyaev 1996). The fire frequency, which is inversely related to the intensity of the fire, can be used to express how site characteristics and fire behaviour are linked. Fire frequencies in a naturally dynamic boreal landscape depend on three major factors.

Firstly, the intimate relationship between soil types, soil moisture and the consequent flammability of vegetation affects the average probability of fire (Pyne 1982; Goldammer and Furyaev 1996). Hence, the fire frequency increases, and the average intensity decreases, along the gradient wet – moist – fresh – dry and from fine-grained soils to coarse soils and from vegetation with inflammable plants like tall herbs to vegetation with dwarf-shrubs and lichens (e g. Zackrisson 1977; Hörnberg 1995). Angelstam and Jasinski (unpublished) found a 23-fold difference in the relative fire frequency between the extremes of this gradient in a naturally dynamic boreal forest landscape in Russia. Secondly, fire frequencies vary in the landscape due to differences in topography. Convex parts burn more frequently than concave (Zackrisson 1977), and south-facing slopes more than north-facing slopes (Högbom 1934). Finally, increasing air humidity reduces fire frequencies to the north (Sannikov and Goldammer 1996), and with increasing altitude (Zackrisson 1977). Similarly, the local climate strongly affects the lightning frequency (Granström 1993) and lightning fire density (Sannikov and Goldammer 1996).

3.2 THREE BOREAL FOREST DISTURBANCE REGIMES

Based on the behaviour of fire in different vegetation types, Angelstam et al. (1993) and Rülcker et al. (1994a and b) developed a conceptual model to explain to practical managers the relationship between the local and regional site conditions and fire in the naturally dynamic landscape. The starting-point in the model was to recognise four major groups of fire frequencies which were given the initials A, S, I and O after the fire frequency classes Absent/Almost never, Seldom, Intermediate and Often. For details see Figure 1 and Angelstam (1997).

On A-sites in naturally dynamic landscapes stand-replacing fires are absent or very uncommon (occurring at intervals >300 years (Hörnberg 1995; Gromtsev, 1996; Angelstam and Jasinski unpublished) and the dominating tree species is the Norway spruce *Picea abies.* Hence the disturbance affecting trees is small scaled and the new succession occur in small gaps (Kuuluvainen 1994). By contrast, on the O-sites disturbances by fire are frequent but the mortality of trees is low, partly because the fire intensity is usually low, and partly because the dominating Scots pine *Pinus sylvestris* is not very sensitive to fire (Sannikov and Goldammer 1996). On I-sites, and partly on S-sites, the effects of fire are more severe to the stand because of higher fire intensifies.

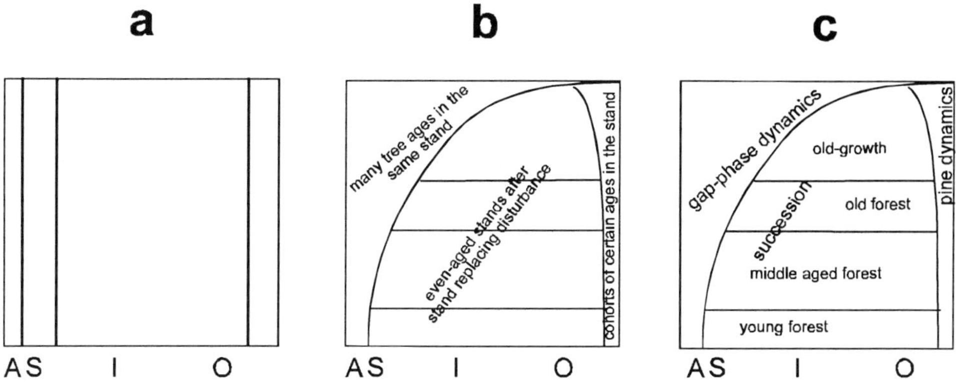

Figure 1. Distribution of disturbance regimes in relation to the forest fire fuel conditions and time after fire disturbance in a natural boreal forest landscape. The horizontal axis represents the relative amount of forest site types and regions where fire is Absent or occurs Almost never (A), via those where fire occurs Seldom (S), those where fire frequencies are Intermediate (1) to those were fire occurs Often (O). -The proportions of A, S, I and O reflect approximately the average distribution of these types of sites and regions in Sweden (Figure la). The distribution of different tree dynamics is shown in Figure lb. The vertical dimension represents the age of forests stands. Ecologically old forests are found both in the old-growth, where conifers begin to die, and in the middle aged forest where the not so long-lived deciduous trees start to die. In the other parts of the figure the variance in tree age is so high that it is not meaningful to present an average age. In Fig. 1c the different disturbance regimes are shown. In the left and upper part of Figure lc, gap phase dynamics dominate and spruce trees of different age are found in the same stand. In the O-area, repeated low intensity fires create characteristic pine forests with tree cohorts having several overlapping age distributions as a result of repeated low-intensity fires followed by recruitment. In the remaining central field, forest stands are usually even-aged within each successional stage after stand-replacing fires.

This figure is thus a simple description of the different components at the stand scale in an average natural boreal/hemiboreal forest landscape, each of which should be represented in a system for describing ecological biodiversity in boreal forest. Individual landscapes may show large variation in site type distribution and lack one or several disturbance types and hence also an ability to host certain sets of species. After the establishment of new tree individuals there is a development in the stand over time. Initially, the amount of dead wood is large. Later, after a deciduous phase with birches *Betula spp.* and aspen *Populus tremula,* the forest may develop old-growth characteristics if it does not burn again for a long time.

A particular area in a given landscape may therefore have escaped a stand-replacing fire for two reasons, either because (1) the site did not burn for a long time because the site, or the region, is not particularly fire-prone (A-, and S-sites) or because (2) the site did not burn by chance (S-, I- and O-sites). Summarising, the boreal forest has three main disturbance regimes (Figure lc) (from Rülcker et al. 1994b):

Gap-phase dynamics in spruce forest. In naturally dynamic landscapes such stands often form corridors, networks or clusters in the wet and moist parts of the landscape. Typically, such sites have a relatively moist and stable microclimate and a continuous supply of dead decaying wood in different stages of decay. The age distribution of trees within a stand is the inverse J-type (Kuuluvainen 1994) and a simple mean age conveys no information of the typical age structure.

Successional development after large scale disturbance. The succession can be initiated by wind, insect outbreaks, fire or human disturbance. Viewed over longer time, the successional stages are ephemeral at a particular site. To persist in the landscape, species therefore must be able to disperse from existing core areas and colonise new sites. Examples of different successional stages are recent burns, successional stages with deciduous trees, old and old-growth forest stands. The trees in a stand are of similar age and the mean age is good descriptor of the forest structure.

Multi-layered pine forests. Dry sites are subjected to frequent low-intensity fires. This produces stands which contain several cohorts of trees. The supply of dead wood in different stages of decay is often continuous.

3.3 MEASURING THE PROPERTIES OF DIFFERENT DISTURBANCE REGIMES

Forest planning has traditionally focused on the stand scale (5-100 ha) by describing the site type, the volume of wood in different age classes and for different tree species. This information is a very good starting point for the assessment of properties that affect biodiversity. However, forest properties should also be assessed in smaller and larger scales, i. e. at the scale of trees and at the scale of landscapes. In Table 1 I list the main properties of the different disturbance regimes in a naturally dynamic forest landscape and divide them between the different scales. I also list some practical methods for assessment.

Several of the required assessment methods, including some bioindicator species (Nitare and Norén 1992), are already being used in practical forest management at the stand level. However, at the level of trees important information is missing, but could easily be collected (i.e. the volume of dead wood in different stages of decay, variation in tree age within a stand and foliage height diversity). At the landscape scale assessment can largely be achieved by analysing the stand scale data with a geographical information system and express the amount, fragmentation and juxtaposition of different stand types in different landscapes (i.e. approx. >10,000 ha each).

4. Determining the sufficient amounts of different properties in boreal forest

Having listed the properties which reflect the composition and structure, in different geographical scales, of naturally dynamic boreal landscapes, the next step is to determine the critical thresholds regarding what amount of the properties found in naturally dynamic landscapes are sufficient for maintaining biodiversity in the managed landscape.

The first task is the identification of species that are unique to the different habitats and ecological properties that can be found in a naturally dynamic boreal forest landscape (see Table 1). If it is possible to translate the species' habitat requirements into criteria based on their environmental demands, these criteria can be used as opposed to the more costly inventories of certain indicator species in the field.

Indicator species are habitat specialists which, if present, indicate that also a range of other species depending on the same or similar properties are present in the landscape. Ideally, the following criteria must be met by a good indicator species:

- Its habitat niche must be narrow and known.
- It should be sensitive to the alteration and/or fragmentation of its habitat. Examples of potential species are those with large area requirements (large size and/or high

TABLE 1. Characteristic boreal forest properties of importance for the maintenance of biodiversity and which therefore need to be assessed in forest ecosystem management, The properties are divided among the three main disturbance regimes and three geographical scales (from Whittaker 1977). At the bottom of the table methods for assessment are suggested.

Disturbance type and tree species composition	at the scale of trees (<1 ha)	at the scale of stands (5-1,000 ha)	at the scale of landscapes (>10,000 ha)
Gap-phase dynamics (Picea abies, Betula spp.)	tree size and age; amount and continuity of dead wood; vertical forest structure	site type; variance in tree age; continuity in tree cover; frequency of flooding or high ground water level	For all forest disturbance types; amount in the landscape in relation to the available site conditions to estimate the potential for restoration of different disturbance regimes
Successional development (Picea abies, Betula spp., Populus tremula, Salix caprea, Sorbus aucuparia)			
• young forest • middle-aged forest • old forest • old-growth forest	burned ground and wood vertical forest structure amount of dead wood amount of dead wood; vertical forest structure; very old trees	site type; tree species composition for age classes; age distribution with high resolution also at higher ages	For all forest disturbance types and all successional stages; amount in the landscape; patch size distribution; juxtaposition of patch types
Pine dynamics (Pinus sylvestris)	burned ground and wood; amount of dead wood; vertical forest structure; very old trees	site type; variance in tree age	
Examples of tools for measurement	estimation of dead wood volume and stage of decay; foliage height diversity; monitoring of indicator species (e.g. fungi)	forest planning data on site, age, tree species; mapping deciduous forest stands; monitoring of indicator species (e.g. lichens)	use of Geographical Information Systems to describe the amount, fragmentation and juxtaposition of stand data in landscapes

trophic position), complex social structure, complex life cycle or life form and/or poor dispersal.
- It must not be too rare.
- It should be easy to detect in most times of the year and to identify unambiguously.
- Finally, to avoid problems related to populations whose reproduction has been affected but not the survival of adults, it should also be possible to determine whether a species that is present on a particular site is transient or permanent and whether it is reproducing.

In the following I review some recent information on potential indicator species whose long-term presence, or preferably whose measurable habitat requirements, could be used to estimate how much of the different properties listed in Table I is sufficient to maintain viable populations of indicator species for the different properties. If viability for all selected indicator species can be assured, then biodiversity will also be maintained.

4.1 GAP-PHASE DYNAMICS

In the absence of stand-replacing disturbance, wet spruce forests have an internal tree dynamics caused by gap formation as individual trees or groups of trees die. Two characteristic properties are the continuity in time at the local scale of dead wood in different stages of decay and the continuity in the tree cover which creates a locally stable moist micro-climate. Remnants of such forests host a large number of red-listed species which are now very rare as a consequence of agricultural development and forest wetland amelioration (Hansson 1997). Properties such as local continuity of a moist climate and dead wood in different stages of decay, degree of soil contact, cause of mortality etc. are difficult to measure directly. Some lichens (e.g. *Usnea longissima* and *Ramalina thrausta)* and fungi (e.g. *Fomitopsis rosea, Fellinus ferruginofuscus),* respectively, are being used as indicators of such conditions (Karström 1992). Long-term studies of the population trends of species specialising in this kind of habitat shows that such stands are often very sensitive to changes in the local microclimate due to removal of adjacent stands. Sjöberg and Ericson (1992) reported that 16 years after the removal of buffer zones of forest surrounding the stands and trees covered with the lichen *Evernia dicaricata*, all individuals of the lichen had gone locally extinct.

4.2 SUCCESSIONAL DEVELOPMENT

4.2.1 Young forest

After a fire there is recently burned wood, a property to which several insect species have adapted (e.g. Wikars 1992). After fire, stands with a high but variable proportion of young deciduous trees usually develop. Two species of grouse specialising on different types of young coniferous (*Tetrao tetrix*) (Swenson and Angelstam 1994) and deciduous young forest (*Bonasa bonasia*) (Åberg et al. 1995) are examples of species whose requirements can be translated to simple measurements available in forest planning maps (i.e. age class and tree species composition).

4.2.2 Middle-aged forest

In the middle aged boreal forest, deciduous trees are becoming biologically old and self-thinning is producing large amount of dead deciduous wood, which are prerequisites for the very demanding woodpeckers *Dendrocopos leucotos and Dendrocopos minor* (Angelstam and Mikusinski 1994; Carlson and Stenberg 1995). The passerine birds *Parus palustris, Sitta europea, Aegithalus caudatus* are less sensitive, but still very rare in intensively managed boreal forest where old deciduous trees are rare (Enoksson et al. 1995). Several species of lichens depending on aspen, sallow *Salix caprea* show similar patterns (Kuusinen 1996). The requirements of all these species can be translated to criteria describing the deciduous component if accurately assessed.

4.2.3 Old forest

For forest birds, several studies show that species depending on, or benefiting from, old forest have shown strongly negative population trends (e.g. Angelstam and Mikusinski 1994; Virkkala et al. 1994). The capercaillie *Tetrao urogallus* is often a symbol of these changes (see Sjöberg 1996). The main factor explaining variations in capercaillie density is the proportion of the landscape covered by old forest (defined as forests which have a good field layer with *Vaccinium myrtillus* and pines of sufficient size in order to carry the weight of foraging capercaillie (Rolstad and Wegge 1987), an easily assessable property using the information in the attribute files of the forest planning maps.

4.2.4 Old-growth forest

A particularly high organismal biodiversity occurs in forests which are 50-100 years older or more than the age at which trees are usually harvested in boreal forest (80 - 120 years depending on the local climate). The lichen *Alectoria sarmentosa* and its associated fauna (Esseen et al. 1996) is a good example. Given sufficient time, several species found in wet spruce forests may also occur in old-growth forest on sites which have escaped fire by chance for a long time. Stand age is therefore an important indicator itself and should be collected with a high resolution also at ages well above the age of tree harvest, i. e to 250 or 300 years.

4.3 PINE DYNAMICS

Because of the systematic way a long history of land use has altered the most productive sites in the boreal landscape, old pine forests still remain at sites with difficult access. While few good indicator species have been defined, the presence of dead wood in different stages of decay and very old trees are probably good indicators on their own.

5. Discussion

A basic foundation for studies of the requirements of viable populations is an understanding of the environments in which these species evolved. The boreal forest is one of the least altered large forest environment in the world and the chances for success should therefore be better than average. To secure the long-term maintenance of biodiversity in taiga forests, it is necessary to change forestry in a direction which imitates natural disturbance regimes such as fire, windfall and flooding closer than today, and avoid management in no-fire refugia. This means not only protection of some habitats but also a differentiation of management that actively produces important structure and maintains natural processes (Fries et al in press) as well as proactive biodiversity management including prescribed burning of forest stands, restoration of the water regimes by closing ditches and active creation of snags during harvest operations.

Formulation of strategic and operative goals for each property should be based both on: a. the composition and structure, in different geographical scales, of naturally dynamic boreal landscapes; and b. the critical thresholds regarding what proportion of the amount of structures found in naturally dynamic landscapes are sufficient for maintaining biodiversity.

Several studies suggest that when continuous habitats break down into remnants there are thresholds in the response (Gardner and O'Neill 1991). Franklin and Forman (1987) modelled landscape dynamics and found that when 70% of the original habitat was left, the first patches became isolated. As fragmentation proceeds, more patches become isolated, and at 30% of the original habitat left, all of the original habitat was in fragments. Continued habitat loss leads to an exponential increase in the distance between patches and true effects of habitat isolation rather than simply a consequence of habitat loss tend not to occur until less than 10-30% of the suitable habitat remained (Andrén 1994). Hence, if the amount of a particular property has been reduced to less than 10%, a further reduction should not be allowed. Conversely, if the reduction has gone even further, then proactive restoration is required.

To conclude, I argue that there is sufficient knowledge to start assessment of forest properties at all relevant scales. By implication biodiversity can thus be measured with sufficient accuracy to initiate the practical application of systems that could later be integrated with the forest planning and management (Angelstam and Pettersson 1997). However, where the land use history is long, assessment and restoration of biodiversity is more difficult. The European boreal forest ranges from highly altered forest landscapes in western Europe to naturally dynamic forests in eastern Europe. Two extremes are the mining district in south central Sweden where logging by clear-felling has been made at least for 5-6 forest generations, and parts of northern Russia where logging of natural landscapes has not yet begun (Angelstam et al. 1995; Angelstam and Majewski 1996). In Sweden the desired threshold value for many properties needs to be approached by restoration from below, i.e. the amount of the property should be allowed to increase, while the desired threshold is being approached by from above in northern Russia, i.e. the property can still be somewhat reduced without risking loss of the species associated with the property. This means that for example the woodland key-habitat approach used in Sweden merely represents attempts to secure the very last remnants (see Angelstam and Pettersson 1997). The classification of forests into three management types (protected forest, restricted management and industrial forests) in Russia is close to an approach which mimics the natural distribution of different disturbance regimes in relation to the site type because the protected forests are to a large extent located in wet spruce forest sites, the forest dynamics of which is difficult to mimic in forest management. Therefore, in defining standards for biodiversity, the European boreal forest should be stratified with respect to the difficulties to maintain and restore biodiversity. For lowland temperate forest in central Europe, which have a very long land use history, mimicking the naturally dynamic forest is a much more difficult goal to reach.

6. Acknowledgements

I thank Lennart Hansson, Kullervo Kuusela, Grzegorz Mikusinski, Hans Ljungkvist, Anna Koffman and two anonymous reviewers for wise comments.

References

Åberg, J., Jansson, G., Swenson, J. and Angelstam, P. 1995. The effect of matrix on the occurrence of hazel grouse (Bonasa bonasia) in isolated habitat fragments. Oecologia. 103: 265-269.

Andrén, H. 1994. Effects of habitat fragmentation on birds and mammals in landscapes with different proportions of suitable habitat: a review. Oikos. 71: 355-366.

Angelstam, P. 1992. Conservation of communities the importance of edges, surroundings and landscape mosaic structure. In: Hansson, L. (ed.). Ecological principles of nature conservation. Elsevier. London. Pp. 9-69.

Angelstam, P. 1996a. Ghost of forest past – natural disturbance regimes as a basis for reconstruction of biologically diverse forests in Europe. In: DeGraaf, R. and Miller, R.I. (eds.). Conservation of faunal diversity in forested landscapes. Chapman and Hall. London. Pp. 287-336.

Angelstam, P. 1996b. Forest landscape management for maintenance of biodiversity – a Swedish perspective. In: Bachmann, P., Kuusela, K., and Uuttera, J. (eds.). Assessment of biodiversity for improved forest management. Proceedings 6. European Forest Institute. Finland. Pp. 69-86.

Angelstam, P. 1997. Landscape analysis as a tool for the scientific management of biodiversity. In: Hansson, L. (ed.). Boreal ecosystems and landscapes – structures, functions and conservation of biodiversity. Ecological Bulletins. In press.

Angelstam, P. and Majewski, P. 1996. Reference landscapes for improved landscape management in Eastern and Western Europe. In: Bachmann, P., Kuusela, K., and Uuttera, J. (eds.). Assessment of biodiversity for improved forest management. Proceedings 6. European Forest Institute. Finland. Pp. 111-123.

Angelstam, P., Majewski, P. and Bondrup-Nielsen, S. 1995. West-east cooperation in Europe for sustainable boreal forests. Water, Air and Soil Pollution. 82: 3-11.

Angelstam, P. and Mikusinski, G. 1994. Woodpecker assemblages in natural and managed boreal and hemiboreal forest – a review. Annales Zoologici Fennici. 31: 157-172.

Angelstam, P. and Pettersson, B. 1996. Principles for landscape planning in Swedish forestry – a review. In: McClain, K.M. and McClary, C.P. (eds.). International Forestry Seminar. Exploring multiple use and ecosystem management: from policy to operational practices. FAO/ECE/ILO Proceedings. Prince George. Natural Resources Canada. Pp. 85-95.

Angelstam, P. and Pettersson, B. 1997. Principles of present Swedish forest biodiversity management. In: Hansson, L. (ed.). Boreal ecosystems and landscapes – structures, functions and conservation of biodiversity. Ecological Bulletins. In press.

Angelstam, P., Rosenberg, P. and Rülcker, C. 1993. Aldrig, sällan, ibland, ofta. Skog och forskning 93(1): 34-41. (In Swedish).

Apps, M.J., Price, D.T. and Wisniewski, J. 1995. Boreal forests and global change. Water, air and soil pollution. 82 (1-2): 191-548.

Berg, Å., Ehnström, B., Gustafsson, L., Hallingbäck, T., Jonsell, M. and Weslien, J. 1994. Threatened plant, animal and fungus species in Swedish forests – distribution and habitat associations. Conservation Biology. 8 (3): 718-731.

Bergeron, Y., Bradshaw, R. and Engelmark, O. 1993. Disturbance dynamics in boreal forest. Opulus press. Uppsala.

Bradshaw, R., Gemmel, P. and Björkman, L. 1994. Development of nature-based silvicultural models in southern Sweden: the scientific background. Forest and Landscape Research. 1: 95-110.

Carlson, A. and Stenberg, I. 1995. Vitryggig hackspett (Dendrocopos leucotos) – biotopval och sårbarhetsanalys, Department of Wildlife Ecology Report 27. Swedish University of Agricultural Sciences. Uppsala. (In Swedish)

DeGraaf, R. and Miller, R.I. (eds.) 1996. Conservation of faunal diversity in forested landscapes. Chapman and Hall. London.

Dudley, N., Gilmour, D. and Jeanrenaud, J.-P. 1996. Forests for life. WWF International and IUCN. Gland.

Engelmark, O. 1984. Forest fires in the Muddus national park (northern Sweden) during the past 600 years. Canadian Journal of Botany. 62: 893-898.

Engelmark, O. 1987. Fire history correlations to forest type and topography in northern Sweden, Annales Botanici Fennici. 24: 317-324.

Enoksson, B., Angelstam, P., Larsson, K. 1995. Deciduous trees and resident birds – the problem of fragmentation within a coniferous forest landscape, Landscape Ecology. 10(5): 267-275.

Esseen, P.-A., Ehnström, B., Ericson, L. and Sjöberg, K. 1992. Boreal forest – the focal habitats of Fennoscandia. In: Hansson, L. (ed.). Ecological principles of nature conservation. Elsevier. London. Pp. 252-325.

Esseen, P.-A., Renhorn, K.-E., and Pettersson, R.B. 1996. Epiphytic lichen biomass in managed and old-growth boreal forests: effect of branch quality. Ecological Applications. 6: 228-238.

Franklin, J. and Forman, R.T.T. 1987. Creating landscape patterns by forest cutting: ecological consequences and principles. Landscape ecology. 1: 1-18.

Fries, C., Johansson, O., Pettersson, B. and Simonsson, P. in press. Silvicultural models to maintain and restore natural stand structures in Swedish boreal forests. Forest Ecology and Management. In press.

Furyaev, V.V. 1996. Pyrological regimes and dynamic of the southern taiga forests in Siberia. In: Goldammer, J. and Furyaev, V.V. (eds.). Fire in ecosystem of boreal Eurasia. Kluwer Academic Publishers. Dordrecht. Pp. 168-185.

Gamlin, L. 1988. Sweden's factory forest. New Scientist. 117: 41-47.

Gardner, R.H. and O'Neill, R.V. 1991. Pattern, process and predictability: the use of neutral models for landscape analysis. In Turner, M.G. and Gardner, R.H. (eds.). Quantitative methods in landscape ecology. Springer. New York. Pp. 289-307.

Goldammer, J. and Furyaev, V.V. (eds.) 1996. Fire in ecosysterm of boreal Eurasia. Kluwer Academic Publishers. Dordrecht.

Granström, A. 1993. Spatial and temporal variation in lightning ignitions in Sweden. Journal of Vegetation Science. 4: 737-744.

Gromtsev, A.N. 1996. Retrospective analysis of natural fire regimes in landscapes of eastern Fennoscandia and their anthropogenic transformation. In: Goldammer, J. and Furyaev, V.V. (eds.). Fire in ecosystem of boreal Eurasia. Kluwer Academic Publishers. Dordrecht. Pp. 45-54.

Hansson, L.(ed.) 1997. Boreal ecosystems and landscapes – structures, functions and conservation of biodiversity. Ecological Bulletins. In press.

Harper, J.L. and Hawksworth, D.L. 1995. Preface. in Hawksworth, D.L. (ed.). Biodiversity – measurement and estimation. Chapman and Hall. London. Pp. 5-12.

Helle, P. and Järvinen, O. 1986. Population trends of North Finnish land birds in relation to their habitat selection and changes in forest structure. Oikos. 46: 107-115.

Heywood, V.H. (ed.) 1995. Global biodiversity assessment. Cambridge University Press. Cambridge.

Hörnberg, G. 1995. Boreal old-growth Picea abies swamp-forests in Sweden – disturbance history, structure and regeneration patterns. Dissertations in Forest Vegetatation ecology 7. Swedish University of Agricultural Sciences.

Högbom, A.G. 1934. Om skogseldar förr och nu och deras roll i skogarnas utvecklingshistoria. (On past and present forest fires and their role in the development of forests.) Almqvist och Wiksells förlag. Stockholm. (In Swedish).

Hunter, M.L. 1990. Wildlife, forests and forestry. Principles of managing forests for biological diversity. Prentice-Hall. New Jersey.

Järvinen, O., Kuusela, K. and Väisänen, R. 1977. Effects of modern forestry on the number of breeding birds in Finland 1945 - 1975. Silva Fennica. 11: 284-294. (In Finnish with English summary.)

Karström, M. 1992. Steget före i det glömda landet. Svensk Botanisk Tidskift. 86: 115-146.

Korpilahti, E. (ed.) 1996. Climate change, biodiversity and boreal forest ecosystems, Silva Fennica. 30(2-3): 86-383.

Kouki, J. 1994. Biodiversity in the Fennoscandian boreal forests: natural variation and its management. Annales Zoologici Fennici. 31: 1-217.

Kuuluvainen, T. 1994. Gap disturbance, ground microtopography, and the regeneration dynamics of boreal coniferous forests in Finland: a review. Annales Zoologici Fennici. 31: 35-51.

Kuusinen, M. 1996. Epiphytic lichen flora and diversity in old-growth boreal forests of Finland. PhD thesis. Publications in Botany from the University of Helsinki 23.

Larsen, J.B. (ed.) 1996. Sustainable forest management. TemaNord 1996: 578. Nordic Council of Ministers. Copenhagen.

MacArthur, R.H. 1965. Patterns of species diversity. Biol. Rev. 430: 510-533.

Mikusinski, G. and Angelstam, P. in press. Economic geography, forest distribution and woodpecker diversity in central Europe. Conservation Biology. In press.

Nitare, J. and Norén, M. 1992. Woodland key-habitats of rare and endangered species will be mapped in a new project of the Swedish National Board of Forestry. Svensk Botanisk Tidskrift. 86: 219-226.

Noss, R.F. 1990. Indicators for monitoring biodiversity: a hierarchical approach. Conservation Biology 4: 355-364.

Pyne, S.J. 1984. Introduction to wildland fire. John Wiley and Sons. New York.

Rassi, P. and Väisänen, R. 1987. Threatened animals and plants in Finland. Government printing centre. Helsinki.

Rolstad, J. and Wegge, P. 1987. Distribution and size of capercaillie leks in relation to old forest fragmentation. Oecologia. Berlin. 72: 389-394.

Rülcker, C., Angelstam, P. and Rosenberg, P. 1994a. Ecology in the forest planning process – a model for large scale forestry. The Forest Research Institute of Sweden. Report 8. (In Swedish with English summary).

Rülcker, C., Angelstam, P. and Rosenberg, P. 1994b. Natural forest fire dynamics can guide conservation and silviculture in boreal forest. SkogForsk Results 2.

Sannikov, S.N. and Goldammer, J.G. 1996. Fire ecology of pine forests of northern Eurasia. In Goldammer, J. and Furyaev, V.V. (eds.). Fire in ecosystems of boreal Eurasia. Kluwer Academic Publishers. Dordrecht. Pp. 151-167.

Sjöberg, K. 1996. Modern forestry and the capercaillie. In DeGraaf, R. and Miller, R.I. (eds.). Conservation of faunal diversity in forested landscapes. Chapman and Hall. London. Pp. 111-135.

Sjöberg, K and Ericson, L. 1992. Forested and open wetlands. In Hansson, L. (ed.). Ecological principles of nature conservation. Elsevier. London. Pp. 326-351.

Stanners, D. and Bourdeau, P. (eds.) 1995. Europe's environment. The Dobris assessment. European Environment Agency. Copenhagen.

Swenson, J.E. and Angelstam, P. 1994. Habitat separation by sympatric forest grouse in Fennoscandia in relation to forest succession. Can. J. Zool. 71: 1303-1310.

Virkkala, R. 1987. Effects of forest management on birds breeding in northern Finland. Annales Zoologici Fennici. 24: 281-294.

Virkkala, R., Rajasarkkä, A., Väisänen, R.A., Vickholm, M. and Virolainen, E. 1994. Conservation value of nature reserves: do hole-nesting birds prefer protected forests in southern Finland. Annales Zoologici Fennici. 31: 173-186.

Whittaker, R.H. 1977. Evolution of species diversity in land communities. Evolutionary biology. 10: 1-67.

Wiens, J.A. 1976. Population responses to patchy environments. Annual Review of Ecology and Systematics. 7: 81-120.

Wikars, L.O. 1992. Forest fires and insects. Entomolgisk Tidskrift. 13(4): 1-12.

Zackrisson, O. 1977. Influence of forest fires on the north Swedish boreal forest. Oikos. 29: 22-32.

SPATIAL ASSESSMENT OF HABITAT SUPPLY AND HARVEST VALUES AS A MEANS OF EVALUATING CONSERVATION STRATEGIES: A CASE STUDY

B.G. DUNSWORTH AND S. M. NORTHWAY
Corporate Forestry, MacMillan Bloedel Limited
65 Front Street, Nanaimo,
B.C. V9R 5H9 British Columbia

Abstract

Changes in harvest scheduling rules can lead to significant changes in total harvest value without significantly affecting the amount of good habitat or the level of fragmentation for a wide range of vertebrate species. In this case study, of the three strategies tested, the most effective strategy for minimizing vertebrate habitat impacts and maximizing harvest value was zoning: half of the landscape unit allocated to provide biodiversity values and excluded from commercial forest development, while allocating the other half to commercial forestry with protection for riparian and soil values.

We developed three harvest scheduling rules representing different levels of constraints on commercial forest activities including the new British Columbia Forest Practices Code regulations. A spatially explicit harvest scheduling model was used to develop a road building and harvest block schedule to maximize financial return while meeting the constraint rules. Similarly, a spatially explicit habitat supply model was used to assess habitat quality and fragmentation.

The Nahmint landscape unit on Vancouver Island was used as a case study to assess the impact of these scheduling rules. The resulting harvest schedules were tested for their level of compliance to the government's biodiversity guidelines, and their performance in providing high quality, unfragmented habitat for six habitat dependency groups of vertebrates. The scheduling rules were also tested for their impact on the value and volume of timber derived from harvesting the forest.

1. Introduction

Biodiversity has become a major issue with respect to forest land management in Canada and British Columbia. In order to understand the current situation, it is important to note the history of the concern for conservation of biodiversity. We begin in 1981 with Ehrlich and Ehrlich's book on species extinction. This and other similar work by Wilson (1984 and 1985) focused public attention on the evolutionary history of tropical rainforests, the vastness of the genetic "library" inherent in the species diversity and the potential loss of species as a result of large scale forest land conversion. Specifically, the causal agents were identified as fragmentation, degradation, and elimination of habitat (Ehrlich 1988).

P. Bachmann et al. (eds.), Assessment of Biodiversity for Improved Forest Planning, 315–329.
European Forest Institute Proceedings no. 18. © 1998 *Kluwer Academic Publishers. Printed in the Netherlands.*

At the same time, these causal factors of species loss were beginning to be conceptually linked to the potential loss of ecological integrity and the provision of ecological services (Ehrlich and Mooney 1983; Wilson 1988; Wilson and Peter 1988). In the context of sustainable development this in turn raised questions about inter-generational and supra-national equity. Thus, the original concern about local species loss was transformed into a global concern about stewardship and land ethic. In this more abstract form the term biodiversity became easily transportable to other regions in the world where there were and continue to be growing concerns over land ethics and stewardship. However, the evolutionary history of these regions, and their resultant diversity of species and risk of species loss, were dissimilar to tropical rainforests.

Conservation strategies focusing on biodiversity in coastal temperate rainforests reflect this history. British Columbia (BC) has recently ushered in a new regulatory environment with the Forest Practices Code of British Columbia Act (BCFPC) (Anon 1994). All of BC's public forests are now managed under the direction of this legislation. While the management objectives defined under previous legislation included provisions for non-timber resources, the BCFPC includes more explicit regulations and guidelines the practice of commercial forestry. These aim to maintain a wide range of resource values by limiting the impact of timber production on related non-consumptive uses.

Currently, in the coastal region of British Columbia, considerable debate centers around what stand and landscape level guidelines are appropriate to conserve biological diversity. The Biodiversity Guidebook (Co-published by BC Ministry of Environment and BC Ministry of Forests 1995) for the BCFPC has rules for seral stage balance, patch size distribution, old growth and interior old growth retention, and forest ecosystem network (FEN) construction. In addition the guidebook recommends stand level rules for wildlife tree patches and objectives for stand structure, vegetation and tree species diversity, and maintenance of coarse woody debris. In short, these rules focus more on forest stewardship than forest habitat and can significantly reduce permitted timber harvest levels and increase harvest costs.

Clearly, stewardship policies that use rule-based approaches are administratively convenient, but compliance has high cost and can give little assurance of performance towards species conservation objectives (Nelson and Shannon 1994; van Kooten 1994; Haley 1996). In order to address conservation of biological diversity, we need to explicitly deal with the problem of species loss, through its causal factors: fragmentation, degradation and elimination of habitat. Dunning et al. (1992) note that landscape variation must be interpreted in a framework of landscape structure, organism response and ecological function. Large scale forest harvesting creates considerable landscape variation and without an understanding of these elements in the context of species responses, static measures of landscape structure have little utility (Kareiva 1987; Kareiva 1990; Morrison et al. 1992; Turner 1989; Wiens 1989).

To assess habitat quality and fragmentation, we require spatially and temporally explicit models which incorporate our best understanding of how landscape structure, organism response, and ecological process interact for the species or habitat dependency groups of concern. Over the last 25 years wildlife research in the Pacific Northwest has developed a considerable body of knowledge about the associative habitat relationships of forest vertebrates (Carey 1989; Hansen et al. 1991; Huff et al. 1992; Meidinger and Pojar 1991; Ralph et al. 1991; Raphael and Marcot 1984). However, we know a lot less about landscape level processes. Forest practices are currently going through a period of rapid change in response to social demands and have the potential to lead to significant landscape level changes. Spatial models are a key tool to help explore possible futures and formulate landscape hypotheses for future testing.

It is also important to have a spatially explicit harvesting model to reflect the spatial aspects of constraints on commercial forest harvesting. Aspatial forest harvesting models have underestimated the effect of spatial constraints (Cox and Sullivan 1995; Yoshimoto and Brodie 1994). The scheduling problem is to assign a future date of harvest to spatially explicit harvest blocks while meeting harvest constraints.

Harvest constraints in BC can be grouped into those protecting soil, riparian, biodiversity, and aesthetic values. These constraints can have the effect of reducing the commercial landbase, affecting the rate of timber harvest, or dispersing the harvest blocks. For different combinations of constraints, a harvest model could find a solution to maximise the net present value of the harvest. This allows for the calculation of the opportunity cost for each set of constraints. Each of these harvest schedules can also be assessed with the habitat supply model for the provision of high quality habitat and fragmentation.

MacMillan Bloedel has linked a spatial habitat supply model with a spatial harvest schedule model in order to explore the interactions of harvest constraint rules and habitat supply. The explicit objectives of this case study were to:

- assess the cost of regulation and the achievement of habitat supply objectives for a variety of harvest scheduling rules on a typical Vancouver Island landscape unit;
- compare habitat supply objectives with the compliance to biodiversity regulations and the extent to which adherence of other resource regulations (visual quality objectives [VQOs], deer and elk reserves, and adjacency rules) contribute to the conservation of habitat.

2. Methods

2.1 LOCATION AND AREA

The Nahmint watershed lies in south-western Vancouver Island and covers approximately 20,000 ha. The headwaters of the valley are at 1600 m in the Gibson-Klitsa plateau, and are drained by the 36 km long Nahmint River to the sea at Alberni Inlet. The valley is a 5 km wide U-shaped glaciated valley. Eight kilometres long, Nahmint Lake is a central feature of the valley.

Much of the bedrock geology of the valley is basalt and pillow lava. The surface geology largely reflects the retreating glaciers of 10,000 years ago, leaving large amounts of morainal material; subsequent weathering has resulted in some colluvial and fluvial deposits.

The climate of the valley is west coast marine. 80% of the 400+ cm of annual rainfall occurs between October and April. January is the coldest month with an average daily minimum of 1°C; July is the warmest with an average daily maximum of 23°C. The landscape unit is broken into four biogeoclimatic variants: CWHvm1 (Submontane Very Wet Maritime), CWHvm2 (Montane Very Wet Maritime), CWHmm1 (Submontane Moist Maritime) Coastal Western Hemlock Variants, and Mhmm1 (Windward Moist Maritime) Mountain Hemlock Variant (Green and Klinka 1994).

2.2 HABITAT SUPPLY MODELING

The habitat supply model incorporates the habitat modeling algorithms developed by Daust (1994) and used in the decision support system, SIMFOR (Daust and Bunnell 1992 and 1994). This raster based model operates using watersheds as landscape units (10,000 ha to 50,000 ha) and a spatial resolution of 1 ha (100 m x 100 m grid). The basic drivers for the model are biogeoclimatic variant (Green and Klinka 1994), site productivity (stand height at 50 years), and forest age. Forest disturbance is derived from the spatially explicit harvest schedule described below. The habitat supply modeling is programmed in ARC/Info AML and consists of three basic components:

2.2.1 Attribute Trajectories

Forest habitat attributes consist of both static and dynamic elements. The static elements include rock, alpine (ice, scrub, and non-treed), and various riparian habitats (river, lake, bog, river-lake, and bog-lake). The dynamic elements include: large trees (>40 cm DBH), small trees (<40 cm DBH), large snags (>40 cm DBH), small snags (<40 cm DBH), large coarse woody debris (>40 cm diameter), small coarse woody debris (<40 cm diameter), canopy gaps, deciduous cover, brush, grass, litter, forest edge, and interior old growth. The propagation of these attributes over time is modeled using attribute trajectories of relative abundance (Figure 1). These are based on expert opinion and validated using MacMillan Bloedel permanent sample plot data (Wells 1996; Wells and Daust 1994). The trajectories are distributed spatially within landscape units based on the spatial distribution of ecological and site productivity units – combinations of biogeoclimatic subzones and site index.

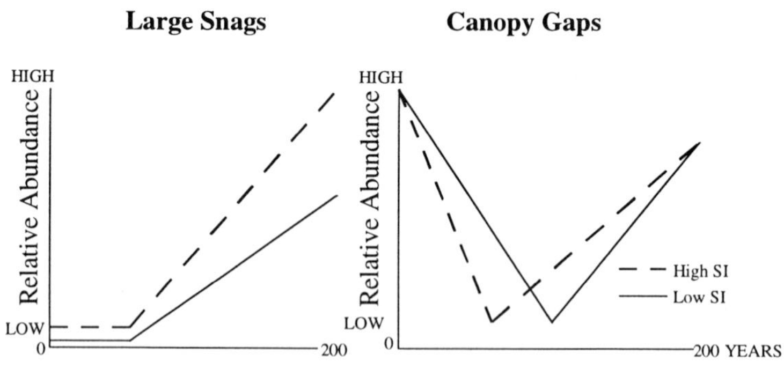

Figure 1. Example schematics of attribute trajectories for the CWHvm1 and CWHmm1 biogeoclimatic variants

2.3.2 Habitat Dependency Groups and Habitat Quality Scores

Species habitat needs are developed using expert opinion and the scientific literature for groups of vertebrate species that have common habitat requirements (Meidinger and

Pojar 1991; Sutherland 1996). This case study used six habitat dependency groups: 1) perching birds (early seral), 2) riparian small mammals, 3) cavity users, 4) elk winter, 5) interior old growth and 6) edge. These groups represent 30 rare and endangered species and 36 representative wildlife species (example, Table 1). The model interface allows the user to explicitly define the species habitat dependency group requirements in terms of an attribute list, minimum and maximum abundance requirements, attribute weight and criticality (essential for survival). The model uses these requirements to evaluate landscapes for habitat quality with the following habitat scoring algorithm for each habitat dependency group:

$$HQS_j = [\Sigma\big((V_{ij} - min_i) / (max_i - min_i)\big) * W_i] / \Sigma W_i \quad (1)$$

where HQS_j = habitat quality score at time j, W_i = weight for attribute i, V_i = value of attribute i, max_i = maximum requirement for attribute i, min_i = minimum requirement for attribute i.

Each hectare is classified for biogeoclimatic variant, forest productivity (SI) and forest age. These are used with the attribute trajectories to derive an attribute value. The attribute values are then scaled, weighted (on the basis of relative importance) and the habitat quality score (HQS) for that time interval computed as a weighted average of all the attribute scores (Equation (1)).

2.3.3 Dispersion and Fragmentation

Species movement through space is modeled as a function of the habitat quality surrounding centers of high quality habitat. The model interface allows the user to

TABLE 1. Example of coastal rare and endangered species and representative wildlife assumed to be associated with model habitat dependency groups.

Dependency Group	Rare and Endangered Species (Common Name)	Representative Wildife Species (Common Name)	
Perching Birds	Tiger Salamander (R)	Great Horned Owl	Western Toad
	Northern Leopard Frog (R)	Saw-whet Owl	Pacific Treefrog
	Sharp Tailed snake (R)	Barred owl	Western Red-backed Salamander
	Gopher Snake (R)	Blue & Ruffed Grouse	Ensatina
	Pine Grosbeak (B)	Band Tailed Pigeon	Common Raven
	Hutton's Vireo (B)	Northern Flicker	Winter Wren
		Hairy Woodpecker	Varied Thrush & Gray Jay
Interior Old Growth	Pacific Giant Salamander (R)	Great Horned Owl	Pileated Woodpecker
	Tiger Salamander (R)	Varied Thrush	Western Toad
	Northern Goshawk (R)	Red Breasted nuthatch	Pacific Treefrog
	Keen's Long-eared Myotis (R)	Chestnut-back Chickadee	Western Red-back Salamander
	Townsend's Big-eared Bat (B)	Black Swift & Marten	Ensatina & NW Salamander

[1] Red- and blue-listed species are designated as R and B, respectively. Based on the British Columbia Conservation Data Center 1994 Listing.
[2] Adapted from Meidinger and Pojar 1991; 106–108.

explicitly define a movement cost function for each habitat dependency group. We have based the movement functions for coastal vertebrates used in this study on literature of forest dependent vertebrate life history and patterns of dispersal (Blair 1940; de Vos 1951; Flyger 1960; Holekamp 1984; Nelson 1993; Trewhella et al. 1988). Annual dispersion potential was converted into circular area and expressed as radial distance. This distance was then assumed to constitute 100% of the species available dispersion through habitat of the highest quality (lowest movement cost). The minimum individual grid cell movement cost was computed using the following equation:

$$\text{Movement Cost per cell (\%)} = 10000 \Big/ \sqrt{(\text{DA} * 10000) / \pi} \tag{2}$$

where DA = annual dispersion area (ha).

Movement costs as functions of habitat quality were then derived using this least cost estimate as the base for high quality habitat and deriving estimates from the literature mentioned above for the relative costs of moving through habitats of poorer quality. Example movement functions for two habitat dependency groups are illustrated in Figure 2.

The model uses these movement functions to interpret landscape fragmentation for each habitat dependency group. Users specify the habitat score thresholds for good quality habitat. Areas of good quality habitat are used as the center of potential fragments, and the model determines if the habitat dependency group could move from one center to any of the others by evaluating the movement cost through the intervening space. High quality habitat that cannot be reached is considered a fragment, and both fragment number and size (the size of the associated high quality habitat) are reported. Figure 3 illustrates the flow of the modeling process from habitat attributes, through the calculation of habitat quality and identification of fragmentation.

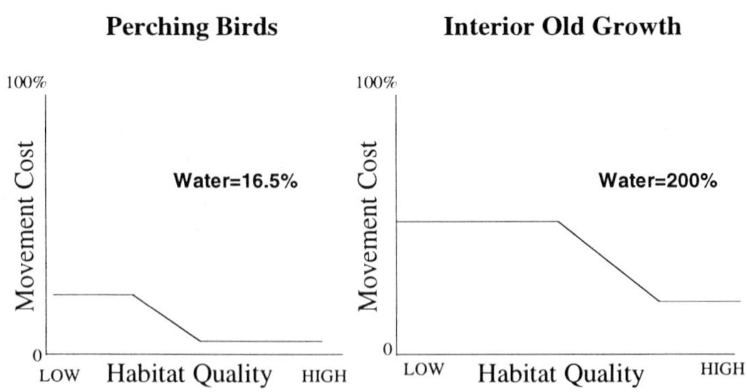

Figure 2. Example schematics of movement function for two habitat dependency groups (see Equation 2)

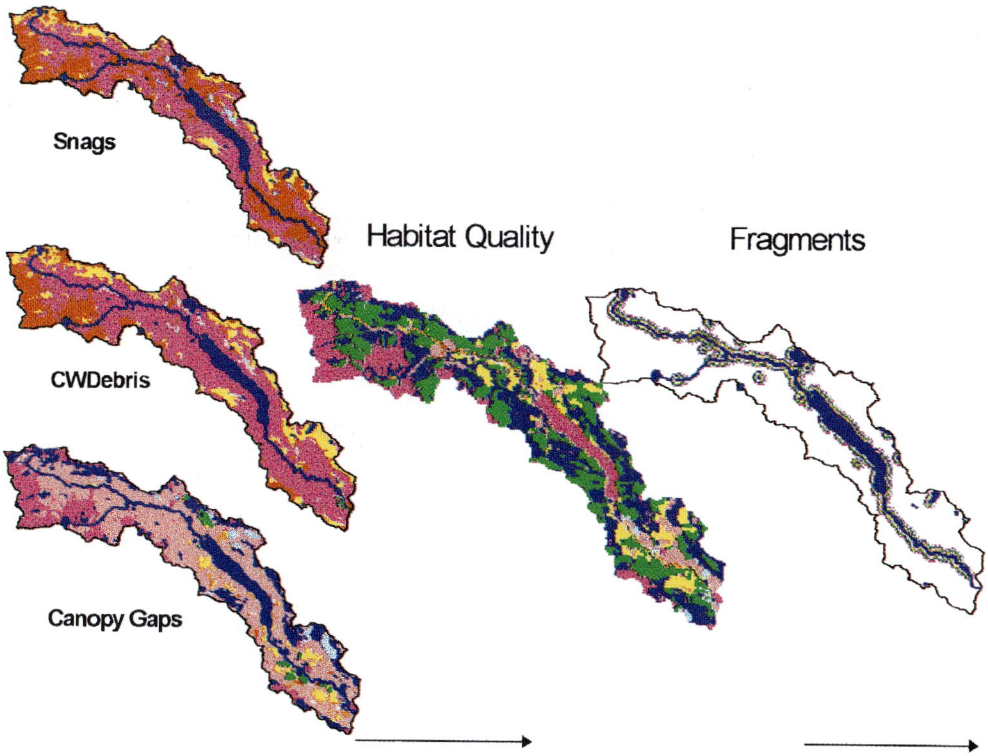

Figure 3. Habitat modeling process

2.4 HARVEST SCHEDULING

The spatially explicit harvest scheduling model determined the timing for the harvest of pre-determined blocks and for the building of pre-determined road segments. Harvest blocks and road segments were assigned to one of the 16 five-year periods making up the 80 years of the plan. The harvest and road schedules maximized net present value (NPV) while harvesting all the available timber under a series of constraint rules.

Three harvest schedules were developed for this case study. Each represented a different set of regulatory constraint rules (Figure 4). All three schedules were constrained to harvest all the available timber in an even-flow of volume over 80 years. The model structure limited each cutblock to one clearcut harvest. Each cutblock was associated with a road segment that had to be built prior to harvest.

The minimum constraint set of rules was designed to protect soil and riparian values. This was accomplished through reduced commercial activity in delineated areas of special sensitivity.

The full constraint set included protection for soil, riparian, aesthetic and biodiversity values. Scenic values were preserved through limiting the alteration of delineated viewscapes and setting a maximum harvest block size. Recreation values were protected through reduced harvest activities in sensitive areas. Biodiversity values were addressed

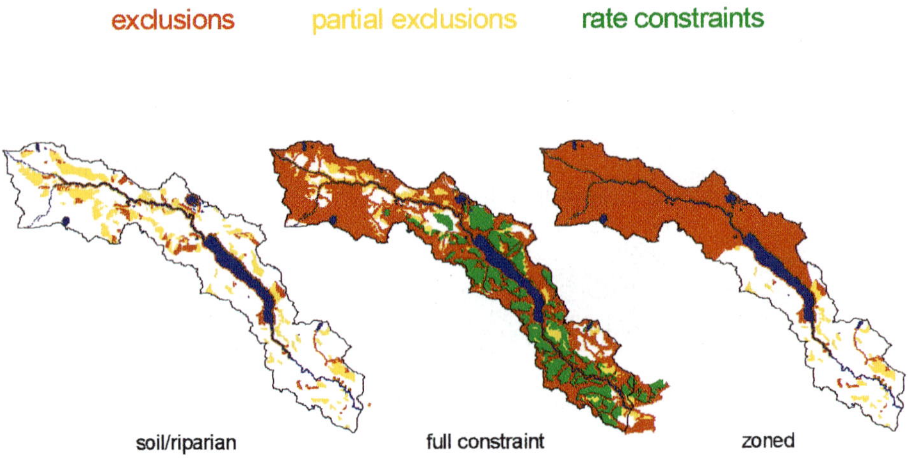

Figure 4. Nahmint harvest constraint options

through FENs. FENs include ridge tops, riparian areas, areas of critical habitat and connecting corridors. Harvesting was excluded from the FENs. Hydrological values and more detailed biodiversity rules were not explicitly dealt with in this case study.

A third set of constraints was used to represent a zoning option. The watershed was divided into approximately equal halves, and the minimum constraint set was applied to the bottom half of the watershed, while harvesting was excluded from the top half.

2.4.1 Financial Measures

Net present value (NPV) at 8% real was used as the measure of benefit derived from the harvest schedule. The NPV was based on the conversion return, calculated as the log values less the harvesting, hauling, and road building costs.

The harvested-log values were based on a 10 year average value from the Vancouver Log Market. Harvesting costs were estimated at $47/m^3$ for conventional methods, $55/m^3$ for longline systems, and $77/m^3$ for helicopter logging. New access road construction was included at $80,000/km. Hauling costs were included at $0.17/m^3/km. A real cost reduction of 2% per annum was included in this study.

2.4.2 Incorporating Constraints

Terrain stability, avalanche, riparian, and difficult regeneration areas were included in the riparian/soil constraint set. Clearcut size restrictions, FENs, deer winter ranges, marbled murrelet reserves, old-growth reserves, VQOs, adjacency/green-up rules and recreation areas made up the full constraint set.

All constraints, but the VQOs, were expressed as specified areas with an associated netdown. Netdown is an area exclusion expressed as a maximum fraction of the area to be harvested. The VQOs were expressed as a cover constraint, with a maximum proportion of the viewscape to be below a specified age. Each potential cutblock was assessed for violation of each constraint. In the creation of the plans, none of the considered constraints were violated.

2.4.3 Optimization Technique

The scheduling problem is easily defined, but difficult to solve. It is a combinatorial problem, and no general algorithms have been found to be successful. Roise (1990) suggested a non-linear programming formulation with adjacency constraints in the penalty function. Random search algorithms have been used by Clements et al. (1990) and Nelson et al. (1988). A biased random search algorithm was tried by O'Hara et al. (1989). Nelson and Brodie (1990) examined mixed integer programming. Yoshimoto et al. (1994) have suggested a random search algorithm coupled with decomposition of the problem into subproblems. Lockwood and Moore (1992) made use of a random search algorithm known as simulated annealing (Kirkpatrick et al. 1983). It is the Lockwood and Moore (1992) implementation of simulated annealing that was used in this paper.

3. Results and Discussion

3.1 BIODIVERSITY RULE COMPLIANCE

Beyond the FENs outlined in Section 2, no other explicit regulatory requirements for biodiversity conservation were modeled. However, under the current BCFPC Biodiversity Guidebook there are additional requirements for age class balance (early seral, mature, and old seral) and interior old growth. We have evaluated the model schedules with respect to their compliance with these items (Table 2). The three schedules provided a variety of levels of compliance, none provided complete compliance. The most persistent violation is the early seral violation in the second period. The early seral age class (<40 years) exceeds 30% in the second period, even under the most constrained schedule. In order to meet the evenflow constraint, this violation could not be avoided as it is an artifact of the historical pattern of logging. The other obvious violation in two of the schedules is the lack of a FEN. This was a consequence of the assumptions for those schedules. Finally, under the least constrained schedule we noted the above violations and mature seral age class (>80 years) violations in the last four periods.

TABLE 2. Biodiversity guidebook compliance violations for each of the case study harvest schedules.[3]

	Years				
Schedule	0	20	40	60	80
Soil/ Riparian	F	FE	FEM	FM	FEM
Full Constraint	-	E	-	-	-
Zoned	F	FE	F	F	F

[3] F–Forest Ecosystem Network; Forest areas: I–Interior Old Growth >9%; O–Old growth (>250 years) >19%; Mature and Old (>80 years) >54%; E–Early seral (<40 years) <23%.

3.2 HABITAT SUPPLY

We evaluated each of the schedules for their provision of high quality habitat and fragmentation, and found surprisingly similar conditions. Cavity users and interior old-growth groups consistently had the highest, maximum good habitat for all schedules at 9,600 ha and 4,600 ha, respectively (Table 3). These levels were attained during the first periods of all schedules. The edge group was the only one that approached these levels near the middle of the soil/riparian schedule. The metrics of maximum and minimum fragment numbers were not as informative for evaluating changes in landscape composition as was habitat quality (i.e., maximum and minimum good habitat) (Table 3).

3.2.1 Fragmentation

There was spatial and temporal variability in fragmentation estimates among the schedules. In general, the soil/riparian schedule created the most fragmentation and the zoning schedules provided the least. Cavity users had sufficient good habitat and dispersion to be able to maintain connectivity in all of the schedules and experienced no fragmentation. The perching bird and elk winter habitat dependency groups, on the other hand, had considerably less good quality habitat than cavity users, were less able to disperse, and were subsequently most impacted by changes in any of the harvest patterns relative to a no harvest baseline.

TABLE 3. Changes in habitat quality and fragmentation in response to different harvest schedules in the Nahmint landscape unit (ha).

Schedule	Habitat Dependency Group	Maximum Good Habitat	Minimum Good Habitat	Maximum Fragment	Minimum Fragment
Soil/ Riparian	Cavity Users	9,605	4,088	1	1
	Riparian Small Mammal	570	388	10	6
	Perching Bird	801	272	4	1
	Elk Winter	131	11	7	2
	Interior Old Growth.	4,647	133	9	4
	Edge	3,944	1,044	11	4
Full Constraint	Cavity Users	9,605	6,702	1	1
	Riparian Small Mammal	572	380	9	6
	Perching Bird	560	183	8	2
	Elk Winter	131	9	9	6
	Interior Old Growth.	4,647	680	10	7
	Edge	3,064	633	11	4
Zoned	Cavity Users	9,605	7,460	1	1
	Riparian Small Mammal	570	423	9	7
	Perching Bird	559	214	6	2
	Elk Winter	131	9	5	1
	Interior Old Growth.	4,647	2,667	9	8
	Edge	2,055	594	8	2
No Harvest	Cavity Users	12,138	9,605	1	1
	Riparian Small Mammal	570	411	8	8
	Perching Bird	559	185	6	2
	Elk Winter	131	9	5	1
	Interior Old Growth.	4,728	4,647	8	7
	Edge	1,951	0	7	0

Differences in fragmentation response were evident across schedules for the perching bird group. They had a negligible increase in maximum good habitat and reduced fragmentation in the soil/riparian schedule compared to the no harvest condition. Alternatively, the riparian small mammals exhibited little change in maximum good habitat and stable levels of fragmentation across schedules. These different "views" of fragmentation reflect the wider dispersal and stronger affinity to early seral conditions modeled in the perching bird group as compared to riparian dependent species. The fragmentation maps allow for a visual assessment of connectivity problems. Figure 5 illustrates habitat quality (red centers) and fragmentation (haloed islands) for three guilds resulting after 80 years of three different harvest schedules.

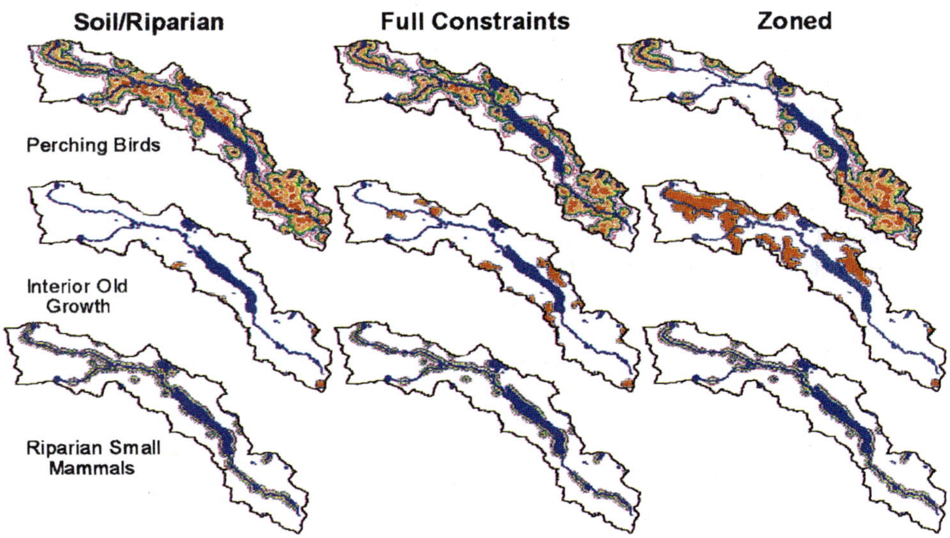

Figure 5. Nahmint habitat assessment by guild and schedule (80 Years)

3.2.2 Habitat Quality

Habitat levels for cavity users and interior old-growth species were notably impacted by harvest schedule. The soil/riparian schedule reduced cavity users minimum good habitat by 40% (2,600 ha) and interior old-growth habitat by 80% (550 ha) from full constraints. The zoning schedule increased the minimum good habitat for cavity users by 3% and interior old-growth habitat by 300% (2,000 ha) over the full constraints (Table 3).

Habitat for the edge group responded in the opposite fashion. The soil/riparian schedule increased the minimum good habitat by 65% (410 ha) over the full constraints. In addition, maximum good habitat increased by 30%. The zoning options decreased minimum good habitat by 6% and decreased maximum good habitat by 33% (1,009 ha) in comparison to the full constraints (Table 3).

Perching birds were the only group to show a positive habitat response to the soil/riparian schedule. Under this schedule they increased their maximum good habitat by 43% (240 ha) and their minimum good habitat by 49% (89 ha) compared to the full constraints (Table 3).

Both the perching bird and edge habitat dependency groups tend to favor early seral conditions and/or conditions associated with high contrast edges between seral stages. These conditions were more prevalent in the soil/riparian schedule due to the higher harvest level and a greater proportion of early seral conditions at any period.

For the provision of good habitat and minimal fragmentation, the most favorable schedule is the zoned schedule. Zoning keeps minimum and maximum habitat quality at or above the full constraints levels (with the exception of the edge habitat). It would provide the highest minimum cavity nesting bird and small mammal riparian habitat, increase interior old-growth habitat minimums by 2,000 ha and decrease edge maximum habitat by 1 000 ha compared to current regulations.

3.3 COMMERCIAL FOREST DEVELOPMENT

We evaluated each of the schedules for their provision of commercial forest activity. The schedules resulted in very different levels of activity and harvest value (Table 4). The inclusion of biodiversity and aesthetic constraints for the fully constrained schedule reduced the forest activity in terms of volume harvested, derived NPV, and short-term conversion return. The zoned schedule resulted in intermediate harvested volume, NPV and short-term conversion returns.

3.3.1. Harvest Volumes

The harvest volumes decreased from the soil/riparian schedule to the full constraint schedule to 56%. Not surprisingly, the zoned schedule resulted in about half the harvest level of the soil/riparian schedule. This was slightly lower than the harvest level from the full constraint schedule.

3.3.2 Harvest Net Present Values

The harvest NPVs decreased with the full constraint schedule to 15% of that from the soil/riparian schedule. The NPVs from the zoned schedule were higher than the full constraint schedule. This is a reflection of cutblock dispersion and the accelerated road building resulting from the full constraints. Thus, even with a similar harvest volume, the accelerated road building of the full constraint schedule resulted in a lower NPV.

3.3.3 Conversion Returns

The average long-term conversion returns were similar for the three schedules. The short-term average conversion return tells quite a different story. The short-term conversion return decreases between the soil/riparian and full constraint. This reflects the consequence of short-term unavailability of high value timber and accelerated road building required under full constraints. The short-term average conversion return also decreases between the soil/riparian and zoned schedules. This is a result of the initial availability of the higher value timber in the top half of the landscape unit. However, over the long term, the historical road building in the bottom results in higher conversion return as the second-growth forest matures.

TABLE 4. Commercial forestry results for each of the case study harvest schedules

Schedule	Logged (ha)	Volume Logged (1,000 m^3)	NPV ($1000 at 8%)	80-Yr average conversion return ($/$m^3$)	1st 5-Yr average conversion return ($/$m^3$)
Soil/ Riparian	9,293	6,060	10,086	17.95	1.58
Full Constraint	5,329	3,370	1,500	17.78	-9.61
Zoned	4,552	2,947	3,579	17.55	-2.14

4. Conclusions

This case study explicitly assesses conservation of biological diversity in forested landscapes by modeling spatial habitat supply and assessing habitat quality and fragmentation for an array of vertebrate species. The harvest scenarios used highlight the following points:

- compliance with biodiversity conservation rules may not achieve habitat objectives;
- zoning is an effective means of achieving habitat and financial objectives.

Our results highlight the difficulty of a rule based approach and the need to define a desirable mix of habitats. Increasing habitat for species that utilize old growth will be at the cost of species that utilize early seral stages. While some have called for targeting the species or habitat mix found under a pattern of natural disturbance, this might only be accomplished with the complete cessation of resource development. Any other desired mixture will necessitate value judgements and will be achieved at the cost of frustrating the desires of some segment of society as well as altering biological communities in "unnatural" ways.

Specifically, we show that the cost of regulation relating to biodiversity and aesthetics can be as high as $9 M for a single landscape unit of which there may be over 125 on Vancouver Island alone. From the standpoint of habitat provision, this reduction in harvest value was associated with an overall increase in good habitat for cavity users and for interior old-growth species, but a reduction of habitat for other species (i.e., edge and perching bird groups). There was also a general increase in fragmentation associated with these regulations.

The zoning scenarios show a better habitat supply situation achieved by concentrating the less constrained harvesting in half of the landscape and excluding harvest from the other half. This resulted in a higher NPV and a more favorable initial cash flow than highly constraining the whole landscape . In this case study, zoning proved better than multiple use.

The results and conclusions from this case study are specific to the Nahmint landscape unit and the scheduling options we tested. Several other options need to be explored. Partial harvests hold the promise of providing an alternative method for commercial forest development that could provide for habitat attributes that are inconsistent with clearcutting methods. A natural disturbance baseline should be developed to illustrate departures in habitat supply and fragmentation for harvest schedule evaluation. Finally, a hydrological model needs to be incorporated into this framework to estimate the impact of harvest patterns on flow, sedimentation, and stream channel stability.

The most important conclusions are that habitat conservation can be costly and that zoning should be examined as an option in providing a mix of commercial forestry and biodiversity values.

References

Anon 1994. Forest Practices Code of British Columbia Act. S.B.C. 1994. C. 41.

Blair, W.F. 1940. Home ranges and populations of the meadow vole in southern Michigan. J. Wild. Management. 4: 149-161.

Carey, A.B. 1989. Wildlife associated with old-growth forests in the PNW. Natural Areas Journal. 9(3): 151-162.

Clements, S.E., Dallain, P.L. and Jamnick, M.S. 1990. An operational, spatially constrained harvest scheduling model. Can. J. For. Res. 20: 1438-1447.

Cox, E.S. and Sullivan, J. 1995. Harvest scheduling with spatial wildlife constraints: an empirical examination of trade offs. J. Env. Man. 43: 333-348.

Daust, D.K. 1994. Biodiversity and land management: from concept to practice. Unpublished M.Sc. thesis. Faculty of Forestry. University of British Columbia. 99 p.

Daust, D.K. and Bunnell, F.L. 1992. Predicting biological diversity on forest land in British Columbia. Northwest Environ. J. 6(2): 191-192.

Daust, D.K. and Bunnell, F.L. 1994. Geographic information systems and forest wildlife: recent developments and future prospects. Proceedings of the XXI IUGB Congress. 1: 348-356.

de Vos, A. 1951. Overflow and dispersal of marten and fisher from wildlife refuges. J. Wild. Mgmt. 15: 164.

Dunning, J.B., Danielson, B.J. and Pulliam, H.R. 1992. Ecological processes that effect populations in complex landscapes. Oikos. 65(1): 169-175.

Ehrlich, P.R. 1988. The loss of diversity: causes and consequences. In: Wilson, E.O. and Peter, F.M. (eds.). Biodiversity. National Academy Press. Washington, DC. Pp. 21-27.

Ehrlich, P.R. and Ehrlich, A.H. 1981. Extinction: The Causes and Consequences of the Elimination of Species. Random House. New York. 305 p.

Ehrlich, P.R. and Mooney, H.A. 1983. Extinction, substitution and ecosystem services. BioScience. 33(4): 248-254.

Flyger, V.F. 1960. Movements and home range of the gray squirrel (*Scuturus carolinensis)* in two Maryland woodlots. Ecology 41: 365-369.

Green, R.N. and Klinka, K. 1994. A field guide to site identification and interpretation for the Vancouver Forest Region. BCMOF Research Branch. Crown Publications. 285 p.

Huff, M.H., Holthausen, R.S. and Aubry, K.B. 1992. Habitat management for the red tree voles in Douglas-fir forests USDA Forest Serv. General Tech. Report PNW-GTR-302. 16 p.

Haley, D. 1996. Paying the piper. Prepared for: Working with the B.C. Forest Practices Code. Insight Information Inc. Vancouver. B.C. 10 p.

Hansen, A.J., Spies, T.A., Swanson, F.J. and Ohmann, J.L. 1991. Conserving biological diversity in managed forests: lessons. Bioscience. 41(6): 382-392.

Holekamp, K.E. 1984. Natal dispersal in Belding's ground squirrel (*Spermophilus beldingi*). Behav. Ecol. Sociobiol. 16: 21.

Kareiva, P. 1987. Habitat fragmentation and the stability of predator-prey interactions. Nature. 326: 388-390.

Kareiva, P. 1990. Population dynamics in spatially complex environments. Phil. Trans. R. Soc. Lond. B 330: 175-190.

Kirkpatrick, S., Gelatt, C.D. Jr. and Vecchi, M.P. 1983. Optimization by simulated annealing. Science. 220: 671–680.

Lockwood, C. and Moore, T. 1992. Harvest scheduling with spatial constraints: a simulated annealing approach. Can. J. For. Res. 23: 468-478.

Meidinger, D. and Pojar, J. (eds.) 1991. Ecosystems of British Columbia. British Columbia. Ministry of Forests. Special Report Series No. 6. 330 p.
Morrison, M.L., Marcot, B.G. and Mannon, R.W. (eds.) 1992. Wildlife-Habitat Relationships, Concepts and Applications. Chapter 3. Habitat Fragmentation. Univ. Wisconsin Press. Pp. 41-97.
Nelson, J.D., Brodie, J.D. and Sessions. J. 1988. Integrating short-term, area-based logging plans with long-term harvest schedules using Monte-Carlo integer programming and linear programming. In: Symp. on Systems analysis in forest resources. USDA For. Serv. Gen. Tech. Rep. RM-161.
Nelson, J.D. and Brodie, J.D. 1990. Comparison of a random search algorithm and mixed integer programming for solving area-based forest plans. Can. J. For. Res. 20: 934-942.
Nelson, J.D. and Shannon, T. 1994. Cost and timber supply assessment of the coastal biodiversity guidelines. Unpublished. Faculty of Forestry. University of British Columbia. 40 p.
Nelson, M.E. 1993. Natal dispersal and gene flow in white-tailed deer in northeastern Minnesota. J. Mammal. 74: 316.
O'Hara, A.J., Faaland, B.H. and Bare, B.B. 1989. Spatially constrained timber harvest scheduling. Can. J. For. Res. 19: 715-724.
Province of BC. 1995. Biodiversity Guidebook. Co-published by BC Ministries of Forests and Environment. ISBN 0–7726–2619–7. 99 p.
Ralph, C.J., Paton, P.W.C. and Taylor, C.A. 1991. Habitat association patterns of breeding birds and small mammals in Douglas-fir/hardwood stands in northwestern California and southwestern Oregon. USDA For. Serv. General Tech. Report PNW-GTR-285. 18 p.
Raphael, M. G. and Marcot. B.G. 1984. Validation of a wildlife-habitat-relationships model: Vertebrates in a Douglas-fir sere. In: Verner, J., Morrison, M.L. and Ralph C.J. (eds.).Wildlife 2000-Modeling Habitat Relationships of Terrestrial vertebrates. Univ. Wisconsin Press. Pp. 129-138.
Roise, J.P. 1990. Multicriteria nonlinear programming for optimal spatial allocation of stands. For. Sci. 36: 487-501.
Sutherland, G. 1996. Habitat dependency groups of vertebrates in coastal western hemlock forests. (In prep.)
Trewhella, W.J., Harris, J.S. and McAllister, F.E. 1988. Dispersal distance, home range size and population density in the red fox: a quantitative analysis. Journal of Applied Ecology. 25:.423.
Turner, M.G. 1989. Landscape ecology: the effect of pattern and process. Ann. Rev. Ecol. Syst. 20: 171-197.
van Kooten, G.C. 1994. Economics of biodiversity and preservation of forestlands in British Columbia. FRDA Publication No. 21. 35 p.
Wells, R. 1996. Forest attribute trajectories in coastal temperate rainforests. (M. Sc. Thesis in prep.)
Wells, R. and Daust, D.K. 1994. A preliminary analysis of stand structure in coastal western hemlock zone forests of various ages. Unpublished report. MacMillan Bloedel Ltd. 25 p.
Wiens, J.A. 1989. Spatial scaling in ecology. Functional Ecology. 3: 385-397.
Wilson, E.O. 1984. Biophilia. Harvard University Press. Cambridge. Mass. 548 p.
Wilson, E.O. 1985. The biological diversity crisis: A challenge to science. Issues Sci. Technol. 2: 245-251.
Wilson, E.O. 1988. The current state of biological diversity. In: Wilson, E.O and Peter, F.M. (eds.). Biodiversity. National Academy Press. Washington D.C. Pp. 3-18.
Wilson, E.O and Peter, F.M. (eds.) 1988. Biodiversity. National Academy Press. Washington D.C. 521 p.
Yoshimoto, A. and Brodie, J.D. 1994. Short and long-term impacts of spatial restrictions on harvest scheduling with reference to riparian zone planning. Can. J. For. Res. 24: 1617-1628.
Yoshimoto, A., Brodie, J.D. and Sessions, J. 1994. A new hueristic to solve spatially constrained long-term harvest scheduling problems. For. Sci. 40: 365-396.

ASSESSING STRUCTURAL DIVERSITY IN MANAGED FORESTS

R. FERRIS-KAAN, A. J. PEACE AND J. W. HUMPHREY[1]
Forestry Commission Research Agency, Alice Holt Lodge, Wrecclesham, Farnham, Surrey, England GU10 4LH
[1]Forestry Commission Research Agency, Northern Research Station, Roslin, Midlothian, Scotland EH25 9SY.

Abstract

Many spatial scales are recognised within multi-objective forest management, but relatively little is known about the effects of current management regimes on biodiversity at a stand level in British forests. This is important, as most management operations are carried out at this level within commercial forests. This paper describes one aspect of a project to assess forest biodiversity at the stand level, namely the description of vertical stand structure. The Project aims to assess levels of biodiversity associated with plantation stands, through the evaluation of new field assessment methods, the collection of baseline data, and an appraisal of the effects of management through the complete range of structural classes generated by clearfelling and replanting silvicultural methods. In other countries, chronosequences have been examined for this purpose.

A three-year project has been initiated to collect baseline ecological data for major plantation forest types in the UK – Sitka spruce (*Picea sitchensis*), Norway spruce (*Picea abies*), Scots pine (*Pinus sylvestris*), Corsican pine (*Pinus nigra* var. *maritima*), and oak (*Quercus robur* and *Quercus petraea*). This information will then be compared with data obtained from analogous semi-natural and high conservation value forests, to help define appropriate target levels of biodiversity that should be achievable within plantations. In addition to generating these baseline data, the project aims to identify key biodiversity indicators for validation and use in future assessment programmes, and the development and monitoring of biodiversity standards. These are standards which are relevant to the site type in question, which are ecologically robust, are clearly understood and acceptable to forest managers and other stakeholders, and which use methodologies which are repeatable and easy to use.

A standardised series of assessments are being undertaken in target sampling areas, comprised of one hectare plots (one within each structure class), and using methods which take account of patterns and processes at larger spatial scales. It is intended that this work be extended to assess landscape scale variation and to incorporate additional forest types and management regimes across Europe, along a forest exploitation and climatic continuum.

This work aims to provide a baseline against which the impacts of changes in management to achieve desired future conditions (targets) can be compared. These targets are defined in terms of structural, compositional and functional attributes of biodiversity, as applied to plantation forest ecosystems. Appropriate levels for these need

P. Bachmann et al. (eds.), Assessment of Biodiversity for Improved Forest Planning, 331–342.
European Forest Institute Proceedings no. 18. © 1998 *Kluwer Academic Publishers. Printed in the Netherlands.*

to take into consideration factors such as pre-afforestation habitat types, physio-chemical site limitations, and the ecological uniqueness of plantation forests in the United Kingdom.

1. Project Background and Objectives

Assessing Biodiversity in Managed Forests is one Project in a broader programme of research being undertaken by the Forestry Commission Research Agency, *The Biodiversity Research Programme* (Hodge et al. this volume). It forms the basis for further research, through the evaluation of field assessment methods and the collection of baseline data. The broad objectives of the Project are:

1. Establishment of chronosequences of different stand ages in major forest types in the UK, under current management of patch clear-felling;

2. Collection of baseline information on levels of biodiversity within and between these chronosequences;

3. Development and validation of biodiversity assessment protocols and *indicators*, with particular reference to the use of the latter as a tool in the rapid and effective assessment of biodiversity in other forests.

1.1 THE IMPORTANCE OF STAND STRUCTURE

Recognising that forest structure can be altered through management, and that species composition alone may not be useful as a biodiversity indicator in forests which are often composed of non-native species, we have concentrated our early research on examining the general relationships between vertical structure and spatial arrangement of vegetation as a means of describing the ecosystem.

In this paper, we present some ideas concerning simple-to-use methods for assessing vertical stand structure, collating these data and their expression. Data are presented for two contrasting Scots pine (*Pinus sylvestris*) chronosequences, one in the lowlands of southern England and the other in the foothills of Scotland.

1.2 CHRONOSEQUENCES AND SITE DEFINITION

Chronosequences have been widely used in forest research (e.g. Pollard 1993), and true chronosequences should have similar soils, climate and site histories (Spies 1991). They should be located on slopes of similar topography, altitude and aspect. Their location in relation to other landscape units should also be comparable. In practice, to satisfy all of these criteria is extremely difficult within the constraints imposed by a working forest.

In selecting sites for this study, a decision was taken to concentrate on only the main commercially grown species. Sampling across the UK was stratified by dividing the country into three broad bioclimatic zones – Uplands, Foothills, and Lowlands – according to a system of *Ecological Site Classification* (Pyatt 1995). These zones are defined by two main variables: *accumulated temperature* (number of day degrees

>5.6°C), and *soil moisture deficit* (mm). Hence, the three zones are as follows:

Lowlands – Warm and Dry, with AT 1375-1925 and SMD 101-180;
Foothills – Cool/warm and moist, with AT 676-1925 and SMD 41-100;
Uplands – Cool/warm and wet, with AT 676-1925 and SMD 0-40.

Since the definition of the zones is more dependent on the wetness element than temperature, a useful and simpler surrogate measure for distinguishing between zones is annual rainfall: >1500mm – Uplands; 800-1500 – Foothills; <800mm – Lowlands.

Initial attempts to define chronosequences solely on the basis of stand age became difficult because of the different species involved, site factors and silvicultural practices. Actual stand age was not as reliable as *Structure Class* in determining the range of stand types required for this study:

Pre-thicket – Ideally a restock site (2nd rotation), comprised of a pure crop (but with some minor natural regeneration of other species) at standard 2m spacing. Crop height should be 2-4m, but variation is likely, and the canopies of adjacent trees should be beginning to interlock in some parts of the site. The ground vegetation should not yet be shaded out.

Mid-rotation – A well-stocked pure stand, not thinned or brashed, with complete canopy closure. Stands should be characterised by reduced light levels under the canopy and little ground vegetation. Crop height should be between 10-20m, with lower branches intact and restricting access.

Mature – This stage should have had at least one thinning, with complete canopy cover of the selected crop species. Crop height should be between 20-25m, with lower branching much reduced. Light levels under the canopy will be higher than in the previous structure class, with more extensive ground vegetation and a developing shrub layer in places. Windthrow should be minimal.

Over-Mature – An older plot, possibly a long-term retention, which has been thinned but has suffered windblow and a degree of canopy degeneration, allowing development of natural regeneration, creation of understorey layers and accumulation of large-sized deadwood. Crop height should be roughly the same as for the mature stage. This stage will be characterised by an incomplete canopy, giving a varied light environment on the forest floor, with extensive ground vegetation development in places.

2. Selection of Assessment Parameters

The multiplicity of definitions for biodiversity make it difficult to decide just what to measure, although there is well documented agreement among forest ecologists that structural and functional diversity should be maintained throughout the forest landscape (e.g. Franklin 1988; Ratcliffe 1995). Within the Biodiversity Research Programme, a working definition of biodiversity has been agreed which encompasses its compositional, structural and functional components. This is a paraphrase of that developed by the United Nations Convention on Biodiversity: "the variability among living organisms from all sources ... and the ecological complexes of which they are part" (UNCED 1992). At this stage in our understanding of plantation forests in the UK, it is not possible to link this to a specific desired future condition or target. Instead, we aim to describe present baseline conditions, while assessing realistic and ecologically meaningful targets under UK conditions.

2.1 VERTICAL STRUCTURE ASSESSMENT

Our method has been developed in the field over two field seasons, 1995-96, and is derived from subjective methods for estimating plant cover in the horizontal plane, familiar to and widely-used by vegetation scientists, e.g. Foliage Height Diversity measures as used in assessments of habitat quality for birds (Erdelen 1984). It involves measuring out a 10m radius from the observer, defining an arc 20m across, and then sub-dividing the canopy into four vertical height bands (Figure 1). Any plant bisecting the arc is recorded, although recognition of the difficulty in assessing this has led us to operate within a 1 metre-wide band (i.e. 0.5 metre either side of the arc). Any part of a plant bisecting this band is included, as long as it is rooted within the 1 hectare study plot. The observer is required to calculate the percent cover of each species within each height band. Until familiar with the method, observers require the use of a clinometer in order to gain a more accurate impression of where the height bands meet.

Within each 1 hectare study plot, sixteen vertical structure assessments are taken, four within each of the 0.25 hectare quadrants. Their location is decided in advance of visiting the site, two assessments due North of the centrepoint of the quadrant (10 metres apart), and two due South. Each series of assessments is carried out by a team of two observers, selected from a pool of six. By combining observers from different geographical locations, it has been possible to ensure a degree of comparability between the results from Upland, Foothill and Lowland forest types.

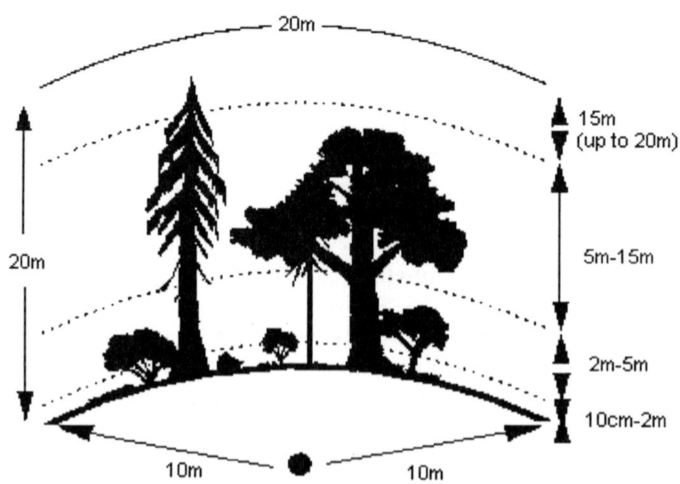

Figure 1. Vertical Structure Assessment: definition of vegetation strata

Observer bias has been minimised through training and field trials have demonstrated that any tendency to over- or under-estimate cover is consistent and predictable, thereby enabling adjustment (E. A. Dickson, pers. comm.). Cover is assessed down to a 5 percent level. For field and shrub layer vegetation, direct validation of visual assessments is possible, and these do not usually present any problems for observers. However, the estimation of cover in the lower and upper canopy layers is much more difficult. Estimates of vertical cover in the lower canopy have been shown to be the most variable between observers, possibly reflecting the inherent variability of this stratum. It is necessary to traverse the 20m arc, making a visual assessment of canopy cover in the horizontal plane directly above the observer.

3. Results

With a large sampling programme across three bioclimatic zones, with five forest types, it has been necessary to condense the details down to one suite of assessment plots for the purposes of this paper. Data are presented for a *Pinus sylvestris* chronosequence in southern England (i.e. the lowland bioclimatic zone), with three plots (one per chronosequence stage) in the New Forest, Hampshire (National Grid References: pre-thicket - SU 253 064, pole-stage - SU 333 053, mature - SU 344 038) and the fourth, over-mature plot in Windsor Forest, Berkshire (SU 884 655). They are all on broadly similar soils, well-drained sands and gravels of a podzolic nature. In addition, results from a chronosequence at Glen Affric, in the foothills zone, are included (pre-thicket - NH 317 363, pole-stage - NH 279 261, mature - NH 273 237, over-mature - NH 238 258). This is within the natural distribution zone for *Pinus sylvestris* in Britain, and its inclusion allows comparisons to be made with a semi-natural Caledonian pinewood stand, as represented by the over-mature plot.

3.1 VEGETATION COMPOSITION

3.1.1 New Forest and Windsor

The pre-thicket plot, comprised of closely-spaced *P. sylvestris* which have regenerated naturally following clearfelling of the previous crop, is very species-poor. The density of trees has limited opportunities for other woody species to regenerate, and the development of a dense mat of *Molinia caerulea* has restricted the growth of other herbs. Particularly notable species which were absent from this plot included bramble (*Rubus fruticosus* agg.) and heather (*Calluna vulgaris*).

In the pole-stage plot, the dominance of bracken (*Pteridium aquilinum*) in the field layer is a characteristic of *P. sylvestris* plantations on podzolic soils in lowland Britain. This species is diagnostic for *Quercus* spp. – *Betula* spp. – *Deschampsia flexuosa* woodland, as recognised by the *National Vegetation Classification* (Rodwell 1991). However, the observed community does not match the semi-natural community due to the absence of *Calluna vulgaris* and bilberry (*Vaccinium myrtillus*), and the under-representation of *Quercus* spp. and *Betula* spp. The build-up of persistent litter and shading from the close canopy has resulted in a ground vegetation characterised by bryophytes, notably *Thuidium tamariscinum* and *Pseudoscleropodium purum*. .

The mature plot is more species-rich than the younger plots, with the open canopy providing greater opportunities for vascular plant species to establish in the sub-canopy layers. *Calluna vulgaris,* although present, is under-represented, and a number of grass species, notably *Molinia caerulea* and tufted hair-grass (*Deschampsia cespitosa*), are more abundant as a consequence of seasonal waterlogging. Natural regeneration of *P. sylvestris* is patchily distributed across the plot and, with occasional shrub cover of other species, may give rise to a two-tiered stand structure. *Pteridium aquilinum* is still locally dominant, presenting a management problem if restoration to heath or wooded-heath is an objective. Although the mature plot has greatest similarity in species composition with *Quercus* spp. - *Betula* spp. - *Deschampsia flexuosa* woodland, the presence of heathland elements is notable.

The over-mature plot, at Windsor, is more diverse, with the ground vegetation present as a mosaic of different communities, with extensive patches of *Calluna vulgaris* and bell heather (*Erica cinerea*), interspersed with areas dominated by *Pteridium aquilinum,* dense clumps of regenerating *P. sylvestris* and large patches of bare ground. In addition to the canopy of *P. sylvestris*, other woody species present include rhododendron (*Rhododendron ponticum*) and sweet chestnut (*Castanea sativa*), the latter having previously been managed by coppicing. These species both produce persistent litter, influencing the distribution patterns of herbaceous ground layer vegetation. The heathland elements in the plant community are more prevalent in the over-mature plot, and the presence of bare ground provides a suitable seedbed for regeneration of *P. sylvestris*.

3.1.2 Glen Affric

For the sites at Glen Affric, the most similar semi-natural community type would be expected to be *Pinus sylvestris - Hylocomium splendens* woodland (Rodwell 1991). Only in the pre-thicket stage plot is this not so, where the community is more similar to an upland heath of the *Calluna vulgaris - Vaccinium myrtillus* type. The pole-stage plot had impoverished field-layer vegetation, with only *Pteridium aquilinum* present to any extent. The shrub-layer in both pole-stage and mature plots was very species-poor, with only occasional *Betula pubescens* in addition to *P. sylvestris*. The field-layer vegetation in the over-mature plot was characterised by the development of hummocks on which *Calluna vulgaris* growth was particularly good, and the cover of vegetation in the field-layer has increased to levels similar to those in the pre-thicket stage plot.

3.2 GENERAL PATTERNS OF VERTICAL STRUCTURE

In spite of the obvious differences between sites, particularly the comparison between lowlands and foothills, there are a number of general patterns in vertical stand structure which can be recognised. All *P. sylvestris* stands, irrespective of age/structure class and origin (whether plantation or semi-natural), are characterised by the presence of horizontal and vertical gaps in all strata. In the case of canopy strata, this can be explained in part by the canopy architecture characteristic of *Pinus* species. *P. sylvestris* in plantation stands does not exhibit a deep crown structure, leaving opportunities for other species to occupy the lower canopy and sub-canopy space. The canopy of *P. sylvestris* stands tends to be uneven, particularly in comparison with similar stands of *Pinus nigra* var. *maritima*, and yet despite this the gaps do not appear to be rapidly exploited by other tree species. Birch (*Betula pendula* and/or *B. pubescens*) is most often the pioneer species which takes advantage of the physical space provided. Self-sown *P. sylvestris* seedlings did not appear to exploit spaces in close proximity to canopy trees in our study plots, but were recorded at greater distance away. This was particularly noticeable in the over-mature plots at Windsor and Glen Affric.

There was considerable variability in vertical structure profiles among the four stand classes at both Glen Affric and the New Forest and Windsor (Table 1). For Glen Affric, the intermediate stand classes – pole-stage and mature-stage – displayed similar ratios of lower canopy cover to field layer cover. These are calculated by summing mean cover values for all four plant groups. For the pole-stage plot: lower canopy cover = 11.87 (10.93 + 0.94), and field layer cover = 8.13, giving a ratio of 1.5; and for the mature plot: lower canopy cover = 5.62 (5.31 + 0.31), and field layer cover = 4.69 (3.13 + 1.56)., giving a ratio of 1.2. This contrasts with the corresponding ratio in both the pre-thicket (no lower canopy present)and over-mature (0.23) stages, in which the field layer cover was much greater and the species composition more diverse. Shrub layer cover was more extensive in the pole-stage plot than any of the others. The New Forest pre-thicket plot was found to differ from the structure which might have been expected from a more open plot in which an extensive field layer could thrive prior to canopy closure. In this study plot, field layer cover has been greatly reduced due to the high density of *P. sylvestris* stems. As an overstocked, naturally regenerated stand, there was little opportunity for ground and field layer vegetation to survive, and the only species to achieve any significant cover was *Molinia caerulea*. Tree height was also greater than would normally be expected for a Pre-thicket plot, with the result that there was higher cover in the lower canopy than in the shrub layer. In this respect, the plot was intermediate between pre-thicket and pole-stage. In both the pole-stage and mature plots at the New Forest, the

TABLE 1. Mean percentage cover for major plant groups in each of four vertical strata, through a chronosequence of structure classes: (a) New Forest/Windsor (b) Glen Affric. Key: UC - Upper Canopy, LC - Lower Canopy, S - Shrub, F - Field. (a) New Forest/Windsor

		Pre-thicket		Pole stage		Mature		Over-mature	
		mean	s.e.	mean	s.e.	mean	s.e.	mean	s.e.
Crop trees	UC	0	-	12.50	1.88	15.31	2.39	3.44	0.99
	LC	19.38	1.70	3.13	0.63	4.69	0.85	1.56	1.27
	S	0	-	0	-	0	-	0.31	0.31
	F	0	-	0	-	0	-	5.00	2.19
Other tree species	UC	0	-	0.63	0.43	2.19	1.64	0.63	0.43
	LC	8.44	2.03	11.25	4.25	2.81	1.44	7.19	1.94
	S	4.69	1.25	4.38	1.70	1.25	0.72	4.06	1.66
	F	0.31	0.31	0	-	0.63	0.63	3.75	1.16
Heather	UC	-	-	-	-	-	-	-	-
	LC	-	-	-	-	-	-	-	-
	S	-	-	-	-	-	-	-	-
	F	0	-	0	-	0	-	0.94	0.50
Other herbs	UC	-	-	-	-	-	-	-	-
	LC	-	-	-	-	-	-	-	-
	S	-	-	-	-	-	-	-	-
	F	0.94	0.50	16.88	3.29	29.38	5.44	16.25	4.88

(b) Glen Affric

		Pre-thicket		Pole stage		Mature		Over-mature	
		mean	s.e.	mean	s.e.	mean	s.e.	mean	s.e.
	UC	0	-	0	-	0	-	0.31	0.31
Crop trees	LC	0	-	10.93	1.53	5.31	1.07	4.69	0.72
	S	4.17	1.04	0.94	0.50	0	-	0.63	0.43
	F	15.42	2.72	0	-	0	-	0.94	0.50
	UC	0	-	0	-	0	-	0	-
Other tree	LC	0	-	0.94	0.68	0.31	0.31	0	-
species	S	2.08	0.74	0	-	0	-	0	-
	F	9.58	2.42	0	-	0	-	0	-
	UC	-	-	-	-	-	-	-	-
Heather	LC	-	-	-	-	-	-	-	-
	S	-	-	-	-	-	-	-	-
	F	13.33	1.42	0	-	3.13	0.90	10.31	2.30
	UC	-	-	-	-	-	-	-	-
Other herbs	LC	-	-	-	-	-	-	-	-
	S	-	-	-	-	-	-	-	-
	F	5.83	3.07	8.13	3.32	1.56	0.99	8.75	1.07

field-layer was dominated by *Pteridium aquilinum*, giving rise to a relatively homogenous sub-canopy structure. In the mature plot the cover of *P. aquilinum* was patchily distributed, and where it did not dominate, other plant species occupied the field-layer. This contrasts with the over-mature plot at Windsor, where the field-layer cover was much more variable. The presence of *Castanea sativa* as a canopy tree in this plot has led to patches of ground with no vegetation cover, only a deep layer of leaf litter. Consequently, there is considerable variation in structure assessments.

The absence of a well-developed shrub-layer in any of the plots may be the result of many interacting factors. In the south of England sites, which have seen *P. sylvestris* planted onto either open heath or replacing semi-natural woodland, the field-layer is characteristically species-poor (Rodwell 1991). Thus, heath shrubs and *Pteridium aquilinum* form the characteristic gap flora, and these may be maintained through the dual action of grazing and, to a lesser extent, burning. However, where the *P. sylvestris* canopy is more complete, these elements may be reduced to a fringe along the edges of rides and recently-cleared areas. Under a more uniform canopy, *Pteridium aquilinum* may dominate the field layer, but has been found to occur in a more patchy distribution with other species as the canopy thins out in the mature and over-mature plots.

Canopy structure shows a consistent pattern for both the New Forest/Windsor and Glen Affric, with little change through the chronosequences. Upper canopy cover, for example, is always low. Although, as the trees grow, more of their foliage enters the canopy layers, self-thinning or management reduces the number of stems, with the result that the profile comprises more gaps. Furthermore, *P. sylvestris* has a characteristically open canopy compared with other conifers, *Picea* spp. for example. Hence, it is unlikely that vertical cover will ever exceed 30-40 percent in the canopy layers. However, mean cover values alone can be misleading, as they conceal valuable ecological information concerning the patterns of cover distribution. For example, although the cover values for upper canopy layer in the over-mature plots at Windsor and Glen Affric are similar, there are significant differences in the way in which that cover is distributed. The structure of the semi-natural *P. sylvestris* at Glen Affric is of very widely-spaced trees with spreading canopies, typically referred to as "granny" pines by virtue of their old age. In contrast, the trees at Windsor are more closely spaced, but they have narrower crowns. They form a noticeable forest canopy, whereas Glen Affric is predominantly open heather moorland with a sparse cover of ageing trees, where associated trees are rarely more than occasional.

4. Discussion

4.1 NEW FOREST AND WINDSOR

The plots at the New Forest and Windsor have been subjected to rather frequent management intervention, with the notable exception of the pre-thicket stage plot. This plot is atypical in that it has arisen from naturally regenerating *P. sylvestris,* which has not been re-spaced, and it has developed into a dense thicket with very impoverished ground- and field-layer vegetation. All other plots in this Lowland chronosequence have been subjected to thinning (pole-stage) and an extra final thinning (mature and over-mature stages). In the case of the over-mature plot at Windsor, the stand density has been deliberately reduced to 30-40 trees per hectare, with the intention of restocking by natural regeneration. As a consequence of management, the plots are relatively homogenous in terms of their structure, although the mature plot shows some spatial variation as a consequence of having been planted as a *Pinus sylvestris - Picea abies* mixture, from which all *P. abies* has been felled.

In common with the Glen Affric plots, the New Forest plots in particular have been subjected to heavy grazing and browsing pressures, both from deer and from the Commoners' ponies which are turned out into the open forest. There is little recruitment into the shrub-layer and, although seedling regeneration is occurring in patches where a seedbed of bare ground is present, the rank vegetation of grasses (unpalatable to deer) appears to prevent successful establishment. This is supported by Kuper (1994), who reports that *P. sylvestris* is unlikely to regenerate in a mixed herb and shrub vegetation, although a few seedlings have been found able to survive the first growing season under such conditions. Furthermore, losses due to herbivory by small mammals, such as bank voles (*Clethrionymys glareolus*) are likely. It is probable that with a reduction in grazing pressure there would be an increase in both the tree species diversity and structural variation , possibly through regeneration of *Sorbus aucuparia* and *Betula* spp., both of which are absent from samples in the three New Forest plots. The heath component of the field-layer vegetation, mainly *Calluna vulgaris*, is under-represented compared with what would be expected from National Vegetation Classification predictions for the site types, and several additional species are absent. The presence of *Pteridium aquilinum* as a constant feature of the understorey further impoverishes the field-layer. Consequently, there is little structure in the sub-canopy layers, which gives rise to a very uneven vertical structure profile.

The over-mature plot at Windsor shows potential for the development of a more diverse sub-canopy structure, largely as a consequence of the management which has taken place. In addition to a reduction in stand density, some vegetation removal through scarification has taken place, in order to provide a seedbed for *P. sylvestris* regeneration.

This has also aided the restoration of the heathland vegetation, with marked areas of *Calluna vulgaris* and *Erica cinerea* regeneration.

There is a marked absence of any additional structural variation resulting from snags and windblown trees, and what little deadwood is present in the plots comprises small diameter brushwood and some harvesting residue. Occasional non-crop trees, notably *Castanea sativa* of former coppice origin, present prior to conifer afforestation at Windsor, have added some structural variation through their deeper crowns. However, *Pinus sylvestris* plantations in the lowlands are characterised by their lack of structural variation. Since we have no "near-natural" baseline for lowland forests with *P. sylvestris* as a major component, it is not possible to set a specific target. Pinewoods elsewhere in Europe show considerable variation in their complexity, reflecting the wide ecological amplitude of *P. sylvestris* (Rodwell and Cooper 1995). The closest pinewood type to plantation stands in lowland Britain is the *Querco-Pinetum* which is favoured by the increased oceanicity of climate towards western Europe. In such a forest type, *Quercus robur* forms a mixed canopy with *P. sylvestris,* and heathy ground cover with species such as *Cytisus scoparius, Deschampsia flexuosa* and *Holcus mollis* (Ellenberg 1982). In the transition to acidophilous oakwoods, *Fagus sylvatica* increases levels of shading and thus reduces *P. sylvestris* to an occasional species. At this stage in our understanding of where plantation stands of *P. sylvestris* fit into the wider European classification of pinewoods, we can make the assumption that increased structural complexity confers some biodiversity benefit. In view of this, it is perhaps important to look for ways in which management could diversify stand structure through mimicking natural processes.

4.2 GLEN AFFRIC

The vertical structure data from Glen Affric presents a number of interesting findings. Firstly, the similarity between the canopy:field layer ratios in the pre-thicket and over-mature plots was unexpected. This is due, in part, to the low canopy cover values in the over-mature plot which has allowed more extensive development of the field-layer vegetation. Low canopy cover is a frequent feature within ancient semi-natural pinewood fragments, and is usually a consequence of timber removal in previous centuries and subsequent lack of regeneration due to over-grazing (Rodwell and Cooper 1995; Bain 1987; Goodier and Bunce 1977). The over-mature plot at Glen Affric is still subject to grazing by red deer (*Cervus elaphus*) and this will restrict the development of the shrub-layer. The occurrence of a shrub-layer in the plantation-origin pre-thicket plot is due to the development of the crop canopy, and the vertical profile does not comprise a number of different strata as is implied by Table 1b.

The over-mature plot, in the area of ancient semi-natural *P. sylvestris* forest, was originally selected as a "baseline" or "benchmark" site against which faunal and floristic characteristics of the plantation plots could be compared, making the assumption that this represented the desired future condition for *P. sylvestris* plantation stands. One of the original selection criteria was the occurrence of a high proportion of ancient trees, snags and fallen deadwood, all classic structural features of ancient semi-natural pinewoods (Reid et al. 1996). Whilst these structural characteristics may well be correlated with other facets of biodiversity (e.g. certain invertebrate groups, Young and Armstrong 1995), a number of structural elements are missing in comparison to what might be termed a "semi-natural pinewood structure". Nixon and Clifford (1995) have suggested that the spatial structure of more "natural" pinewood remnants in Scotland is characterised by groves of juvenile and middle aged trees in a matrix of widely spaced mature individuals, and the authors cite work in natural conifer forests in Scandinavia (see Engelmark et al. 1994, in: Nixon and Clifford 1995) which illustrates this spatial structure. Recent Forestry Commission work has highlighted the paucity of tree species in many native pinewoods (C.J. Nixon, pers. comm.). In instances where grazing pressure is relieved, the resultant mix of natural regeneration contains a high proportion of broadleaved species such as rowan (*Sorbus aucuparia* L.), birch (*Betula pubescens* L.), and willow (*Salix* spp.).

However, to recognise these newly regenerating stands, with their more diverse structure and composition as our "biodiversity bench-marks" poses new difficulties. These stands should rightly be considered as pioneer or early successional stands, and they will lack many of the elements of old growth" perceived to confer high biodiversity status on such forest stands. In addition, many of these stands owe their present composition and structure to past human interference. Natural disturbance is also a complicating factor. The Scottish pinewoods form an oceanic outlier of Eurasian boreal forests (Worrell 1996), but across their Scottish range they occur within a number of greatly contrasting climatic regimes, from the drier East, through to the windier and wetter West (Rodwell and Cooper 1995). Consequently, the prevailing natural disturbance regime changes across this climatic continuum. It is considered that stand-replacement disturbances due to fire have been more common in the eastern pinewoods, whereas in the West, small patch-disturbance due to windthrow is more likely (Nixon and Clifford 1995). Thus, "natural" even-aged stands regenerating after fire may be more likely in the East, and intimate mixed-aged stands more common in the west (e.g. Glen Affric). It has been suggested that the role of small gap replacement in pinewood dynamics may have been underestimated, and could be more ubiquitous than previously thought (Worrell 1996).

It is important to realise that the vertical structure profile approach to assessing biodiversity does have a number of limitations. First among these is the problem of representing spatial arrangement of total cover values for each of the strata, in such a way that a distinction is made between aggregated and dispersed cover. The use of some expression of evenness may help to overcome this, and we are seeking to refine the calculations accordingly. However, the real value of the current approach is that it is relatively simple and can be added into existing forest survey techniques used by foresters. There are numerous ways in which silvicultural intervention can alter forest stand structure, and rather fewer in which species composition can be altered within the constraints of multi-objective forestry, and so it is important to have developed reliable and repeatable techniques for assessing vertical structure. We continue to examine associated datasets, for invertebrates, songbirds, and small mammal diversity, in order to look for useful, consistent relationships to structural criteria.

For the purposes of comparing biodiversity among native and plantation stands of *P. sylvestris*, one solution might be to build upon the approach of Young and Armstrong (1995), and select native stands which encompass a full range of structural types. In addition, there would be benefits from applying this methodology in other countries, in forest stands managed by different silvicultural methods, since harmonisation of assessment techniques will allow freer exchange of data and expertise. An alternative method would be to establish public expectations for a mix of desired future conditions, or desired levels of ecosystem services provided. In either case, forest structure could be manipulated to meet these requirements.

5. Acknowledgements

Thanks are due to George Gate, Communications Branch, Alice Holt, for producing figures, and to Andrew Brunt, Ellie Dickson, Rebecca Lander, Suzanna Meekins, and Antonia Nichol, for assistance in field assessments.

References

Bain, C. 1987. Native Pinewoods in Scotland: A Review. The Royal Society for the Protection of Birds. Edinburgh.

Ellenberg, H. 1982. Vegetation Ecology of Central Europe. 4th Edition. Cambridge University Press. Cambridge.

Engelmark, O., Kullman, L. and Bergeron, Y. 1994. Fire and age structure of Scots pine and Norway spruce in northern Sweden during the past 700 years. New Phytologist. 126: 363-368.

Erdelen, M. 1984. Bird communities and vegetation structure: I. Correlations and comparisons of simple diversity indices. Oecologia. Berlin. 61: 277-284.

Franklin, J.F. 1988. Structural and Functional Diversity in Temperate Forests. In: Wilson, E.O. (ed.). Biodiversity. National Academy Press. Washington DC. Pp. 166-175.

Goodier, R. and Bunce, R.G.H. 1977. The native pinewoods of Scotland: the current state of the resource. In: Bunce, R.G.H. and Jeffers, J.N.R. (eds.). Native Pinewoods of Scotland. Proceedings of the Aviemore Symposium. Institute of Terrestrial Ecology. Cambridge. Pp. 78-87.

Hodge, S.J., Patterson, G.S., and McIntosh, R. this volume. The approach of the UK Forestry Commission to the conservation of forest biodiversity.

Kuper, J.H. 1994. Sustainable development of Scots pine forests. Wageningen Agricultural University Papers 94(2). Wageningen Agricultural University. The Netherlands.

Nixon, C.J. and Clifford, T. 1995. The age structure of native pinewood remnants. In: Aldhous, J.R. (ed.). Our Pinewood Heritage. Proceedings of a Conference. Inverness 20-22 October 1994. Forestry Commission. The Royal Society for the Protection of Birds. Scottish Natural Heritage. Pp. 177-185.

Pollard, D.W.F. 1993. An introduction to the Forest Ecosystem Dynamics Program. In: Marshall, V. (ed.). Proceedings of the Forest Ecosystem Dynamics Workshop. 10-11 February 1993. Pacific Forestry Centre. Forestry Canada B.C. Ministry of Forests. Victoria B.C. Pp. 1-3.

Pyatt, D.G. 1995. An Ecological Site Classification for Forestry in Great Britain. Forestry Commission Research Information Note 260. The Forestry Commission. Edinburgh.

Ratcliffe, P.R. 1995. Ecological Diversity in Managed Forests. In: Ferris-Kaan, R. (ed.). Managing Forests for Biodiversity. Forestry Commission Technical Paper 8. Forestry Commission. Edinburgh. Pp. 3-7.

Reid, C.M., Foggo, A. and Speight, M. 1996. Deadwood in the Caledonian pine forest. Forestry. 69: 275-279.

Rodwell, J.S. 1991. British Plant Communities. Vol.1. Woodlands & Scrub. Cambridge University Press.

Rodwell, J.S. and Cooper, E.A. 1995. Scottish pinewoods in a European context. In: Aldhous, J.R. (ed.). Our Pinewood Heritage. Proceedings of a Conference. Inverness 20-22 October 1994. Forestry Commission. The Royal Society for the Protection of Birds. Scottish Natural Heritage. Pp. 4-22.

Spies, T.A. 1991. Plant Species Diversity and Occurrence in Young, Mature, and Old-Growth Douglas-Fir Stands in western Oregon and Washington. In: Ruggiero, L.F., Aubry, K.B., Carey, A.B. and Huff, M.H. (eds.). Wildlife and Vegetation of Unmanaged Douglas-Fir Forests. General Technical Report PNW-GTR-285. May 1991. USDA Forest Service. Pacific Northwest Research Station. Portland. Oregon. Pp. 111-121

UNCED 1992. Statement of Forest Principles. The United Nations Conference on Environment and Development. Rio de Janeiro. Brazil.

Worrell, R. 1996. The Boreal Forests of Scotland. Forestry Commission Technical Paper 14. The Forestry Commission. Edinburgh.

Young, M.R. and Armstrong, G. 1995. The effect of age, stand density and variability on insect communities in native pine woodlands. In: Aldhous, J.R. (ed.). Our Pinewood Heritage. Proceedings of a Conference. Inverness 20-22 October 1994. Forestry Commission. The Royal Society for the Protection of Birds. Scottish Natural Heritage. Pp. 206-221

DIVERSITY OF FOREST LICHENS IN SWITZERLAND

A Pilot Study in the Central Plateau and the Pre-Alps

M. DIETRICH, C. SCHEIDEGGER
Swiss Federal Institute for Forest, Snow and Landscape Research
CH-8903 Birmensdorf, Switzerland

Abstract

As a symbiosis the lichens are among the most threatened organisms in Central Europe and other industrialised regions of the world. Their conservation requires basic information on diversity, distribution and dynamics. Therefore, a national inventory of epiphytic lichens has been carried out in Switzerland. The proposed fixed-area plot sampling makes it possible to assess simple and reproducible values on lichen diversity for large areas, as well as at the stand level. On the 69 forest observation plots of the pilot study a total of 202 lichen species was observed. On average there were 23 species per plot. In forest areas, three main factors are thought to control lichen diversity: climate, air pollution and forest management. For each factor variables were tested to get an idea of their impact on species richness of epiphytic lichens. Analyses which include further variables of ecology and forest management would increase the possibilities of establishing cause-effect relationships.

1. Introduction

The quantification of biological diversity is an important objective of the assessment of non-timber resources in forest surveys (Köhl et al. 1995). Until now, indices of tree diversity have often been the only measurement of biological diversity in forest inventories. Nevertheless, compared to the overall biological diversity in most European forests, richness of tree species is poor. Because of the special nature of lichens, knowledge of their diversity has great importance for environmental studies. Lichens are a mutualistic symbiosis between a fungal and an algal or cyanobacterial partner. Water uptake as well as uptake of gaseous substances has no selective mechanisms. Therefore, on the one hand air pollution is considered to be a major threat to lichens, and on the other hand, lichens are very good bioindicators of atmospheric pollution (Herzig and Urech 1991). Further, as the physiology of the stenoceous lichens only functions effectively within a narrow range of ecologically stable conditions, they are very sensitive to any alteration or destruction of their habitats, including consequences of forest management. Epiphytic lichens are known to be good indicators of a long ecological continuity, present in old growth forests (Rose 1992). As a consequence of these facts, it is not surprising that lichens are among the most threatened organisms in Central Europe and other industrialised regions of the world.

P. Bachmann et al. (eds.), Assessment of Biodiversity for Improved Forest Planning, 343–349.
European Forest Institute Proceedings no. 18. © 1998 *Kluwer Academic Publishers. Printed in the Netherlands.*

The diversity of lichens in forests is normally not apparent, because it is mainly dominated by inconspicuous crustose species; the more conspicuous foliose and fruticose lichens are sparse (Dietrich and Scheidegger 1997). In Switzerland, floristic studies on epiphytic lichens have until recently mainly taken the form of overall investigations of the flora of small and species-rich forest areas (Camenzind and Wildi 1991; Dietrich 1991; Groner 1990; Scheidegger et al. 1991). The diversity of lichens reported in such studies is usually considerably biased by an unequal consideration of the organisms or the habitats. The first bias is often encountered when species difficult to identify are neglected in a survey, such as sorediate crustose lichens (Dietrich and Scheidegger 1996*a*). The second may lead to considerable errors when habitats of species with a narrow ecology are inaccurately represented.

This pilot study was aimed at contributing to the development of a representative survey of epiphytic lichens at a regional and national level. The main objective of the survey is to provide representative and reproducible data on lichen frequency for the establishment of an official, regionalised Red List to secure the survival of the threatened species. In addition, standardised sampling in forest and non-forest areas guarantees the assessment and monitoring of epiphytic lichen diversity at the level of tree species, at the plot level and at the level of different strata. Furthermore, indicator values of light, temperature, continentality, humidity, reaction, mineral nutrients and especially of toxitolerance will be calculated in order to monitor environmental changes including air pollution. In this paper we present some results on the diversity of forest lichens, including analyses of possible impacts.

2. Methods

In 1993 and 1994 data sampling was carried out on the fixed-area plots of the Swiss Forest Inventory (EAFV and BFL 1988; Zingg 1988). The plot unit was a 500 m^2 circle, as defined on a horizontal plane. The intersection points of the 1 x 1 km grid of the Swiss co-ordinate system were taken as plot centres. All the observation sites were situated in the two production regions Central Plateau and Pre-Alps (Dietrich and Scheidegger 1996a). Stratification was performed in terms of region, vegetation formation (forest and non-forest), and vegetation belt (colline-submontane, lower and upper montane, and sub-alpine). Definitions of the strata followed the Swiss Forest Inventory (EAFV and BFL 1988). In selecting the plots care was taken to ensure that at least ten were chosen for each combination of stratification criteria of which there were 4 in the Central Plateau and 8 in the Pre-Alps. Within each combination plots were selected arbitrarily, taking into account ease of access, but with no special attention to potentially rich habitats. Here, for data analyses three forest strata were defined: Central Plateau (colline-submontane and lower montane vegetation belt), lower Pre-Alps (colline-submontane and lower montane) and upper Pre-Alps (upper montane and sub-alpine).

On each plot every tree with a diameter at breast height (dbh) $\geq$12 cm was normally sampled, though with a maximum of three individuals per tree species and dbh-class of 12 cm. The entire epiphytic lichen flora was recorded as cover percentage of the stem investigated (0-170 cm).

Lichen diversity was interpreted based on species richness per plot, per tree species and per tree stem. To demonstrate possible cause-effect relationships of species richness, simple correlation was used for testing the altitude, the number of years (classified) since the last harvest (EAFV and BFL 1988) and the atmospheric pollution, whereas the degree of atmospheric pollution was measured in terms of the indicator value of toxitolerance (Wirth 1992). For characterisation at the plot level we always used the lowest value indicated by the lichen species. The importance of the tree composition was analysed in terms of tree diversity (number of species) and in terms of dominating tree species.

3. Results

A total of 570 trunks from 21 different tree species was investigated on 69 observation plots in forests. *Picea abies* (264 individuals on 58 plots) was by far the most frequent species sampled, followed by *Fagus sylvatica* (114; 31), *Abies alba* (61; 19) and *Fraxinus excelsior*. Tree species with low frequency revealed on average the highest number of lichen species: *Pinus mugo arborea* 11,9 (n = 18), *Acer pseudoplatanus* 11,7 (18), *Salix alba* 11,3 (8). From all investigated trees only 8 had no lichens on the lower tree trunk. The maximum lichen number per trunk was 35 (Dietrich and Scheidegger 1996b).

With no special attention to lichen-rich forest areas, the sampling method revealed a total of 202 different species. On average 23 lichen species were observed per plot. Only two plots in the Plateau had an observed lichen diversity lower than 7 (Figure 1a). The highest species number was 49, found in the upper Pre-Alps (Figure 1c).

TABLE 1. Altitude, average number of lichen species, tree species and average lowest value of toxitolerance for the three strata (n = sample size)

	Altitude of plots (m)	Species richness	Number of different tree species	Toxitolerance	n
Central Plateau	430-750	14.8 ± 7.8	2.52 ± 1.68	3.71 ± 0.76	21
Lower Pre-Alps	495-983	19.9 ± 6.9	3.13 ± 1.27	2.88 ± 0.67	24
Upper Pre-Alps	947-1636	33.2 ± 9.9	2.13 + 0.83	1.54 ± 0.58	24

Toxitolerance: 1 = very low; 2 = quite low; 4 = medium; 6 = moderatetly high; 9 = very high

One significant factor controlling richness of lichen species over large areas was the altitude. Correlation between the number of lichen species per plot and the altitude of the plot revealed the value of $r = 0.60$ ($p > 0.99$). As a consequence, compared to the upper Pre-Alps, the observed average species richness in the Central Plateau and the lower Pre-Alps was poor (Table 1). However, in these strata high plot values for species richness were not excluded (Figure 1a and b). Furthermore, several taxa from the Red List of macrolichens of Switzerland (Clerc et al. 1990) were only recorded in the upper Pre-Alps.

Based on the observed lichen species, the calculated average lowest indicator value of toxitolerance was highest in the Central Plateau and lowest in the upper Pre-Alps (Table 1). Negative correlation between values for toxitolerance and sampled lichen diversity per plot is given by $r = -0.74$ ($p > 0.99$), indicating the marked impact of the degree of atmospheric pollution on the lichen diversity.

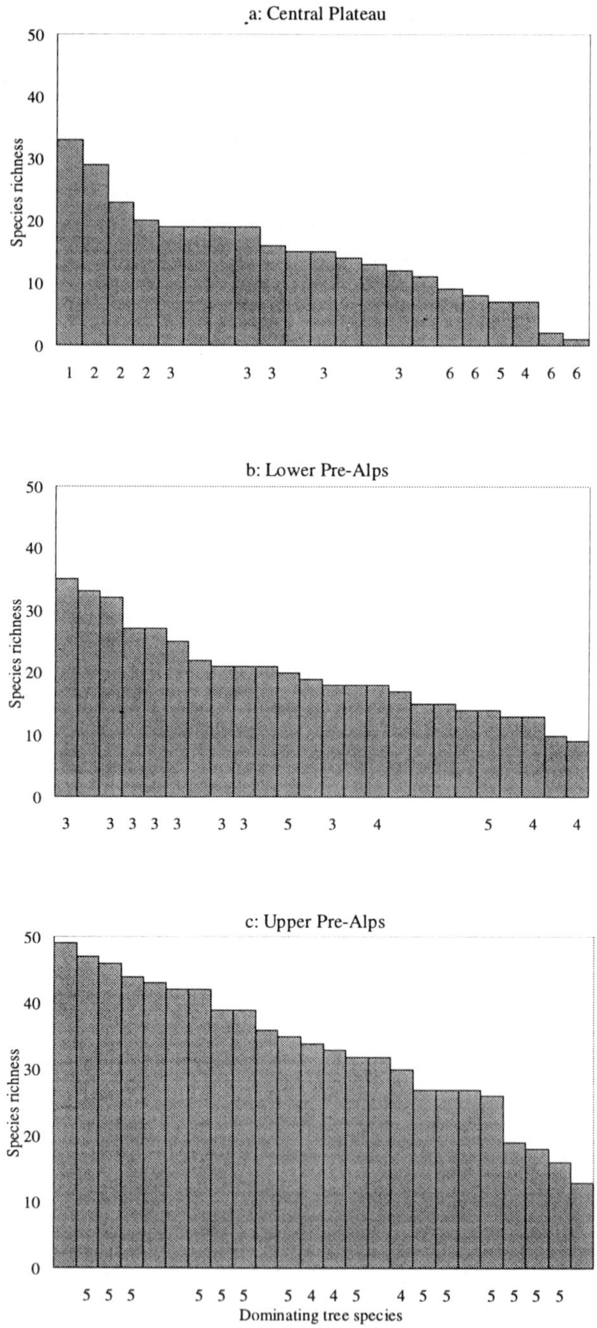

Figure 1. Ranked lichen diversity on the plots of each stratum. Dominating tree species are indicated for the plots: 1 = *Salix alba*; 2 = *Fraxinus excelsior*; 3 = *Fagus sylvatica*; 4 = *Picea abies* and *Abies alba*; 5 = *Picea abies*; 6 = *Picea abies* (plantation); blank = more than one tree species

Forest management influences the lichen diversity in terms of alteration and destruction of habitats. As no data on the intensity of the last harvest were available, classified values for the number of years since the last harvest were tested as one aspect of forest management. They showed a positive correlation to species richness of lichens at the lower trunk ($r = 0.48$; $p > 0.99$). This demonstrates the importance of a long ecological continuity for the establishment of a diverse lichen flora.

In general every tree species is characterised by its particular chemical and physical features of the bark, offering different conditions to epiphytic lichens with different ecological amplitudes. Nevertheless, although the average tree diversity per plot was lowest in the upper Pre-Alps (Tab. 1), the observed lichen diversity was distinctly highest. In fact, correlation between tree diversity and lichen diversity is given by $r = 0.05$ ($p = 0.29$).

In contrast to the plots of the upper Pre-Alps, the natural composition of tree species in Central Plateau and lower Pre-Alps has often been changed in favour of *Picea abies*. Plots in these two strata being dominated by *Picea abies* showed a low lichen diversity throughout (Fig. 1a and b). In contrast, the stands which were dominated by tree species living in their natural habitat, such as *Fagus sylvatica*, *Fraxinus excelsior* or *Salix alba*, revealed the highest number of lichen species in Central Plateau and lower Pre-Alps.

Picea abies, which is in the study area naturally restricted to the upper Pre-Alps, is as well the most frequent tree species in the Central Plateau (EAFV and BFL 1988). There, the lichen diversity per stem of *Picea abies* was extremely low (3.5 ± 0.7; $n = 24$), this not only in the plantations of even-aged trees. A higher variation in tree age only resulted in a slightly higher value (4.1 ± 0.6; 22). In comparable situations the stands of *Fagus sylvatica* had a distinctly higher species number per stem (9.2 ± 1.1; 22). In comparison, *Picea abies* living in its natural habitat favoured a rich lichen flora. In the upper Pre-Alps, the observed lichen diversity per stem of *Picea abies* on the plots dominated by this tree species was 15 ± 1.3 ($n = 87$).

4. Discussion

About 800 epiphytic lichen species are expected to grow in Switzerland. Therefrom 202 were sampled on the 69 forest observation plots of this pilot study. With an average of 23 species per plot, and only 8 trees with no species on the lower tree trunk, epiphytic lichens were shown to be a weighty component of species diversity in forests. The fixed-area plot sampling made it possible to assess simple and reproducible values of lichen diversity for large areas, as well as for the stand level. For the national survey proportional random allocation of the sampling units to the strata will be performed. Stratification according to region, vegetation formation and vegetation belt was shown to be an appropriate tool for the detection of patterns of lichen diversity (Dietrich and Scheidegger 1997).

In general, three main factors are thought to control lichen diversity in forests: climate, air pollution and forest management. Concerning climate, mainly the amount of precipitation is responsible for different lichen floras (Schauer 1965; Buschardt 1979). Air pollution not only changes the flora, but also decreases species richness (Herzig and Urech 1991; van Dobben 1993). Sustainability in forest management is of great consequence for the conservation of forest lichens (Rose 1992; Rosentreter 1995).

The quantification of possible effects of environment using only actual diversity data is difficult. In large, heterogeneous populations true effects may be swamped by apparent effects (Schreuder 1995). In this study, altitude was shown to be the most apparent effect on species richness. Altitude is not having impact on climate alone, but also on air pollution (traffic, industry, heating, agriculture) and possibilities of forest management (accessibility, slope, tree selection). Therefore, the distinctive correlation between lichen diversity and calculated toxitolerance, as well as between lichen diversity and the number of years since the last harvest, may obviously be caused by the altitude. Correlation between the altitude and the toxitolerance is given by $r = -0.73$ ($p > 0.99$), between the altitude and the number of years since the last harvest by $r = 0.33$ ($p > 0.99$). The calculated correlation between the number of sampled tree species and the number of observed lichen species was not evident. Again, the altitude influences the tree diversity ($r = -0,30$; $p > 0.95$). In general, a higher substrate diversity results in a higher richness of lichen species (Lücking 1994; Kuusinen 1994). Here, the impact of dominant tree species on the lichen diversity is assumed to be greater than the overall diversity of trees, demonstrating the significance of the selection and cultivation of tree species.

Analyses which include further variables of ecology and forest management would increase the possibilities of establishing cause-effect relationships concerning lichen diversity, at least when repeated in certain time periods. Thereby, based on the lichen flora additional indicator values, such as for light, temperature, continentality, humidity, reaction and mineral nutrients (Wirth 1992), could be of importance (Dietrich and Scheidegger 1996b).

5. Outlook

For the national survey of diversity, distribution and dynamics of epiphytic lichens, sampling intensity per plot will be increased by including all other parts of the trees, as well as shrubs and dwarf shrubs. The size of the random sample will be 826. As a subset of the second Swiss Forest Inventory 237 plots are going to be located in forest areas. Designed as long-term ecological observation the survey may be repeated in certain time periods, and as such will allow the monitoring of lichen frequency, diversity and distribution, as well as of derived parameters of ecology. In addition, the sampling design will enable further analyses to be done, combining the lichen data with the data of the second Swiss Forest Inventory.

6. Acknowledgements

We thank the Swiss National Science Foundation (Project No. 31- 32567.91) and the BUWAL (Federal Office of Environment, Forests and Landscape; WSL Project No. 6.94.900) for financial support.

References

Buschardt, A. 1979. Zur Flechtenflora der inneralpinen Trockentäler. Cramer. Vaduz.

Camenzind, R. and Wildi, E. 1991. Die epiphytische Flechtenflora des Gurnigel-Gantrischgebietes (BE), Botanica Helvetica 101, 183-197.

Dietrich, M. 1991. Die Flechtenflora des Merliwaldes, Giswil/OW (Zentralschweiz). Botanica Helvetica. 101: 167-182.

Dietrich, M. and Scheidegger, C. 1996a. The importance of sorediate crustose lichens in the epiphytic lichen flora of the Swiss Plateau and the Pre-Alps. Lichenologist. 28: 245-256.

Dietrich, M. and Scheidegger, C. 1996b. Diversität und Zeigerwerte von epiphytischen Flechten der häufigsten Baumarten: Ein methodischer Ansatz zur Beurteilung von Umweltveränderungen im Wald und im Freiland. Botanica Helvetica. 106: 85-102.

Dietrich, M. and Scheidegger, C. 1997. Frequency, diversity and ecological strategies of epiphytic lichens in the Swiss Central Plateau and the Pre-Alps. Lichenologist 29.

van Dobben, H.F. 1993. Vegetation as monitor for deposition of nitrogen and acidity, Doctoral Thesis, University Utrecht.

EAFV (Eidgenössische Anstalt für das forstliche Versuchswesen) and BFL (Bundesamt für Forstwesen und Landschaftsschutz) (eds.) 1988. Schweizerisches Landesforstinventar: Ergebnisse der Erstaufnahmen 1982-1986. Berichte der Eidgenössischen Anstalt für das forstliche Versuchswesen. 305: 1-375.

Groner, U. 1990. Die epiphytischen Makroflechten im Bödmerenwaldgebiet, Muotatal SZ. Berichte der Schwyzerische Naturforschenden Gesellschaft. 9: 77-93.

Herzig, R. and Urech, M. 1991. Flechten als Bioindikatoren: Integriertes biologisches Messsystem der Luftverschmutzung für das Schweizer Mittelland. Cramer. Berlin.

Lücking, R. 1994. Foliikole Flechten und ihre Mikrohabitatspräferenzen in einem tropischen Regenwald in Costa Rica. Doctoral Thesis. University Ulm.

Köhl, M., Bachmann, P., Brassel, P. and Preto, G. (eds.) 1995. The Monte Verità conference on forest surveys designs. 'Simplicity versus efficiency' and assessment of non-timber resources. Swiss Federal Institute for Forest, Snow and Landscape Research (WSL/FNP). Birmensdorf.

Kuusinen, M. 1994. Epiphytic lichen flora and diversity on *Salix caprea* in old-growth southern and middle boreal forests of Finland. Annales Botanici Fennici. 31: 77-92.

Rose, F. 1992. Temperate forest management: its effects on bryophyte and lichen floras and habitats. In: Bates, J.W. and Farmer, A. (eds.). Bryophytes and lichens in a changing environment. Clarendon Press. Oxford. Pp. 211-233.

Rosentreter, R. 1995. Lichen diversity in managed forests of the Pacific Northwest, USA. In: Scheidegger, C., Wolseley, P.A. and Thor, G. (eds.). Conservation biology of lichenised fungi. Mitteilungen der Eidgenössischen Forschungsanstalt für Wald, Schnee und Landschaft. 70: 103-124.

Schauer, T. 1965. Ozeanische Flechten im Nordalpenraum. Portugaliae Acta Biologica. (B) 8: 17-229.

Scheidegger, C., Dietrich, M., Frey, M., Keller, C., Kuhn, N. and Wildi, E. 1991. Zur Waldflechtenflora des westlichen Aargauer Mittellandes und ihrem Wandel seit 1960. Mitteilungen der Aargauischen Naturforschenden Gesellschaft. 23: 175-191.

Schreuder, H.T. 1995. Simplicity versus efficiency in sampling designs and estimation. In: Köhl, M., Bachmann, P., Brassel, P. and Preto, G. The Monte Verità conference on forest survey designs: 'Simplicity versus efficiency' and assessment of non-timber resources. Swiss Federal Institute for Forest, Snow and Landscape Research. Birmensdorf. Pp. 56-63.

Wirth, V. 1992. Zeigerwerte von Flechten. Scripta Geobotanica. 18: 215-237.

Zingg, A. 1988. Schweizerisches Landesforstinventar: Anleitung für die Erstaufnahme. Berichte der Eidgenössischen Anstalt für das forstliche Versuchswesen. 304: 5-117.

Integrating Forest Biodiversity Issues in Forest Management Planning

SUSTAINABILITY AND BIODIVERSITY – FOREST HISTORICAL NOTES ON TWO MAIN CONCERNS OF ENVIRONMENTAL UTILISATION

ANTON SCHULER
Professor of Forest History
Department of Forest and Wood Research
Swiss Federal Institute of Technology
ETH-Zentrum
CH-8092 Zürich

Abstract

The concept of 'sustainability', one of the most important principles in 18th century forestry of the German speaking world and uniquely concerned with the sustainability of timber production, was joined in the course of the 19th century by the concept of the 'role of the forest in nature'. In the 20th century, there followed the forest political theory of the different forest functions and, finally, the modern sustainability concepts.

'Biodiversity' is not recorded as such in forest historical records but can be made accessible by using suitable 'tools' of historical research, some of which may yet have to be developed. By concentrating on 'sustained forestry' 19th century forestry (dis)qualified and largely banned various forest utilisation that were not restricted to timber production and frequently applied within small areas as 'harmful' minor utilisation. The result was not only a loss of a great variety of locally adapted legal forms but also of biodiversity.

1. Introduction

Sustainability and biodiversity are two concepts of paramount importance in the discussion on environment, environmental utilisation and nature conservation, and are now literally on everybody's tongue. From a forest historical point of view, however, they are far apart, in fact, some 250 years – at least as regards their appearance in relevant documents. The German term for sustainability made its first appearance in the very first forestry textbook in German, published in 1713, and consequently underwent many changes as to definition, context, and meaning. The term of 'Biodiversität' (biodiversity), however, appears to be new. It does not occur either in the 'Waldbauliche Terminologie' by Brünig and Mayer (1980) or in the German dictionary 'Der Grosse Duden', for example in its 1989 edition, which "lists not only the main vocabulary of the German language but also technical and specialised language terms insofar as they are of importance to the general public". Does the fact that the word 'biodiversity' was not listed mean that it was still so little known in 1989, i.e. less than 10 years ago? As forest

P. Bachmann et al. (eds.), Assessment of Biodiversity for Improved Forest Planning, 353–360.
European Forest Institute Proceedings no. 18. © 1998 *Kluwer Academic Publishers. Printed in the Netherlands.*

historians, we have to live with the fact that it is pointless to look for that particular term in older or more recent documents. A different approach is needed.

2. Sustainability

According to the philologist and ethnologist Kurt Kehr (1993) 'sustainability' is a technical word used in forestry, but not a technical term in the strict sense of the word as its meaning is not finally defined and fixed. It is a keyword the meaning of which has undergone changes since its introduction into forestry in the early 18th century, not in the positivistic sense of an "improvement" but as an adaptation to changes in science. It is clearly a term from the field of social ethics, and it is not surprising, therefore, that there are many differences of form and content depending on social and regional usage – and this not only because of translation problems.

Such varied and partly also contradictory aspects of sustainability call for limitations. Rather than presenting here a comprehensive history of sustainability, we will follow the concept and idea of sustainability over a long period of time, looking at further relationships, for example with economic and social history.

The time before about 1800 has frequently been called the 'age of timber' but also the 'age of destructive deforestation' (Radkau and Schäfer 1987). Forest utilisation was characterised by a varied demand for products for daily use, including all kinds of timber as well as other 'forest products'. As a main resource, timber was probably most important as fuelwood, particularly in connection with the constantly growing demand for energy or raw materials for trade and early industrial purposes, but also because of population growth, especially in urban areas. Because of the poorly developed transport systems this great demand for various forms of forest products often resulted in a lack of timber, at least locally.

In this situation adequate timber supplies had to be permanently safeguarded. A logical consequence was the establishment of independent forestry, which both brought about and presupposed appropriate scientific teaching and research. In the very first textbook in the German language devoted solely to forestry, the word 'nachhaltend' (sustained) made its first recorded appearance, namely in '*Sylvicultura oeconomica*, Anweisung zur wilden Baum-Zucht' by Hans Carl von Carlowitz, published 1713 in Leipzig. The book follows the tradition and manner of the baroque 'Hausväter' (husbandry) literature which itself goes back to works of the Renaissance and the Graeco-Roman agricultural writers, rediscovered, and in the 16th and 17th century, also translated and published. Motivation for '*Sylvicultura oeconomica*' was the generally increasing overall lack of timber which was to be counteracted with practical measures. Clear-cut areas were to be reforested to increase timber production so that private homes as well as the building trade, breweries – the latter mentioned explicitly –, mining and ironworks could be provided with fuelwood. Continuous and regular supplies of timber were to become the main task of the emerging forest industry.

The word 'nachhaltend' itself is mentioned in chapter 7 of '*Sylvicultura oeconomica*' which deals with the necessity and usefulness of timber utilisation: In order to ensure permanent timber supplies everything possible should be done for a "continuous, stable and sustained (nachhaltende) utilisation" as this was of essential importance for the country. The '*Sylvicultura oeconomica*' does not only represent the beginnings of forestry literature in the German language but it also introduces the concept of sustainability, here used in its adjectival form of 'sustained'. The concept of sustainability is from then on permanently and increasingly present.

'Onomatologia forestalis-piscatorio-venatoria', published 60 years later, cautions "to cut timber sustainably", and goes on to say that there was more to it than one first thought (Onomatologia 1773; Zürcher 1965). The division of the forest into different annual cuts was not sufficient as nature did not necessarily follow our design ("Dessein"). Is it too daring to see here some sort of reference to ecology and biodiversity?

In Switzerland such a division of the forest into annual cuts was also the centre of the regulation on forest treatment by the Zürich Authorities who, in the forest mandate of the same year of 1773 (Waldungsmandat 1773), requested their subjects to treat forests more "forest-conform", perhaps more sustainably, than so far. Timber should no longer be cut so untidily and as single trees but in regular cuts, taking into consideration the entire turnover time ("from beginning to end") and planned ("eingerichtet") according to the conditions of each individual forest. Forest planning is thus actually mentioned in writing but not the term 'sustainability' though its meaning was undoubtedly implied.

Some 40 years later, forest utilisation was still not in a "proper relationship" with "*sustained* timber yield". In most cases it was necessary to limit timber utilisation. According to Hirzel, head forester, and his 'Instruction' it was therefore the most important but also the most difficult task of foresters to reduce timber cuts appropriately, i.e. to keep them on a sustained level (Instruction 1808).

In the first half of the 19th century the concept of sustainability was considerably extended, as can be seen from the works of the Swiss forest pioneer, Karl Albrecht Kasthofer, who, in his early work (Kasthofer 1818), had considered the difference between the amount of timber cut and the amount of timber growing as sufficient to measure sustainability. In 1846, Kasthofer limited his assessment no longer to the amount of timber cut, spatially and temporally, but also took into consideration site and growth capacity of the stand (Kasthofer 1846). Sustainable use was ensured only when site conditions did not deteriorate.

Also in the first half of the 19th century and in view of the increasing realisation that large clear-cuts in previously non-accessible or hardly accessible pre-alpine forests were harmful, a growing number of people pointed out the relationship between excessive exploitation and flooding. This is particularly clear from the 1849 Memorandum on mountain deforestation ('Ueber die Entwaldung der Gebirge') by Xavier Marchand who emphasised that in the sense of an (extended) sustainability "the Government did not only have the right, but in fact had the duty to protect the work of previous generations and the hopes and the safety of future generations from the whims of a single generation. It was the duty of the Government to safeguard the forests." This is not only a reference to the main function of the forests, to supply timber, but it also concerns another important function, the protection function, as well as the effects on the local climate, particularly on the precipitation pattern, attributed to the forest (Schuler 1995). The problem of the regulating effects of forests on water drainage and the relationship between clear-cutting (over-exploitation of natural resources) and increased erosion were of great importance for future development. This is particularly obvious when Marchand severely takes to task those who "merely see the forest as a producer of timber, are aware of the growth potential of a tree and consequently appreciate its dead body only, in other words all those economists who have never got further than the rudimentary stage in their studies of natural science" (Marchand 1849). Another 'Leitmotiv' therefore joins 'sustainability', namely the part played by the forest in nature.

In 1856, the Swiss Forestry Society requested the Swiss Federal Council to commission a survey on the state of the forests and torrents in the mountains (Bericht 1856). The Society thus adopted Marchand's argument that the forest had other important parts to play in nature than merely producing timber. Mention was made that, due to the non-sustained treatment of mountain forests, the communes, as forest owners, also suffered loss of revenue. In addition, these regions with their extended grounds, in the long run only suitable for timber production, could no longer fulfil properly their national task, namely supplying fuelwood and building timber to the Plateau regions, i.e. regions devoted to arable farming and industry. Forests utilised and tended on a sustained basis, as was the ideal of the Forestry Society, were not only meant to ensure income and welfare (by providing sufficient timber) but would also fulfil almost automatically all other demands on the forests: protective and regulating effects in connection with air and water, protection against natural hazards, against noise and insolation, conservation of soils and their fertility and much more. In fact, the 'Kielwassertheorie' ('wake theory', i.e. the theory that the social functions of forests are automatically provided in the 'wake' of a proper production forestry) is here anticipated, a theory that was to reach its peak about a hundred years later with Viktor Dieterich and his forest function theory (Dieterich 1953). In the petition of the Swiss Forestry Society, submitted in 1856, emphasis was, however, on two points in connection with sustainability. Namely, on the one hand, sustained timber supplies at a time when a possible substitution of timber as an energy factor owing to the new transport possibilities was only just beginning to emerge, and, on the other hand, on safeguarding the protection functions (erosion control and protection against natural hazards) owing to a sufficiently large forested area.

As a next step, article 24 of the revised Constitution of 1874 was instituted giving the Federal Government the right to control hydraulic engineering and forestry in the mountains, and, in 1876, the Swiss Federal Forest Police Law came into force. Sustainability was thus introduced at least indirectly at the level of the Constitution and the Federal Law; and a sufficient forest coverage was definitely ensured by the legal requirement of maintaining or, in case of insufficient forest coverage, enlarging it, and by observing the principle of sustainability by introducing compulsory management plans for public forest owners. The main element of these management plans and, in general, the only one subject to approval was the stipulation of a sustainable allowable cut and the control of sustainable utilisation.

The transition from schematic silviculture to so-called near-to-nature silviculture, introduced at the turn of the century by Arnold Engler, was of fundamental importance for further development. It followed after a starting phase of assessing the forests and working out management plans and after the application of the Swiss Federal Forest Police Law to the whole of Switzerland. The further development must also be seen in connection with new methods of forest management, in particular the control method. Without it a continuous control of forest development and observation of sustainability would hardly have been possible. The origins of the control methods can be found in the treatment of selection ('plenter') forests where temporal and spatial control is considerably more complicated than with the so-called 'Altersklassenwald' (compartment forest). The doctrine of near-to-nature silviculture provided the principle of sustainability with a new aspect. According to Leibundgut sustainability has much to do with a "positive attitude towards forestry" (Leibundgut 1945). "In my opinion this attitude does not only mean the conservation of the forested area as indicated by the wording of the Federal Law, nor is it restricted to a similar timber harvest or financial return every year. The term 'sustainable' corresponds to the positive attitude towards forestry as demonstrated by a strong farming community who made liberation from the false clear-cutting doctrine possible."

Towards the end of the 20th century sustainability is increasingly considered to be a norm for mankind to deal with ecosystems rather than a scientific characteristic of the forest or a production figure dictated by forest planners (Kehr 1993). Sustainability, finally, has become a "behavioural norm for the dealings of humans with ecosystems. From having been a limitation, the conservation of ecological base becomes the actual goal" (Glück 1994; Bernasconi 1996). Conservation of the ecological base – does this not mean conservation or promotion of biodiversity?

3. Biodiversity

The varied forms of forest utilisation of the past, so often evoking almost nostalgic memories of the good old times with cows grazing in the woods, lopped trees and leaves for bedding, show a landscape and environment that, though it may be under human influence, appears to be intact also with regard to biodiversity. Most written forestry sources list these utilisation as harmful or detrimental that should have no place or should be abolished in 'regulated' forest utilisation, in 'proper' forestry. Such ideas can be found in the forest regulations and forest laws accompanying the development of forestry and forest policy from the 17th century until the present. The dilemma of modern forest history stems from the fact that until recently forest history has normally considered the increasing amount of forest legislation and the extension of purely forest-oriented utilisation as progress. Questions on environmental history, perhaps in connection with biodiversity, or on social history, have not been asked and dealt with, or only exceptionally. For the modern approach, oriented towards environmental and social history, not only the timber-producing large coppice with standards and high forests are of importance but also the small and scattered farm forests as well as the many forms of brushwoods and coppice forests. These were partly used for agricultural purposes, and it was frequently not quite clear or it was irrelevant whether they were part of the forests or the open country (Radkau and Schäfer 1987).

In addition to the 'usual' minor utilisation such as forest grazing, leaf cutting, collection of tanbark, hunting, or honey production, there existed many other forms of utilisation. There are often no records of them nor are they mentioned in the forest history literature because they were of importance to an underprivileged class only. A reference to the importance of such varied forms of utilisation for the rural population can be found in the margin of an official letter on a Bernese forest dispute of 1757 (Stuber 1996): "Why does one deliberately talk of timber only when it is a question of gaining all benefits of the forest?" Benefits, at that time meaning the various uses still made of the forest. Economic and social historians have begun to approach this problem: "We have no idea to what extent, right into the 19th century, the poorer families literally depended on gathering for their existence, with the forest being the most important place: fruit, berries, mushrooms, buckwheat, fir needles and herbs for tea, leaves for bedding, fodder for small livestock, etc." (Braun, 1984). Why should this not also be of interest to forest history which, after all, is not a timber production history?

The introduction of regulated forestry is closely connected with the Agrarian reform at the end of the 18th century. The large communal grazing on pastures and in forests was considered to be the main reason for the slowness induced by 'non-regulated' agriculture (Radkau and Schäfer 1987). The Agrarian reform freed agriculture 'at last' from forest utilisation. It also brought forestry the long wished for autonomy. However, those unable to fend for themselves legally or otherwise or at most as beggars, etc., were left out in the cold. Privatisation of the farming economy and the beginning of the regulation of utilisation rights close the circle to the previously mentioned attempts at abolishing special forms of utilisation (Stuber 1996). The poor and other inhabitants who had no normal rights of utilisation found it particularly difficult to get forest products, though they had partly (and temporarily) restricted rights on the village common or the co-operative common. These rights were frequently limited to the 'Schachen' forests, second-class softwood forests, which have now gained new importance as wetland forests in the discussion on nature conservation and biodiversity.

There are few forestry records available for the assessment of past biodiversity. Valuable data can be gained, also for previously mentioned minor utilisation, from the forerunners of the actual management plans, the "visitation reports", available, for instance, in the canton of Zurich since the early 19th century. They contain information on rights, utilisation possibilities and their extent. And they point out the necessity of abolishing these rights as they were not conducive to proper sustainable forestry. Undoubtedly, such claims reflect the current interpretation of the concept of sustainable forestry, since then aimed at permanent and regular timber production.

In this connection, old forms of utilisation, some still existing today, are of great interest as they document the most diverse utilisation needs which would hardly have been possible without a large 'biodiversity', which was itself, however, a consequence of these diverse utilisation needs. As long as these rights of utilisation existed, they almost automatically excluded uniformity of utilisation and therefore of the forest stands. The emphasis on timber production, for a long time of first importance because of the demands for sustained forest utilisation, encouraged redemption of the old rights and privileges.

Remains of joint utilisation of agriculture and forestry can be found in the 'Pâturages boisés' (meadows interspersed with forest trees) of the Swiss Jura with its special landscape characteristics. They have been mentioned in the forest legislation and thus are part of the forested area although grazing as the probably most important minor forest utilisation is still possible there.

Some old forms of ownership such as plant and forest "superfizies" continue to exist in different parts of Switzerland (Ganzoni 1954; Liver 1965). They have only survived because clarification and redemption would obviously be quite complicated as they are not normal easements but forms of so-called shared ownership (Schuler 1996). As early as 1838 Johannes Bohl, then cantonal head forester in the Grisons, described in his Instructions 'Anleitung zur Verbesserung des Bündnerischen Waldwesens' the ownership forms in many parts of his canton as "complicated and uncertain". In the lower and higher Alpine meadows there were a large number of forests with grounds and grazing belonging to one person, the timber, however, to another. In many places one had only a very vague idea as to who the owner might be (Bohl 1838). As an excellent forester Bohl suspected, of course, that this was the reason for the sad conditions of the forests and that the ownership question would have to be tackled and clarified immediately. The fact that also the latest cantonal forest law stipulates redemption of the forest 'superfizies' shows that this was perhaps just a little more difficult than he may have thought it to be at the time.

Many of the old legal forms have obstinately survived despite all efforts at clarifying and usually redeeming them in favour of sustained uniform timber production. The question now is whether, in so doing, we have not destroyed conditions that would otherwise have encouraged biodiversity? What has suffered undoubtedly is the "diversity" of solutions of, and approaches to, environmental utilisation which were adapted to the local conditions and socio-economic environment and therefore showed a diverse development pattern.

4. Conclusion

The history of sustainability shows that the measures developed and undertaken from the narrow perspective of regulated forestry's introductory phase eventually led to a sharp decline in biodiversity. Due to the circumstance that traditional forest history's point of reference was almost solely restricted to regulated forestry, the history of biodiversity (amongst many other aspects of the development of forest utilisation) cannot be successfully investigated without the methods of a forest history oriented towards the approaches of Environmental History, methods which are not limited to the perceptional restrictions of regulated forestry.

(Translated by Rosmarie Louis)

References

Bericht an den Bundesrath über das Forstwesen in der Schweiz, vom schweizerischen Forstverein. 7 July 1856. P. l.

Bernasconi, A. 1996. Von der Nachhaltigkeit zu nachhaltigen Systemen. Forstliche Planung als Grundlage nachhaltiger Waldbewirtschaftung. Dissertation ETH No. 11 195. Beih. No. 76. Swiss Federal Institute of Technology.

Braun, R. 1984. Das ausgehende Ancien Régime in der Schweiz. Aufriss einer Sozial- und Wirtschaftsgeschichte des 18. Jahrhunderts. Göttingen and Zürich.

Brülhart, A. (1993) Nachhaltigkeit als Jubiläumsthema des Schweizerischen Forstvereins. Schweiz. Z. Forstwesen. 144: 79.

Brünig, E., Mayer H. 1980. Waldbauliche Terminologie. Fachwörter der forstlichen Produktion. Universität für Bodenkultur. Institut für Waldbau. Vienna.

Dietrich, V. 1953. Forstwirtschaftspolitik. Eine Einführung. Hamburg and Berlin.

Duden Deutsches Universalwörterbuch 1989. Herausgegeben und bearbeitet vom Wiss. Rat unter der Mitarbeit der Dudenredaktion unter Leitung von G. Drosdowski. 2. völlig neu bearbeitete und stark erweiterte Auflage. Mannheim, Zürich, Wien.

Ganzoni, R.Z. 1954. Beitrag zur Kenntnis des Waldeigentums in Graubünden unter besonderer Berücksichtigung des Engadins, des Bergells, des Puschlavs und des Münstertales. Diss. iur. Universität Bern. Beiheft Nr. 4 zum Bündnerwald, Chur.

Glück, P. 1994. Entstehung eines internationalen Waldregimes. Centralbl. ges. Forstwesen, 111: 75-92.

Hasel, K. 1971. Waldwirtschaft und Umwelt. Eine Einführung in die forstwirtschaftspolitischen Probleme der Industriegesellschaft. Hamburg and Berlin.

Instruction über die Wissenschaften und Grundsätze, worauf eine regelmässige Behandlung und Verwaltung des Forstwesens beruhet. In Hinsicht des Forstwesens von dem Canton Zürich zur Anleitung der Erwählten Forstzöglingen entworfen und bearbeitet von Cantons Forstinspector Hirzel, dann denselben vorgelesen und erklärt vom Octobre 1808 bis Merz 1809. Manuscript in the Zurich Cantonal Archives.

Kasthofer, K. 1818. Bemerkungen über die Wälder und Alpen des Bernerischen Hochgebirgs. Ein Beitrag zur Bestimmung der Vegetationsgrenze schweizerischer Holzarten, des Einflusses der Waldungen auf die Kultur des Hochgebirgs, des Verhältnisses der Forstwirthschaft zur Landwirthschaft und der Bedinge für Verbesserung der Alpenwirthschaft. 2. Edition. Aarau.

Kasthofer, K. 1846. Kurzer und gemeinfasslicher Unterricht in der Naturgeschichte der nützlichsten einheimischen Waldbäume, in der Schlagführung zur Förderung der natürlichen Wiederbesaamung der Wälder, in der Bestimmung der nachhaltigen Holzbenutzung und in der Waldsaat und Waldpflanzung. Genf.

Kehr, K. 1993. Nachhaltig denken. Zum sprachgeschichtlichen Hintergrund und zur Bedeutungsentwicklung des forstlichen Begriffes der "Nachhaltigkeit". Schweiz. Z. Forstwes. 144: 595-605.

Leibundgut, H. 1945. Die Zukunftsaufgaben unserer Waldwirtschaft. Schweizerischer Bund für Naturschutz. Vol. 1. Basel.

Leibundgut, H. 1975. Wirkungen des Waldes auf die Umwelt des Menschen. Wir und die Umwelt. Erlenbach. Zürich.

Liver, P. 1965. Zur Geschichte und Dogmatik des Eigentums an Bäumen auf fremdem Boden in der Schweiz. In: Festschrift Karl Siegfried Bader. Zürich. Pp. 281-300.

Marchand, A. [X.] 1849. Ueber die Entwaldung der Gebirge. Denkschrift an die Direktion des Innern des Kantons Bern. Jenni. Bern. 59 p.

Onomatologia Forestalis-Piscatorio-Venatoria oder vollständiges Forst-, Fisch- und Jagdlexikon 1772/1773. Frankfurt a. M. Leipzig (Supplement Stuttgart 1780).

Peters, W. 1984. Die Nachhaltigkeit als Grundsatz der Forstwirtschaft, ihre Verankerung in der Gesetzgebung und ihre Bedeutung in der Praxis. Die Verhältnisse in der Bundesrepublik Deutschland im Vergleich mit einigen Industrie- und Entwicklungsländern. Diss. University of Hamburg.

Radkau, J. and Schäfer, I. 1987. Holz. Ein Naturstoff in der Technikgeschichte. Reihe: "Kulturgeschichte der Naturwissenschaften und der Technik". Reinbek bei Hamburg.

Schuler, A. 1993. Das Prinzip der Nachhaltigkeit und der Aufbau der schweizerischen Forstwirtschaft. Schweiz. Z. Forstwes. 144: 263-269.

Schuler, A. (ed.) 1995. Nachhaltigkeit als Grundlage menschlichen Handelns. Seminar anlässlich des 150-Jahr-Jubiläums des Schweizerischen Forstvereins vom 8. - 11. September 1993 in Interlaken. Beih. Nr. 75 Schweiz. Z. Forstwes. Zürich.

Schuler, A. 1995. Naturgefahren und die Bedeutung der Wälder im "Haushalt der Natur". Schweiz. Z. Forstwes. 146: 629-640.

Schuler, A. in press. Von der vielfältigen Nutzung des Allmendwaldes zur Multifunktionalität heutiger Wälder. Bemerkungen aus schweizerischer Sicht. News of Forest History.

Stuber, M. 1996. 'Wir halten eine fette Mahlzeit, denn mit dem Ei verzehren wir die Henne'. Konzepte nachhaltiger Waldnutzung im Kanton Bern, 1750 bis 1880. Diss. phil.hist.. University of Bern.

Zürcher, U. 1965. Die Idee der Nachhaltigkeit unter spezieller Berücksichtigung der Gesichtspunkte der Forsteinrichtung. Diss. no. 3688. ETH Zürich.

AN APPRAISAL OF BIOLOGICAL DIVERSITY 'STANDARDS' FOR FOREST PLANTATIONS

IAN F. SPELLERBERG
Department of Resource Management, Lincoln University
Canterbury, New Zealand

JOHN W.D. SAWYER
Department of Conservation
Bowen Street
Wellington, New Zealand.

Abstract

A literature review of forest management practices for biological diversity in plantation forests was undertaken. The review revealed many ways of increasing levels of Biological Diversity but there were very few examples of standards being used. Here, the concept of 'standard' is explored and the role of standards in management of forest plantations for biological diversity is appraised. Recognition of different perceptions of standards and flexibility in standards are suggested. A theoretical framework for biological diversity standards which recognises spatial and temporal components is suggested. It is concluded there should be more than just aims and objectives for management of plantations forests for different levels of Biological Diversity; the use of standards could improve forest management practices. We suggest that the concept of standards has wide applications in biological conservation.

1. Introduction

Many countries in Europe, Scandinavia and North America have adopted a policy of multi-use forestry for plantation forests. By way of contrast, this concept has not been adopted in some countries such as New Zealand. The concept embraces the production of timber products while at the same time allowing for management of the forests for recreation, education, amenity, and wildlife. Increasing pressures on land-use and the growing impact of afforestation seemed to have brought about a practice of multi-use forestry which is acceptable by both the forest industry and the wider community.

The management of forest plantations for wildlife is not new (Leopold 1939) and neither is the conservation of forest (tree) genetic resources (Palmberg-Lerche 1993). However, management of forest plantations for certain kinds and levels of Biological Diversity has only more recently been advocated (Hunter 1990). The 1992 Convention on Biological Diversity has prompted discussion about management of plantation forests for not only for certain species of wildlife but also for Biological Diversity. The concept

P. Bachmann et al. (eds.), Assessment of Biodiversity for Improved Forest Planning, 361–365.
European Forest Institute Proceedings no. 18. © 1998 *Kluwer Academic Publishers. Printed in the Netherlands.*

of Biological Diversity has been used in policy to ensure that variety becomes part of conservation; the Convention refers specifically to variety within species, between species and of ecosystems.

Biological Diversity is found at different levels of organisation within three components, taxonomic, genetic and ecological (Angermeier and Karr 1994). The quality and quantity of Biological Diversity at different levels of organisation may be difficult to measure except at the species level. It is perhaps not surprising therefore that biological diversity is sometimes equated with species. There seems to be general support for the view that it is important to specify levels of Biological Diversity (Harper and Hawksworth 1994).

This brief essay explores the concept of standards as a way of ensuring that aims and objectives are both agreed in policy and put into practice. We appraise the use of standards with respect to management of forest plantations for Biological Diversity. Examples of standards have been described in detail elsewhere (Spellerberg and Sawyer 1996).

2. Methods

A literature search was undertaken using library databases and those of research establishments (e.g. the Nature Conservancy in North America, the USDA Forest Service). Key words included plantation, forestry and forest but jointly with wildlife, biological diversity, biodiversity and plants and animals. Over two hundred letters were written to forest research stations in North America, Scandinavia, Australasia and Europe. Site visits were undertaken to plantation forests in England, Scotland and Wales. A bibliographic database was established using Cardbox-Plus software and nearly 700 references were accumulated.

A synthesis was undertaken of the methods (reported in the literature) used for increasing levels of biological diversity in plantation forests. The results of this have been reported in detail elsewhere (Spellerberg and Sawyer 1993).

The literature search revealed examples of the use of standards in forest management practice. Consequently, we decided to explore the concept of standards and their applications in management of forest plantations for different levels and components of biological diversity.

3. The concept of standards in conservation

A standard may be defined as ‘an acceptable or approved example of something against which others are judged or measured’ or as ‘a level of excellence of quality’. It is something which can be measured or quantified.

Management for conservation of wildlife or for various levels of biological diversity must be based on aims and objectives. The overall aim could be to increase Biological Diversity in a plantation and the objective could be increased levels in a particular locality or increased levels of a particular kind of biological diversity. Examples of increasing levels of Biological Diversity include increasing numbers of certain species of wildlife or increasing the diversity of wildlife at a particular locality. The adoption of certain agreed standards could help to achieve the aims and objectives. The use of standards are valuable for two main reasons:

(i) they provide a measurable goal;
(ii) within the concept of multi-use forestry, standards enable forest managers to determine the effectiveness (in both monetary terms and in Biological Diversity terms) of their forest management practices.

We suggest that the second reason is particularly valuable and more so where monitoring programmes to appraise forest management practices have been implemented. Without standards, the success or failure of the forest management practices could at the worst be disputed.

4. Putting standards into context

The question of who sets the standards will inevitably be asked. If management of forests for biological diversity is undertaken as part of a policy for multi-use forests, then it would seem useful to facilitate wide consultation with those who use the forest or the forest products. For multi-use forests we suggest not only a consultation process but also a partnership approach involving forest managers and those who make use of the forest for amenity, education, ecological research, recreation and other purposes.

The perceptions of standards in a general sense may vary widely. For example perceptions of standards for forests may include:

(i) those which satisfy the recreational visitor to the forest;
(ii) those which maintain ecosystem integrity ('integrity' is used here as outlined by Angermeier and Karr 1994);
(iii) those which satisfy the informed naturalist;
(iv) those which are as near as possible to the 'natural' forest condition.

The use of the term 'natural' in connection with forest wildlife and forest communities has been well discussed (see for example Peterken 1987; Hansen et al. 1991). What is believed to be natural is one thing but can it be achieved is something else. We believe that while establishing standards in theory should include the concept of 'naturalness', it is also important to be realistic. In constructing a policy for standards we suggest that it is important to consider the following:

(i) what is possible;
(ii) what is legally required;
(iii) what is locally, regionally and globally significant;
(iv) what is vulnerable.

There may be geographical and edaphic factors which constrain the establishment of standards for levels of biological diversity and there will almost certainly be constraints imposed by silvicultural practices.

In some countries there may be legal requirements and constraints affecting what standards can be achieved in practice. Some forest plantations include designated sites of conservation, geological, cultural or heritage interest.

The significance of designated sites will vary from what is considered significant at different geographical scales; from international to local. For example at an international scale, wetland sites designated under the Ramsar Convention). Landscape values may also be important to consider at both an international level and at a local level.

Finally, the vulnerability and fragility of the forest may impose constraints on what standards are put into practice. Some landforms or ecosystems such as the lowland heathlands of southern England (many of which are found within afforested areas) are vulnerable to physical and chemical perturbations. Traditional forest practices such as removal of dead wood for fuel could impose constraints on standards (dead and decaying wood have long been identified as being an important habitat in woodland and forest ecosystems (Elton 1966)).

We suggest that whatever standards are agreed, they can never be definitive. The biotic system is forever changing and so also is the forest plantation. Therefore standards for levels of biological diversity have to be reappraised as conditions and circumstances change over time. As well as a temporal basis for appraising standards there is also a spatial basis. In some instances, for example, desirable standards may reach far beyond the forest boundary. We suggest that it is important not only to recognise a temporal and spatial basis for standards but also variability. A defined optimum for a standard is not possible and therefore in defining standards we are really defining a level of variability about a mean or fixed point.

5. Standards for forest practice and for forest wildlife

Standards have long been used as part of forest management practices (these are reviewed extensively in Spellerberg and Sawyer (1993)). For example in the UK Nature Conservation, Water, and Landscape Guidelines have been adopted as stewardship principles for the Forest Enterprise (Forestry Commission 1991). Watercourse management standards for the UK have been in existence for many years (Mills 1980). In North America, Best Management Practices are used (Rieker 1991) and the guidelines and standards are communicated through a newsletter, the *Forest Research News*. In Europe and in North America, minimum standards have sometimes been adopted in connection with wildlife objectives. For example the minimum volumes of dead wood to be provided for nature conservation have been established.

6. Quantifying standards for levels of biological Diversity in forest plantations

We have suggested an approach which takes into consideration spatial scales and the three ecological components of composition, structure and function (Spellerberg and Sawyer 1996). The spatial scales could range from a forest stand to a plantation and even the plantation in the wider landscape. At each scale and within each ecological component we suggest that objectives, standards and indicators be established.

For example, the objective, the standards and the indicator could be as follows for composition at a forest stand level:

(i) The objective is to accommodate site-specific keystone species;
(ii) The standard is the presence (e.g. nesting or foraging) of a certain population level;
(iii) The indicator are population counts of the keystone species.

The structural component at a stand level could be foliage height diversity and the functional component at a stand level could be habitat heterogeneity or forest patch mosaics within the forest stand (and movement of wildlife between those patches).

We conclude that management of forest plantations to achieve Biological Diversity objectives should be based on agreed standards. The theory of standards and practice in forestry could also be applied in the wider context of ecological restoration and biological conservation in general. This conclusion seems timely in view of current discussion about ecological integrity and environmental health. State of the environment reporting is here to stay and should (indeed, must) include standards applicable to levels of biological diversity whether that be taxonomic, genetic or ecological.

References

Angermeier, P.L. and Karr, J.R. 1994. Biological integrity versus biological diversity as policy directives. BioScience. 44: 690-697.

Elton, C.S. 1966. The pattern of animal communities. Methuen. London.

Forestry Commission 1991. Forests and Water: guidelines. Forestry Commission. UK.

Hansen, A.J., Spies, T.A., Swanson, F.J. and Ohmann, J.L. 1991. Conserving biodiversity in managed forests. BioScience. 41: 382-392.

Harper, J.L. and Hawksworth, D.L. 1994. Biodiversity: measurement and estimation. Philosophical Transactions of the Royal Society. London. B. 345: 5-12.

Hunter, M. 1990. Wildlife, forest and forestry: principles of managing for biological diversity. Englewood Cliffs. N.J. Prentice Hall.

Leopold, A. 1939. A biotic view of the land. Journal of Forestry. 37: 727-730.

Mills, D.H. 1980. The management of forest streams. Forestry Commission Leaflet No. 78. HMSO.

Palmberg-Lerche, C. 1993. The conservation of forest genetic resources: issues and programmes. Diversity. 9: 26-29.

Peterken, G.F. 1987. Natural features in the management of upland conifer forests. Proceedings of the Royal Society of Edinburgh. 93B: 223-234.

Rieker, H. 1991. Best management practices for upland and wetland forests. The Institute of Food and Agriculture Sciences. University of Florida. Florida Cooperative Extension Services. FRC 8.

Spellerberg, I.F. and Sawyer, J.W.D. 1993. Biodiversity in plantation forests: increasing levels and maintaining standards. Unpublished report to the Forest Authority. UK. P. 209.

Spellerberg, I.F. and Sawyer, J.W.D. 1996. Standards for biodiversity: a proposal based on biodiversity standards for forest plantations. Biodiversity. 5: 447-459.

A PROPOSAL TO COMBINE WOOD-PRODUCTION AND BIODIVERSITY MANAGEMENT BY A LANDSCAPE ECOLOGICAL APPROACH IN BOREAL EUROPE

KULLERVO KUUSELA
European Forest Institute
Torikatu 34
80100 Joensuu, Finland

Abstract

The goals of forest management have become more and more diverse in the past few decades. In addition to the traditional wood and other production functions, the aims of forest management now include also such aspects as the cultural values of forests and maintaining a desired level of biodiversity.

The landscape ecological approach may help in achieving the objectives forest management is currently facing. It is a hierarchical system of climatic vegetation zones and landscape units, forest types, stand development stages and species communities.

This paper attempts to outline a system which includes a desirable biodiversity as an aim in the boreal coniferous wood-production forests where wild fires have been eliminated and have lost their role as the most important maintainer of biodiversity. The system is constructed by logical reasoning using the available information in the frame of references of the landscape ecological approaches used in the Russian and Canadian forest classification systems illustrated in Appendices 1 and 2.

1. Introduction

Forest as a renewable natural resource was long used for producing commodities and immaterial benefits. Only measures taken were harvesting and consuming. When the human population and its needs increased and reached the level at which harvesting exceeded the sustainable productivity of natural forests, methods and measures had to be developed and employed for growing and harvesting wood and other benefits on a sustainable basis.

The aim of traditional forest management regimes is to study, develop and use resource assessments and planned growing and producing methods by which sustainable and progressive production of multifunctional forests can be achieved and maintained. Production is sustainable when removals of wood do not exceed the level of regeneration. In progressive management, methods and measures are used to develop forests towards desirable conditions where the sites' biological potential is in full use, and the quantity and quality of the multifunctional product mix is as high as possible as well as economically profitable.

P. Bachmann et al. (eds.), Assessment of Biodiversity for Improved Forest Planning, 367–380.
European Forest Institute Proceedings no. 18. © 1998 *Kluwer Academic Publishers. Printed in the Netherlands.*

Globally, forests are decreasing and their composition is changing in such a way that certain organism species are increasingly threatened by extinction, and the biodiversity of organism communities is decreasing. This has called for new measures to preserve and maintain the biodiversity of forests:

- Establishing nature conservation forests protected from all anthropogenic disturbances.
- Supplementing biodiversity management to traditional forest management regimes.

The aim of this paper is to describe the landscape ecological approach studied and developed in the countries of the boreal coniferous zone (e.g. Corns and Annas 1986; Gromtsev 1993; Ekosistemy landschaftov... 1995) and to propose how biodiversity can be included in forest management by production diversity regimes.

2. Levels of landscape ecological approach

2.1 CLIMATIC VEGETATION ZONES

Climate is the principal environmental factor to which organisms, i.e. plants, animals and microbes, have to adapt. It is described by such parameters as the annual solar radiation on the site, mean annual air temperature and precipitation, the length of the growing season, the sum of the effective temperature during growing season etc. Consequently, climatic vegetation zones can be determined and described by the compositions of climatic climax plant communities adapted to particular climates (Eyre 1968; Kuusela 1994). From the landscape ecology point of view, four principal vegetation zones can be recognised in Europe:

Tundra, where tundra grasses and sedges with an understorey of mosses and lichens dominate in a cold climate, supplemented by dwarf shrubs, and where stunted trees can grow on sheltered sites. Tundra is divided into two subzones:

- latitudinal or arctic tundra
- elevation or mountain tundra on sites located higher than the average elevation above sea level in a particular principal zone.

Boreal coniferous forest (taiga in Russia), where conifers dominate in climax stands due to their strong ability to compete with broad-leaved trees. The latter ones are pioneers which regenerate on sites after disturbances and dominate in the first stages of serial successions.

Summergreen broad-leaved forest, where trees such as oak and beech dominate in climax stands due to their strong ability to compete with conifers in a climate warmer than in boreal zone.

Mixed evergreen forest in the Mediterranean climate.

Ecotones can be found between the principal zones. They are transitional zones where the tree species composition in climax stands gradates from one principal zone to another.

Between tundra and closed boreal forest there is ecotonal forest tundra where tundra plants dominate in ground layers, and stunted small trees grow both isolated and in small groups.

Between boreal coniferous and summergreen broad-leaved forest there is ecotonal mixed forest where coniferous and broad-leaved trees can more or less equally compete in climax stands.

Boreal coniferous zone is divided into northern, middle and southern sub-zones (taigas). The ability of broad-leaved trees to compete with conifers increases from north to south. The summergreen broad-leaved trees proper such as oak, ash, elm, etc. can grow either isolated or in small groups in the southern sub-zone.

2.2 LANDSCAPE AND ITS TYPES

Landscape (geographical landscape) has two components: geolandscape and biolandscape. When combined, they form geobiolandscape.

Geolandscape is composed of the inorganic (non-living) elements such as climate; bedrock and soil parent materials and their birth and development history; the forms of land's surface (relief); surface- and ground-water regimes; and water courses.

Biolandscape is composed of living organisms, principally of plants, and their communities.

Geobiolandscape is includes these two components and their interactions, as well as their physical and chemical processes. Together they cause and maintain soil processes and soil types such as podzols in the boreal zone; microclimates inside vegetation cover; wetland vegetation's, especially mires in which plant necromass decomposes incompletely into peat in anaerobic conditions under groundwater table; and peat lands.

Landscape type is a relatively large area. Inside its boundaries the geocomponents are similar to the extent agreed and differ from those of the neighbouring areas which represent the types of their own.

Natural landscape and its types have been and currently remain outside anthropogenic influences.

Cultural landscape and its types have been and are currently under anthropogenic influences. They are created and maintained by the combination of human activities and natural forces.

2.3 FOREST ECOSYSTEM AS VEGETATION COMMUNITY

Forest ecosystem (biocomplex) is a functional unit composed of the interacting organism communities, their geoenvironment and biological, physical, and chemical processes (Figure 1).

Community is composed of species living in a habitat. Because photosynthesising green plants (flora) are the primary producers, ecosystem units can be described as plant communities. They have also animal and microbe species (fauna) adapted to them. Consequently, each plant or vegetation community is a bio-unit comprising fauna species.

The abstract concept of an a association of vegetation communities is a grouping of communities on a wider area including similar habitats, community dynamics and geoconditions.

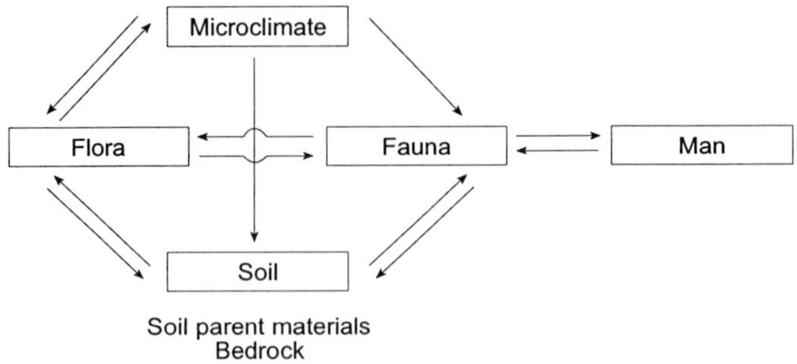

Figure 1. Ecosystem (biocomplex)

Organisms, i.e. species and their individuals, collaborate geocomponents, and physical and chemical processes in developing soil, its processes and types. Soil, the living element of a site, is located on soil parent materials and bedrock. It is composed of inorganic and organic necromass particles, roots of plants, fungi and bacteria. Soil is an essential part of the ecosystem's nutrient cycle in which plants use nutrients, and necromass is broken down by decomposers into nutrients for green plants.

Trees, the tallest and most dominating plants in forest ecosystems, form a microclimate inside their biomass. The microclimate mitigates the extreme values of the macroclimate. It has its own oxygen, carbon dioxide and moisture regimes which offer habitats for species unable to endure the extremes of the macroclimate.

By its nature, forest ecosystem is a dynamic entity of interacting components and processes. In addition to internal factors, the dynamics are caused by geofactors such as the variations of the macroclimate, rising of land from the sea, wind and water erosion, changes in water courses, landslides, earth quakes, volcanic eruptions, wild fires and storms.

3. Systematics of forest types as vegetation associations

3.1 FORMATIONS

In the hierarchy of vegetation communities and associations presented by Eyre (1968), the highest concept is a formation type (life form). A formation type, capable of occupying a particular region, is composed of species communities and their associations adapted to the climate of the particular region.

Boreal coniferous forests, summergreen broad-leaved forests, and evergreen mixed forests are the principal forest formation types in Europe. Formation types are considered to equal vegetation zones.

Boreal forests are divided into subformations. They are determined and described by the tree species dominating in the climatic climax stages of forest plant communities.

- Pine formation, dominated by pine (*Pinus silvestris*), occurs on rocky and stony sites, on dry sand, and on gravel and moraine sites, where its competitive power is stronger compared to other species, and where in natural conditions wild fires occur by relatively short intervals. Pine dominates also on nutritionally poor peat lands.

- Spruce (*Picea abies*) formation occurs on moist, nutrient rich mineral sites and peat lands where the peat layer is relatively thin and water is flowing.
- Birch (*Betula*) formation occurs on forest tundra, particularly on the shores of water courses.
- Black alder (*Alnus glutinosa*) formation occurs on limited areas of temporarily flooded shores of seas, lakes and other water courses in southern boreal subzone.
- Mixed pine and spruce formations are recognised in the Russian forest typology (Jakovlev and Voronova, 1959).

3.2 FOREST TYPES AND THEIR GROUPS

Forest type is an association (an abstract concept) of forest plant communities in which the original composition of plants is adapted to geoconditions. It is an association of plant sociologically similar communities.

Complex type is a group of forest types which have defined common characteristics.

Each forest type represents a particular site quality – bonitet – on which the growing of trees and the production of wood can be predicted, and which are significantly different from those of other forest types.

For the purposes of growing wood, the bonitet of a forest type can be estimated more accurately by the relationship of stand height over stand age. This relationship predicts stand bonitet which is the combination of site bonitet and the growing potential of a particular tree species and their genetic quality.

Within each formation, forest types are studied and described by ground-plant associations, i.e. by the number and composition of plants other than trees, paying special attention to the plants indicating the nutrient fertility and water regime of the site. Indicator plants require a minimum fertility to survive and propagate.

In addition to nutrient fertility and water regime, the site bonitet is dependent on the thickness of soil parent materials on bedrock and the stoniness of the soil.

The original and basic forest types are vegetation associations which develop in natural conditions without major disturbances until they have reached, or are close to reaching, climax stages. In reality, forest types are most often in the form of serial succession started by natural and cultural disturbances. They are called derivative types (see Chapter 4).

3.3 FOREST TYPES AS HABITATS FOR ORGANISMS

Biodiversity is composed of the absolute and relative number of species and their individuals, i.e. of the number of biological units living in their habitats. Enlarged concept comprises also the genetic, structural, functional, pattern, process, spatial, and temporal diversity of biological units and their habitats.

Habitat is a concrete space where the individuals of one or more species naturally or usually live or are found. A habitat can be either natural or cultural (see Chapter 2.2).

The scheme of habitat development is illustrated in Figure 2. It starts in climatic-edaphic (inorganic, lifeless) stage which is the habitat (geohabitat) for pioneer organisms. Geohabitat can be land rising from the sea, or a site after a landslide or volcanic eruption. Pioneer organisms and geofactors prepare community habitats for new species. At the end of this development, there is a set of geobiohabitats in each of which live organisms adapted and specialised to it. They compose the overall biodiversity.

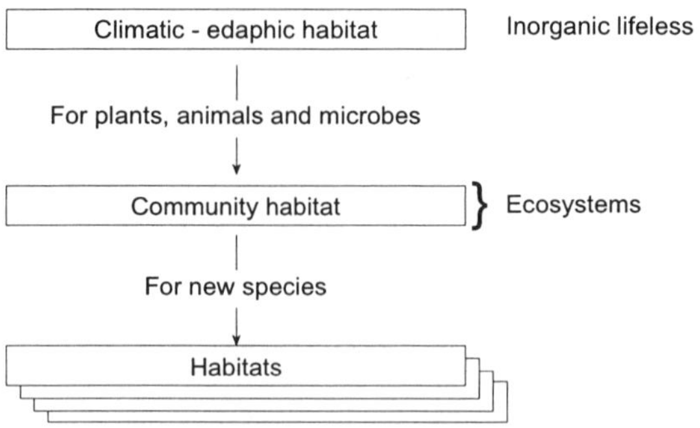

Figure 2. Development of habitat diversity and its totality.

4. Dynamics of forest communities

4.1 NATURAL SUCCESSIONS AND CLIMAXES

The last continental glazier covered the northern half of the British Isles, Scandinavia, Finland, large areas of the north-western Russia, Baltic countries and the northern central Europe. It started to melt and retreat towards northwest about 15,000 years ago. Exposed inorganic land was the geohabitat for pioneer organisms to start successional priseres towards climatic climax communities (Figure 3).

In the first stage, the pioneer community is composed of algae and lichens which can grow even on bare rock. They start to build up primitive soils from inorganic and organic particles. Then come mosses and turf-forming grasses. Woody scrubs and finally trees come in the consequent stages. Pioneer trees, such as birch spreading by light seeds transported by wind, are often among the first colonisers. The number of species increases. Forest-soil profiles are developed by podzolization processes. Microclimates develop within the vegetation, especially under tree crowns.

In the final climatic climax stage, the deep-rooted, long-living climax trees, their shade tolerating seedlings, and ground-vegetation species dominate. The climax stage is more or less in equilibrium with geofactors. The equilibrium, however, is a dynamic one. The community renews itself and possibly develops slowly into a different community.

Priseral successions and climaxes described above represent communities on mineral sites where water regime is favourable for trees. On sites where relief and soil parent materials stagnate the water and rise the ground-water table, wetlands and peat-forming mires cover large areas in the cool and humid boreal climate. They are dominated by ground plants such as sphagnum-masses, wetland grasses, and sedges and dwarf-shrubs.

Arresting factors, e.g. temporarily impeded drainage, may change the direction of priseral successions into subclimaxes. When the arresting factor is removed, the succession develops toward a true climax.

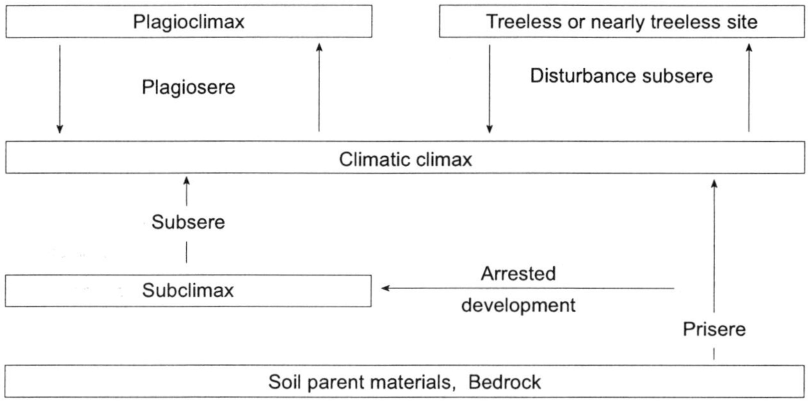

Figure 3. Successional seres and climaxes of forest plant communities.

The dynamics of boreal disturbance ecology from climatic climax to treeless or nearly treeless regeneration site and back to climax is schematically presented in Figure 4 (Kuusela 1990; Hannelius and Kuusela 1995). The disturbance, in most cases wild fire, as well as storm or insect damage, kills all or most trees, opening the site for pioneer plants (Stage 1). Fire and direct sun radiation accelerate the decomposition of necromass and raw humus, increasing the amount of nutrients which plants can use, and decreasing the acidity of humus by 1 to 2 pH-units. Also the amount of water available for plants is increased. Ground vegetation and tree seedlings flourish. They and decomposing necromass components represent the highest level of biodiversity (Stage 2).

Birches (*Betula pendula* and *B. pubescens*), aspen (*Populus tremula*) and grey alder (*Alnus incana*) are true pioneer trees on most of the mineral sites. Exceptions are the most barren sands, gravels and rocky lands where pine (*Pinus silvestris*) is the pioneer as well as the true or apparent climax tree.

In the following stages of 3 to 6, the number of trees per hectare and the genetic diversity of the stands decrease. At the same time the nutrients tied into humus, thus unavailable for plants, and the acidity of humus increase, and biological productivity and diversity decrease.

During these successional stages the climax vegetation, in the case of trees, spruce (*Picea abies*), and to minor extent Siberian fir (*Abies sibirica*) in the eastern part of the Russian taiga in Europe, invade by shade tolerant seedlings under pioneer trees and rise to dominance in mature and over-mature climax stands. Spruce creates cool and humic microclimates and accelerates the development of raw humus. Biological productivity and diversity are lowest in Stage 6.

In natural conditions wild fires recur approximately once in 50 to 100 years on assorted sands and gravels, and in 150 to 200 years on moist loams and clays. Disturbances can change every one of the stages 2 to 6 into regeneration areas. If there were no wild fires, spruce would also dominate on most of the sands and gravels where pine is an apparent climax tree.

When a climax stand has been outside major regenerative disturbances long enough, the stand degenerates, root and stem rot increase and old trees start to die one by one and in small groups. The stand regenerates itself in the openings left by dead trees, mostly by

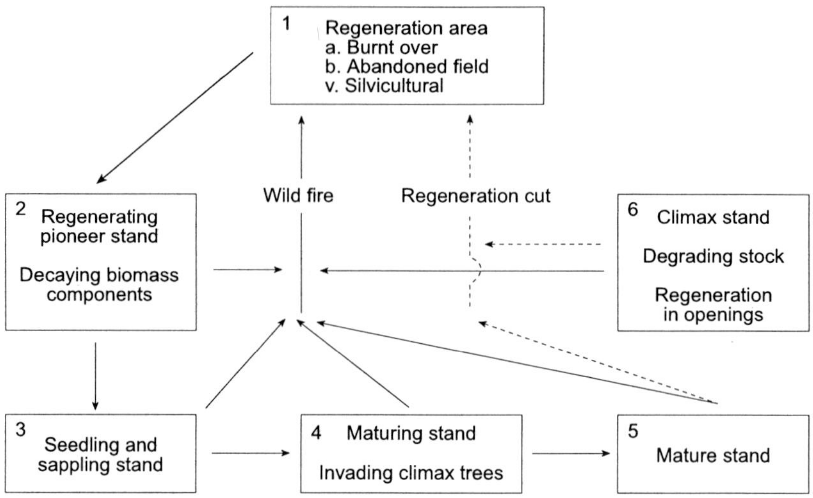

Figure 4. Major regeneration cycle 1-6, ⟶ in boreal disturbance (fire) ecology. Minor cycle occurs inside stage 6. Silvicultural regeneration occurs in the stage 5 or in 6 when natural forest is opened for production. Natural disturbance can start regeneration in each stage.
– Cultural - - -➤ Natural - - -➤ Cultural

shade-tolerant spruce seedlings and also by the vegetative sprouts of broad-leaved trees. This is called a minor regeneration cycle, composed of mini-successions. Nutrients tied into raw humus, unattainable for plants, however, are not released to the extent as in the major regeneration cycle. Biological productivity remains relatively low.

As a part of diversity dynamics, fauna species follow in time and replace those successional stages to which they are adapted and specialised.

4.2 CULTURAL SUCCESSIONS AND CLIMAXES

Wild fire as a natural regenerator has been eliminated in the Nordic Countries and their occurrence and areas have decreased in the Russian taiga in Europe. Natural disturbances have been replaced by human activities such as wood harvesting, slash-and-burn (shifting) cultivation, pasturing in forests, and by silvicultural harvesting and wood growing.

Forest communities started by human populations are called plagioseres and plagioclimaxes (Fig. 3). Continuous cultural measures maintain plagioclimaxes such as forest pastures and broad-leaved and pine stands on sites where spruce is the climax tree. Despite the cultural communities, natural factors are trying to develop plagiocommunities towards climaxes. E.g. in most of silviculturally maintained birch and pine stands there is some spruce undergrowth. If silvicultural measures and fire fighting were eliminated, spruce would gradually rise to dominance.

In cultural forests, the major regeneration cycle is composed of Stages 1 to 5, then returning to Stage 1 (Figure 4). From the point of view of the overall biodiversity, the proportion of broad-leaved trees is much higher in cultural forests than in natural ones.

This has been demonstrated by the results of repeated forest inventories. For example, in the middle taiga of the Russian Karelia, some 100 to 150 years ago broad-leaved trees dominated on about 10% of forest land while their dominance is currently 20 to 30% (Gromtsev 1993).

In Finland, the dominance broad-leaved species was at its peak when the period of burn-and-slash cultivation and pasturing in forest ended. After that their proportion decreased because of the extensive use of broad-leaved trees for fuel. When fuel wood consumption was decreased to its current level in the 1970s, their amount started to increase.

5. Overall habitat diversity

The overall habitat diversity is composed of all the stages of the major and minor regeneration cycles of boreal disturbance ecology (Figure 4). Each stage represents its own and specific kind of macrohabitat with flora and fauna adapted to it. If any of them is missing or occurs in relatively small area, overall diversity decreases. A great overall diversity is maintained by rotating the major cycle.

An important observation is that natural forest, especially if protected from wild fire, is the result of "one treatment regime". Therefore its habitat diversity is lower than of cultural forest which is the result of variable silvicultural regimes and supplemented by nature conservation reserves.

Macrohabitats of cultural forests are stands or compartments which have boundaries and areas on forest maps. In addition, there are microhabitats such as small wet places, springs and their surroundings, shores of water courses, solitary or groups of rare tree species and old, partly decayed trees etc. Their areas are so small that their boundaries and areas are not marked on forest maps. They are, however, essential components of the overall diversity.

6. Biodiversity management

Biodiversity described by the number of species and their individuals is a too wide and complicated field of specific knowledge to be mastered by the professional forest managers. A feasible way to include biodiversity in forest management is to use habitats, i.e. forest and stand types, their successions and derivates as diversity units.

Biodiversity management can be developed and employed as a part of forest management in a particular geographical area the following steps:

1. Vegetation zone or possible subzone is determined and described.
2. Landscape type (landscape ecological region) is determined and described.
3. Formation types by edaphic types are determined, described and marked on forest maps.
4. Forest types and possible treeless vegetation types are determined, described and marked on forest maps.
5. Derivative forest types, i.e. the current stands and their tree-species composition and age structures, which are the results of natural and cultural disturbances, are determined, described and marked on forest maps.

6. Microhabitats and habitats of rare, "to be specially protected" species are recognised and registered in stand descriptions.
7. Survey results are analysed and the condition of overall habitat diversity is described taking into account the natural dynamics of species compositions in the particular geobioconditions.
8. A desirable (target) composition of macro- and microhabitats is determined and described, including possible strict natural reserves.
9. The actual and desirable overall habitat diversity are compared with each others and the need to improve the actual one is recognised.
10. A plan as a part of general forestry plan to develop the current actual overall habitat diversity towards the desirable one is prepared and executed in connection of other forestry measures.

References

Corns, J.G.W. and Annas, R.M. 1986. Field guide to forest ecosystems of West-Central Alberta. Northern Forestry Centre. Canadian Forestry Service.

Ekosistemy landshaftov zapada severnoij taijgi (stryktyra, dinamika) 1995. Rossijskaja Akademija Nauk. Kareljskij Naytshnyj Tsentp. Institut Lesa. Petrozavodsk.

Eyre, S.R. 1968. Vegetation and soils. A World Picture. Edward Arnold. London.

Gromtsev, A.N. 1993. Landshaftnye zakonomernosti struktury I dinamiki spednetaeÓhnyh sosnovyh lesov Karelii. Rossijskaja Akademija Nauk. Kareljskij Naytshnyj Tsentp. Institut Lesa. Petrozavodsk.

Hannelius, S. and Kuusela, K. 1995. Finland the country of evergreen forest. Forssa Printing House Ltd.

Jakolev, F.S. and Voronova 1959. Tipy lesov Karelii I ih prirodnoe rajonirovanie. Gosydarstvennoe Izdatelstvo Kareljskoj ASSR. Petrozavodsk.

Kuusela, K. 1990. The dynamics of boreal coniferous forests. Finnish National Fund for Research Development. SITRA 112. Helsinki.

Kuusela, K. 1994. Forest resources in Europe 1950-1990. European Forest Institute. Research Report 1. Cambridge University Press.

Appendix 1

CANADIAN FOREST LANDSCAPE CLASSIFICATION SYSTEM

Source: Field Guide to Forest Ecosystem of West-Central Alberta. 1986. I.G:W. Corns and R.M. Annos. Northern Forestry Centre. Canadian Forest Service.

This system is adapted from the biogeoclimatic system developed in British Columbia by Krajina (1965 and 1969) and it parallels the scheme used by the Canadian Committee of Ecological Land Classification, the habitat type of the United States Forest Service (Pfister et al. 1977) and the operational group of the ecosystem classification developed in Ontario (Jones et al. 1983). The system is based on two types of units:

1. Biogeoclimatic, or zonal (large units).
2. Ecosystem, or site (small-scale units).

1. The large bioclimatic units identify vegetation and soil development patterns resulting from similar macroclimates. They are described hierarchically in terms of ecoregions, subecoregions, and their phases.

   1.1. Ecoregion: A geographical area that has a distinctive, mature ecosystem and specific edaphic variation as a result of a given regional macroclimate.
   1.2. Eco-subregion: A subdivision of an ecoregion that is controlled by a macroclimate but exhibits less variability in soils and vegetation than the ecoregion.
   1.3. Phase: Inside ecoregions and eco-subregions, identifies extensive area of ecosystems which is for topographic or topoedaphic reasons a typical of the regional climate. E.g. an extensive area of frost pockets in mountainous terrain. The phase is not a formal taxonomic unit but may be identified and mapped for management or descriptive purposes.

2. The small-scale ecosystem, or site, units group together similar sites on the basis of soil, vegetation, and physiographic properties. These are the ecosystem types and ecosystem associations and their phases.

   2.1. Ecosystem type: An abstract taxonomic class defined as all land areas belonging to the same soil family with the potential of supporting vegetation belonging to the same plant association.
   2.2. Ecosystem association: An abstract taxonomic unit above the ecosystem type and within an ecoregion. It is defined as all land areas with the potential of supporting plant communities with similar successional development belonging to the same plant association. More than one soil family may be represented.
   2.3. Phase of ecosystem associations: Is recognised in order to facilitate more-precise management interpretations and are distinguished on the bases of differences in slope class, aspect, soil, climate, or some other feature of soil or vegetation. It is not a formal taxonomic category.

2.4. Facies: A form of ecosystem association by a given tree species. E.g. in the *Picea glauca/Viburnum/Aralia* ecosystem association on similar sites may support pure stands of aspen, white spruce, paper birch, or pine, depending on the stand origin. Thus this ecosystem association may have four different facies according to the dominant tree species.

2.5. Biogeocoenose: The smallest ecosystem recognised in the field, and determined by the extent of the plant community and the associated polypedon.

Units described in the Field Guide mentioned above are the ecoregion, ecosystem association (forest type), and in few cases the ecosystem association phases.

Example

Ecoregion: Lower Boreal Cordilleran (LBC)

Populus tremuloides communities dominate zonal sites with mixtures of *Picea glauca* and *Pinus contorta.*

Ecosystems associations:

1. PL-Sb/*Ledum*/Lichen (lodgepole pine - black spruce)/Labrador tea/Lichen.
2. Sw/*Hylocomium* (white spruce/feather moss).
3. PL-Sb/*Ledum*/*Pleurozium* (lodgepole pine - black spruce)/Labrador tea/feather moss.
4. Sw/*Viburnum*/*Rubus pubescens*, all facies (white spruce/mooseberry/dewberry).
5. Sw/*Viburnum*/*Aralia*, spruce facies (white spruce/mooseberry/wild sarsaparilla).

Each ecosystem association is described by tree-species and ground vegetation compositions; environmental parameters (climate, soil, drainage and land form); productivity (potential yield of stem wood); successional relationships; similar associations, comments; management interpolation etc.

Appendix 2

General description of the classification system of the climatic vegetation zones, forest-typological areas, forest formations and forest types in Russian Karelia. (Jakovlev, F.S. and Voronova, V.S. 1959. Karelia's forest types and their natural regions. In Russian.)

I. Vegetation zone: Boreal coniferous forest, taiga.

- I.1. Subzone: Northern taiga.
- I.2. Subzone Middle taiga.

II. Forest typological, i.e. landscape ecological regions.

- I.2. In middle taiga (greenmoss coniferous forests).
  Regions:
- II.1. Watershed region Jänisjoki-Suojoki, greenmoss forests dominated by pine.
  Total of 7 regions.
- 1.1. Subregion Jänisjärvi-Kualisma: moraine lands and glazier eskers,
  moderately hilly, peatbogs.
  Total of 2 subregions.

III. Forest formations.

1. Pine dominance
   Pine-spruce dominance
   Spruce dominance
   Formations dominated by birch, aspen or grey alder are successions towards climax formations dominated by conifers.

IV. Forest type complexes in middle taiga, I.1.

1. Pine dominance.
   - 1.1. Lichen, one type
   - 1.2. Greenmosses, two types
   - 1.3. *Sphagnum*, one type.

2. Pine-spruce dominance
   - 2.1. Moss-lichen stony, one type
   - 2.2. Grass-greenmoss, two types
   - 2.3. *Polytrichum*, one type
   - 2.4. *Sphagnum*, one type.

3. Spruce-dominance
   - 3.1. Greenmoss, three types
   - 3.2. Peatbog-grass, one type
   - 3.3. *Polytrichum*, one type
   - 3.4. *Sphagnum*, one type.

V. Examples of forest types and their descriptions.

IV.1.2. Greenmoss, pine dominance

Vaccinium vitis-idae: Pine 10 (in tenths); bonitet IV; *Vaccinium vitis-idae, Vaccinium myrtillus, Calluna vulgaris, Majanthemum bifolium, Deschampsia flexuosa, Rubus saxatilis, Convallaria majalis*, etc.; greenmosses: *Pleurozium, Dicranum, Hylocomium*; morain hills, ridges and slopes; podzol illuvial-ferrous, sand-gravel and esker gravel, thin, fresh.

Vaccinium myrtillus: Pine 9, birch 1, and spruce; bonitet IV; *Vaccinium myrtillus, Vaccinium vitis-idae, Majanthemum bifolium, Pyrola* minor; greenmosses: *Hylocomium, Pleurozium*; hills, lover slopes; depression on level ground; podzol illuvial-ferrous, sand, medium thickness, moist.

IV.1.2. Greenmoss, spruce dominance

Vaccinium myrtillus: Spruce 9, birch 1, pine and aspen; bonitet IV; *Vaccinium myrtillus, Vaccinium vitis-idae, Trientalis europea, Majanthemum bifolium, Pyrola rotundifolia, Pyrola secunda, Calamagrostis arundinacea, Convallaria majalis, Geranium silvaticum, Dryopteris spinulosa, Lastrea dryopteris, Carex digitata*; greenmosses: *Hylocomium, Rhytidiadelphus, Pleurozium*; lover slopes of hills, hills; level grounds; ground podzol illuvial-ferrous, mixed sand, silt-sand, from fresh to moist.

Oxalis acetosella: Spruce 10, pine, birch and aspen; bonitet III; *Vaccinium myrtillus, Vaccinium vitis-idae, Matteuccia struthiopteris, Carex digitata, Oxalis acetosella, Dryopteris spinulosa, Pyrola rotundifolia, Pyrola secunda, Geranium silvaticum*; greenmosses: *Hylocomium, Rhytidiadelphus*; lover slopes, depressions on level ground; strongly podzolized, illuvial-ferrous, silt-sand, from fresh to moist.

DIVERSITY OF GROUND-LAYER VEGETATION IN BEECH FOREST

Comparison of semi-natural and managed beech stands in Northern Hungary

T. STANDOVÁR
Department of Plant Taxonomy and Ecology, L. Eötvös University
Ludovika tér 2., H-1083 Budapest, Hungary

Abstract

The comparison of six beech (*Fagus sylvatica*) stands showed that one semi-natural plot possesses more vascular species than five managed plots together.

The causes for higher diversity were studied using the vegetation and site data collected from the 576 systematically distributed 0.5 by 0.5 m quadrates in the 1.44 ha semi-natural plot.

It was shown that vascular species richness and total cover are strongly related to site heterogeneity expressed in terms of steepness, stoniness, etc.

The analyses of data, collected from a 30 by 30 m sub-plot with relatively uniform site properties, indicated that even within such a small area, vascular species richness and total cover vary not only according to site heterogeneity, but also according to the spatial pattern of small canopy gaps.

Results support the assumption that the creation of large even-aged stands can substantially reduce natural diversity of vegetation in temperate deciduous forests by homogenizing the habitat mosaic produced by natural small-scale gap dynamics.

1. Introduction

As the title of this conference suggests, nowadays the assessment of biodiversity has been a stressing demand for improving forest planning and management practices. This need is a result of increasing public awareness and of international commitments of most governments (Rio and Helsinki processes).

In an ideal situation, foresters should only ask ecologists to give "the answers" to the questions involved in this problem (what is biodiversity, how we could measure it, what are the effects of management on diversity, etc.). Unfortunately there are no ready answers for all such questions.

Recently, the importance of biodiversity in forests has been emphasised in hundreds of publications (e.g. Bachmann et al. 1996; Boyle and Boontawee 1995). The scientific challenge is to find ways of quantifying different levels of diversity. It is also important to set up our standards and baselines with which we compare existing diversity (Standovár 1996; Pielou 1995). Peterken (1996) formulated the following basic approach: natural

P. Bachmann et al. (eds.), Assessment of Biodiversity for Improved Forest Planning, 381–388.
European Forest Institute Proceedings no. 18. © 1998 *Kluwer Academic Publishers. Printed in the Netherlands.*

woodland provides a reference point for managed woodland, particularly when the objectives of management include nature conservation.

This approach calls for studies in which as natural as possible stands are studied and compared with managed stands. In this way both ecological relationships and the effects of management can be studied. Because total richness can not be measured, at least for practical purposes, some surrogate measures of biodiversity also need to be developed.

This paper shows some preliminary results of a project with similar motivations, namely: 1) a strong ecological interest in canopy understorey relationships in forest vegetation at the stand scale; 2) the realisation of the need for studying the effects of management induced structural changes on ground-layer vegetation; 3) the wish to contribute to the validation of possible surrogate measures (indicators) of forest biodiversity.

The richness of ground-layer vegetation was studied in one nearly natural stand and five even-aged homogeneous beech stands of different ages. The primary objective of this study was to determine if the higher diversity found in the ground-layer of the semi-natural stand is caused by the more heterogeneous site, by its more complex canopy structure, or both. Most of the semi-natural stands in Hungary (and probably elsewhere) have survived because of the inaccessibility of the area. This means that they grow on extreme, often steep and rocky sites with exceptional heterogeneity. So, the aim has been to test if the lack of management has caused any additional increase of diversity by allowing a more complex canopy structure to develop.

2. Materials and Methods

2.1. STUDY AREA

This study was carried out on the northeastern slopes of the Kékes, which is the highest point in Hungary (1,014 m above sea level). Mean annual temperature is 5.7 °C and annual precipitation averages 784 mm. The bedrock of the area is andesite. Since the topography is extremely steep, differently eroded shallow forms of illuviated brown forest soil occur, scree being also characteristic.

All study plots are located at elevations above 700 meters, where zonal vegetation is montane beech wood (*Aconito-Fagetum*). On rocky sites maple-ash-lime mixed woods (*Phyllitidi-Aceretum subcarpaticum*) occur (Kovács 1975).

The sample plot for this study representing the semi-natural stand structure is located within the Kékes North Forest Reserve. According to the forest management plan data the stand is 160 years old, but in reality a mosaic of different forest developmental phases is characteristic of the area. Trees older than 200 years occur together with many other age classes.

Control plots (1 ha each) in managed stands are even-aged nearly pure beech stands of about 35, 45, 50, 85 and 95 years of age.

2.2. DATA COLLECTION

In the semi-natural stand a 120 by 120 m plot with heterogeneous stand structure was studied. Vegetation was systematically sampled in 0.5 by 0.5 m quadrates that were set out on a grid at 5 meter intervals. Seven variables were estimated to describe the overall characteristics in each of the 576 quadrates. They were aspect, steepness of slope, the proportion of 3 possible substrates (stone, soil, dead wood), percentage cover of bryophytes and vascular plants (abbreviated as ASP, SLOPE, STONE, SOIL, WOOD, BRYO, VASCU respectively). In addition, all vascular species were recorded. Their dominance was estimated in percentages.

Just to have a rough estimate of richness, a less intensive preliminary vegetation survey was carried out in the managed plots. Line intercept method (Manly 1992) was applied for vascular species. They were recorded along a 500 meter long randomly laid line with 1 cm resolution. In addition, species that did not intercept the line were also recorded within each 1 ha plot.

2.3. NUMERICAL METHODS

Species-dominance diversity was expressed by calculating Shannon's diversity index. T-statistics were calculated for Shannon diversities to compare pairs of plots. Diversity ordering was carried out by applying Rényi's function (Rényi 1961). All these calculations were performed by using the NuCoSa and DivOrd program packages (Tóthmérész 1993a and 1993b).

Hierarchical numerical classification was used to distinguish quadrate groups with similar physical characteristics (PTypes) and similar vegetation (VTypes). Weighted group average method with Euclidean distance function was applied (Greig-Smith 1983). Five variables (ASP, SLOPE, STONE, SOIL, WOOD) were used.

Spearman's rank correlation analyses were used to study correlation between physical site characteristics (SLOPE, STONE, SOIL, WOOD, PType membership) and simple vegetation characteristics (BRYO, VASCU, number of vascular species denoted as VASSPNO).

The relationships between physical quadrate types and certain vegetation characteristics were studied by using Kruskal-Wallis test (**H** statistics), the non-parametric analogue of classical analysis of variance (Conover 1980). The analyses were performed by using the STATISTICA software package (StatSoft 1983-1994).

3. Results and Discussion

3.1. SPECIES DIVERSITY IN MANAGED AND SEMI-NATURAL PLOTS

Altogether 57 species were recorded in the six plots. As these basic results show (Table 1), the 1.44 ha plot representing the semi-natural stand had the highest vascular species richness (48). Of the 9 species that were not found in the 576 quadrates (covering 1% of the plot) there are 7 species that were recognised within the 1.44 ha semi-natural plot. In spite of the convincing contrast in species richness, pair-wise T-tests did not show significant difference of Shannon's diversity between the semi-natural and managed stands in any case. However, diversity ordering by Rényi's function showed that the semi-natural stand possessed higher diversity than the others, except for one case (Figure 1).

TABLE 1. A comparison of species richness, diversity and evennes of the 6 sample plots. (S1 - total number of vascular species in the 1 ha managed plots, S2 - number of vascular species intercepting the 500 m line, H - Shannon's diversity, E - evennes). Shannon diversity was calculated using S1 and a cover value of 0.0002 for the species that did not intercept the sample line.

Sample	S1	S2	H	E
Semi-natural	48	-	3.6551	0.6544
Managed 35 years	9	5	2.0554	0.8852
Managed 40 years	22	11	2.5681	0.7423
Managed 45 years	22	18	1.8778	0.4503
Managed 85 years	20	10	2.7699	0.8338
Managed 95 years	18	13	3.3907	0.9163

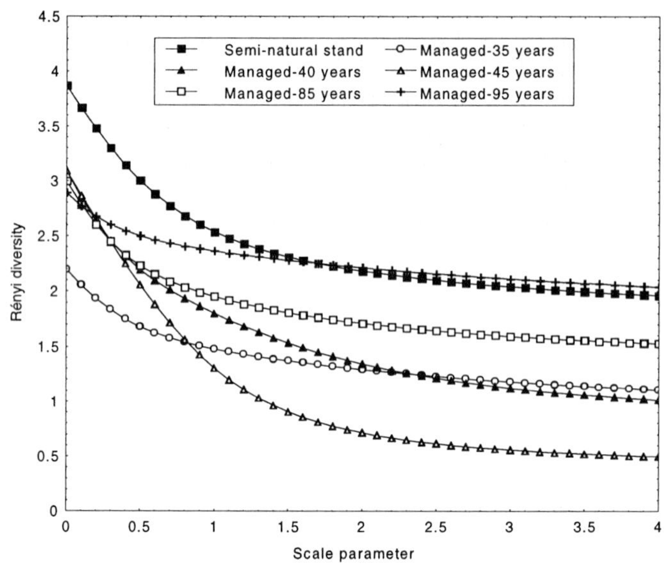

Figure 1. Diversity ordering of the 6 data sets.

3.2. EFFECTS OF SITE HETEROGENEITY VERSUS CANOPY COMPLEXITY

Even these preliminary results indicate that the semi-natural stand holds higher level of vascular species diversity than managed ones. The next step was to determine if the effects of site heterogeneity on species richness can be separated from that of canopy complexity.

The results obtained by hierarchical classification of the 576 quadrates were treated at the six-cluster-level. Allocated results for the 120 by 120 m area are shown in Figure 2. Table 2 shows average quadrate characteristics for the 6 clusters. Numbers in bold indicate important characteristics of these "physical quadrate types". Types 1, 3 and 4 represent steep sites with little soil substrate, type 2 is also rocky but less steep. Only types 5 and 6 are characterised by high proportion of soil substrate and they are also moderately steep.

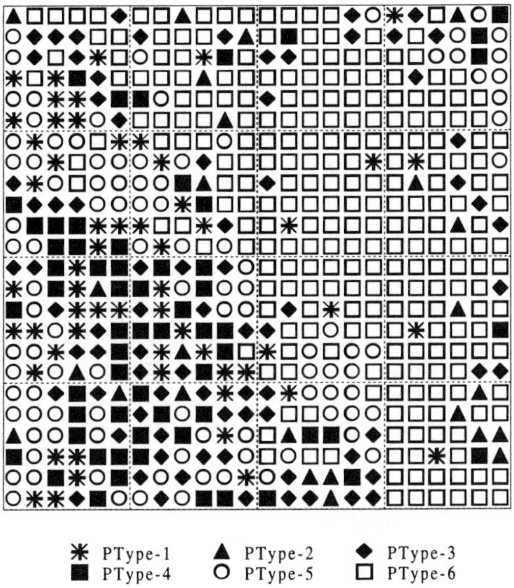

Figure 2. Distribution of "physical quadrate types" within the 120 by 120 meter plot.

TABLE 2. Mean of 6 characteristics in the "physical quadrate types" (PType) distinguished by numerical classification of 576 quadrates by variables ASP, SLOPE, STONE, SOIL, WOOD. (SLOPE=100 means scree without explicit steepness)

PType	SLOPE	VASCU	BRYO	STONE	SOIL	WOOD	Valid N
1	52.8	23.3	13.2	9.8	36.1	**54.8**	58
2	18.2	14.5	22.8	**68.0**	26.6	1.4	25
3	**100.0**	24.0	21.9	46.2	51.9	2.8	79
4	**100.0**	21.2	48.4	**83.0**	15.6	1.5	61
5	22.3	21.0	2.4	9.4	**86.6**	3.9	101
6	24.9	13.9	2.2	7.3	**89.5**	3.2	252
All types	45.2	18.3	11.9	23.9	67.9	8.2	576

The results of rank correlation analyses served test to whether simple vegetation characteristics (BRYO, VASCU, VASSPNO) were correlated with site properties (SLOPE, STONE, SOIL, WOOD, PType membership). In most cases significant correlation was found (Table 3).

TABLE 3. Significant rank correlation found between vegetation and site characteristics

Variable pairs	Kendall Tau	Z	p-level
VASCU & PType	-.092140	-3.3070	.000943
VASCU & STONE	.102105	3.6647	.000248
VASCU & SOIL	-.095192	-3.4165	.000634
VASSPNO & PType	-.073699	-2.6451	.008166
VASSPNO & SOIL	-.069939	-2.5102	.012067
BRYO & PType	-.417554	-14.9864	.000001
BRYO & SLOPE	.415488	14.9122	.000001
BRYO & STONE	.694353	24.9210	.000001
BRYO & SOIL	-.606808	-21.7789	.000001

This could easily indicate that richness of ground layer vegetation is not related to canopy structures, but rather, merely to spatial variations in site quality. The situation is confused, because canopy structures themselves are related to site steepness and stoniness. In this particular case it means that larger gaps can be found in the patches where site conditions are harsh (steep with little soil substrate).

Since the detailed tree by tree survey of stand structure has not been accomplished for the whole plot yet, only the bottom right (cf. Figure 2) 30 by 30 meter plot was studied in more detail to test if there is any "site-independent" effect. This plot is relatively uniform in terms of PTypes, most quadrates belong to PType-6, the one with the least harsh site conditions. A photograph of the canopy was taken perpendicularly for all the 36 quadrates to illustrate canopy gap patterns. Unfortunately these photographs can not be presented here, so canopy data are used only for supporting assumptions.

The 36 quadrates in the plot were grouped into 2 clusters (VTypes 1 and 2) by numerical classification using vegetation data (percentage dominance of vascular species). The spatial distribution of VTypes is shown in Figure 3 overlaid by that of PTypes. Of the 29 quadrates of PType-6, 13 belonged to VType-1, 16 to VType-2, i.e. there is no relationship between vegetation and site types. However, as the Kruskal-Wallis test showed, there is a significant dependence between VType membership and species richness and cover (cf. Table 4).

Even the above results themselves indicate that VType membership is not site-related. This view is corroborated if one analyses the species' performance within this small (30 by 30 m) plot. The photographs showed that the quadrates in VType-1 had dense tree canopy above, whereas quadrates of VType-2 are located within or near to small canopy gaps. These non-documented facts are in harmony with vegetation data, treated as

TABLE 4. Mean and standard deviation of cover and species richness of vascular plants (VASCU, VASSPNO) in the two vegetation types.

Groups	Mean		Standard deviation		N
	VASCU	VASSPNO	VASCU	VASSPNO	
VType-1	6.77	2.55	6.56	2.33	18
VType-2	55.27	5.16	20.75	2.30	18
All Groups	31.02	3.86	28.89	2.64	36

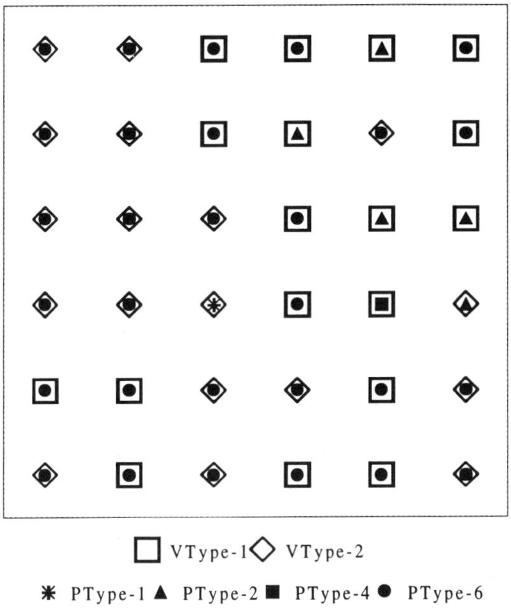

Figure 3. Joint distribution of "physical quadrate types" (PType) and vegetation types (VType) within the 30 by 30 meter plot.

"indirect canopy data", as follows: there are fewer species and much less cover in the herb layer of VType-1 (cf. Tab. 4). There are some species (*Oxalis acetosella, Chelidonium majus, Urtica dioica, Geranium robertianum, Stachys sylvatica*) that are practically confined to VType-2 (gaps or dead wood). Others (*Galium odoratum, Fagus sylvatica, Dryopteris filix-mas, Viola sylvestris, Mercurialis perennis, Pulmonaria obscura*) react with a significant increase in dominance in the gaps. Seedlings of *Fraxinus excelsior, Acer pseudoplatanus, A. platanoides* occur only in VType-2.

4. Conclusions

As these preliminary results suggest, even a small semi-natural stand can host a higher level of plant species diversity than 5 uniform stands of different ages. They also imply that under spatially heterogeneous canopy, small areas can contain patches providing habitats for species with different requirements. These patches are in proximity, so dispersal is not likely to limit the long-term survival of species.

In temperate deciduous forests the natural disturbance regime would be small-scale gap dynamics. Hence it is questionable that the use of large clear-felling followed by natural regeneration can be justified as sustainable management considering the Rio and Helsinki commitments.

Further research is needed to elucidate the effects of forest management on biodiversity. As a contribution, within this project we aim at conducting more detailed comparative studies on canopy-understory interactions in semi-natural and managed stands. We plan to carry out comparative studies of not only species diversity, but also investigate compositional and structural diversity. Other groups of organisms (bryophytes, fungi, certain insect groups) are also in the focus of our interest.

In this way it is hoped to get a better understanding of important stand structural characteristics that are closely linked with overall forest biodiversity.

5. Acknowledgements

Firstly I am grateful to Réka Aszalós and Péter Ódor, my students, who have stimulated this project from the outset, taken part in the field work and hopefully will become full-hearted co-workers with their expertise. Many thanks are due to Ilona Paszterkó, Erzsébet Szurdoki, László Gálhidy and Zsolt Skribanek for their kind help in the field work. I am also grateful to Péter Aszalós who helped this work by programming our PSION data capture. This work was partially financed by grants given by the Hungarian Scientific Foundation (MHB "A Magyar Tudományért" Alapítvány, No. 61/95/II) and the National Science Foundation (OTKA, No. F21300). These grants are greatly acknowledged together with the scholarship given by the European Forest Institute that made my participation possible.

References

Bachmann, P., Kuusela, K. and Uuttera, J. (eds.) 1996. Assessment of Biodiversity for Improved Forest Management. EFI Proceedings No. 6. European Forest Institute. Joensuu.

Boyle, T.J.B. and Boontawee, B. (eds.) 1995. Measuring and Monitoring Biodiversity in Tropical and Temperate Forests Proceedings of a IUFRO Symposium held at Chiang Mai. Thailand. August 27th-September 2nd, 1994. CIFOR. Bogor.

Conover, W.J. 1980. Practical Nonparametric Statistics 2nd ed. John Wiley & Sons.

Kovács, M. 1975. Beziehung zwischen Vegetation und Boden. Akadémiai Kiadó. Budapest.

Manly, B.F.J. 1992. The design and analysis of research studies. Cambridge University Press. Cambridge.

Peterken, G. 1996. Natural Woodland: Ecology and Conservation in Northern Temperate Regions. Cambridge University Press. Cambridge.

Pielou, E.C. 1995. Biodiversity versus old-style diversity: measuring biodiversity for conservation. In: Boyle, T.J.B. and Boontawee, B. (eds.). Measuring and Monitoring Biodiversity in Tropical and Temperate Forests. Proceedings of a IUFRO Symposium held at Chiang Mai. Thailand. August 27th-September 2nd, 1994. CIFOR. Bogor.

Rényi, A. 1961. On measure of entropy and information. In: Neyman, J. (ed.). Proceedings of the 4th Berkeley Symposium on Mathematical Statistics and Probability. Vol. I. University of California Press. Berkeley. Pp. 547-561.

Standovár, T. 1996. Aspects of Diversity in Forest Vegetation. In: Bachmann, P., Kuusela, K. and Uuttera, J. (eds.). Assessment of Biodiversity for Improved Forest Management. EFI Proceedings No. 6. European Forest Institute. Joensuu.

StatSoft 1983-1994. STATISTICA for Windows. Vol. I-III. StatSoft Inc. Tulsa.

Tóthmérész, B. 1993a. DivOrd 1.50: A Program for Diversity Ordering. Tiscia. 27: 33-44.

Tóthmérész, B. 1993b. NuCoSa 1.0: Number cruncher for community studies and other ecological applications. Abstracta Botanica 17(1-2): 283-287.

ASSESSING EFFECTS OF FOREST MANAGEMENT ON BIODIVERSITY IN THE PACIFIC NORTHWEST, USA

Comparison of management options with expert panels

EDWARD E. STARKEY
U.S. Department of the Interior
National Biological Service
Forest and Rangeland Ecosystem Science Center
Corvallis, Oregon 97331, USA

Abstract

For at least the last twenty years, management of federal forests in the Pacific Northwest region of the USA has been at the center of intense controversy. In 1993, President Clinton created the Forest Ecosystem Management Assessment Team (FEMAT) and asked it to identify management alternatives that would provide habitat for viable populations of spotted owls, marbled murrelets, salmon and other species associated with old-growth forests. The team was instructed to develop alternatives which provided a medium to high probability of ensuring the viability of these species.

For seven options, the team used expert panels to assess the likelihood of maintaining habitat for more than 1,000 terrestrial plant and animal taxa that are closely associated with late-successional forests, as well as for at-risk fish species and stocks. The likelihood of maintaining connected, viable late-successional ecosystems was also evaluated. Collectively, these assessments may represent the most extensive evaluation of biological risk ever undertaken in support of natural resource decision-making.

In this paper, the FEMAT process is presented from the perspective of a team member. The assessment process by which alternatives were evaluated by expert panels is described and reviewed, and lessons learned are discussed.

1. Introduction

For at least the last 20 years, management of federal forests in the Pacific Northwest region of the USA has been at the center of intense controversy. The problem began soon after World War II when intensive forest management was first applied to federal forests. Forest managers adopted management techniques which were generally based on European methods. Forests were clear-cut harvested , regenerated with a single species, thinned, and harvested after a relatively short interval (40-80 years). Snags and woody debris were commonly removed. Managers assumed that this type of management would not negatively affect other resources such as water, soils, fish and wildlife.

P. Bachmann et al. (eds.), Assessment of Biodiversity for Improved Forest Planning, 389–396.
European Forest Institute Proceedings no. 18. © 1998 *Kluwer Academic Publishers. Printed in the Netherlands.*

During the 1970s and 1980s, scientific studies increasingly suggested that high yield, short rotation forestry would not support many of the organisms associated with older forests. By then, most of the forests of the region had been heavily logged and consisted of a highly fragmented mosaic of recent clearcuts, young plantations and thinned stands interspersed with uncut natural stands. More than 80% of prelogging old-growth forests had been harvested by the early 1980s (Booth 1991).

Criticism of federal forest management policies increased following the finding that northern spotted owls *(Strix occidentalis caurina)* were uncommon in younger, second-growth forests, and were closely associated with older forests (Forsman et al. 1977 and 1984). This charismatic species provided a focal point for much of the debate that followed, and served as a surrogate for the broader issue of biodiversity preservation in very old forests.

In 1989, the northern spotted owl was listed as a threatened species under the Endangered Species Act of 1973 and the Seattle Audubon Society sued the Forest Service for failing to adopt forest management plans which would comply with the National Environmental Policy Act of 1969, the Endangered Species Act, and the National Forest Management Act as revised in 1982. Subsequently, a federal judge granted an injunction against timber harvest on most federal lands which contained spotted owl habitat. The marbled murrelet *(Brachyramphus marmoratus),* a seabird which nests in old-growth forests, was added to the threatened species list in 1990. In 1991 and 1992, additional legal rulings further constrained federal agencies from harvesting timber.

In 1992, management of northern spotted owls and federal forests of the Pacific Northwest became an issue in the Presidential election campaign. During a visit to Oregon, candidate Clinton promised to return, if elected President, and attempt to resolve the issue. On April 2, 1993, President Clinton returned to Portland to attend a forest conference, and to discuss key issues. Following the conference, at President Clinton's request, the Forest Ecosystem Management Assessment Team (FEMAT) was created and asked to identify management alternatives that provide for habitat conditions that would support viable populations of species associated with old-growth forests of western Oregon and Washington, and northwestern California. Because of legal concerns and deadlines, the team was asked to complete the process within 90 days.

In this paper, the FEMAT process is presented from the perspective of a team member. The assessment process by which alternatives were evaluated is described and reviewed. Selected results are presented as well as lessons learned during the assessment process.

2. Goals and Objectives

The team's goal was to identify management alternatives which would "attain the greatest economic and social contributions from the forests" and comply with "applicable laws and regulations". The team was asked to develop alternatives with a "medium to very high probability of ensuring the viability of species" and which met the following objectives (Thomas 1994):

(a) Maintenance and/or restoration of habitat conditions for the northern spotted owl and the marbled murrelet that will provide for viability of each species, well distributed within their current ranges on federal lands,

(b) Maintenance and/or restoration of habitat conditions to support viable populations, well distributed across their current range, of species known to be closely associated with old-growth forest conditions,

(c) Maintenance and/or restoration of spawning and rearing habitat on federal lands to support recovery and maintenance of viable populations of anadromous fish species and stock and other fish species and stocks considered "sensitive" or "at risk", and

(d) Maintenance and/or creation of a connected or interactive old-growth forest ecosystem on federal lands within the region.

3. The Team

The team consisted of Economic, Social, Terrestrial, Aquatic, and Spatial Analysis Groups as well as numerous support personnel. Approximately 100 scientists and as many as 500 technicians, administrative and support staff were involved in the effort. Terrestrial and Aquatic Groups included 24 and 18 ecologists, respectively.

Terrestrial and Aquatic Ecology Groups contributed to the development of various management options and were largely responsible for formulation of management standards and guidelines to be applied under these options. An additional major responsibility included the assessment of the likely effects of management options on plant and animal species commonly associated with old-growth forests.

4. The Assessment Process

For 10 options, the Terrestrial Ecology Group evaluated the likelihood of maintaining sufficient habitat, well distributed across federal lands, to provide for continued viability of northern spotted owls and marbled murrelets. For seven of these options, similar assessments were completed by the Terrestrial and Aquatic Groups for more than 1,000 plant and animal species closely associated with old-growth forests (Meslow et al. 1994). These groups also assessed the likelihood of maintaining functional, interacting, late successional and old-growth forest ecosystems. However, only assessments of individual species or groups are discussed in this paper.

The Terrestrial and Aquatic Ecology Groups faced a number of challenges in evaluating management options. For most options, the assessment process included more than 1,000 taxa thought to be associated with late-successional or old-growth forests. For nearly all of these taxa, habitat relationships are only poorly understood and there is inadequate information upon which to base predictive models. Furthermore, the team was asked to complete their work in 90 days, and thus faced a serious time constraint.

Therefore, expert panels were convened to provide professional judgments of habitat quality under various options. More than 70 experts were selected from universities, the private sector, and government agencies. Panels consisting of 3-5 scientists were provided all available information on options to be evaluated. They were asked to project the likelihood of achieving specific habitat outcomes on federal lands during the next 100 years. Those projected outcomes were:

Outcome A. Habitat is of sufficient quality, distribution, and abundance to allow the species population to stabilize, well distributed across federal lands. This outcome would provide compliance with the National Forest Management Act.

Outcome B. Habitat is of sufficient quality, distribution, and abundance to allow the species population to stabilize, but with sufficient gaps in the historic species distribution on federal land. These gaps cause some limitation in interactions among local populations.

Outcome C. Habitat only allows continued species existence in refugia, with strong limitations on interactions among local populations.

Outcome D. Habitat conditions result in species extirpation from federal land. This outcome would clearly not comply with the National Forest Management Act and, for listed species , the Endangered Species Act.

In an attempt to quantify scientific and personal uncertainties, each panelist was given 100 likelihood points to assign to any combination of outcomes for each taxon and option. Panelists could express confidence in a particular outcome by placing all or most points in a single alternative or they could express high levels of uncertainty by distributing points across outcomes. Panelists were told that likelihood's should not be considered to be probabilities in the statistical sense, instead, they represented degrees of personal belief in particular outcomes. Scores for all panelists were combined into a single average for each outcome (Table 1). No additional statistical analyses were performed.

5. Assessment Results and Discussion

Although panelists evaluated seven Management Options, the following discussion focuses on three options which illustrate the full range of timber harvest levels which would have been permitted under those options evaluated (Tables 1 and 2). Option 1 would have allowed only a very small amount of timber harvest (0.1 billion board feet) and was generally the most restrictive option considered. Option 9 would have provided for an intermediate level of timber harvest (1.2 billion board feet) and harvest under Option 7 (1.8 billion board feet) would have been greater than that permitted by any of the other options. Option 7 incorporated the most recent forest management plans of federal agencies, and assessment results therefore reflect the panelists projections of continued implementation of these plans. Only selected taxa are discussed to illustrate results of the assessment process.

Much of the assessment analysis focused on the likelihood of Outcome A (*well distributed habitat*) for taxa and options. This Outcome was emphasized because it would likely meet the requirements of the National Forest Management Act. Additionally, options were compared to identify those which reached an 80% likelihood of achieving this outcome. Selection of this analysis threshold was based on professional judgment that an 80% likelihood would provide a "high probability of ensuring the viability of species" (Meslow et al. 1994). Outcome D was also of concern because extirpation of species was considered to be unacceptable.

TABLE 1. Projected future likelihoods of habitat outcomes[a] for Northern Spotted Owls under selected Forest Management Options

Option	Outcome[a]			
	A	B	C	D
1 (very little harvest)	89	10	1	0
7 (greatest harvest)	71	25	4	0
9 (intermediate harvest)	83	18	0	0

[a]Habitat outcomes: A=well distributed, B=gaps, C=refugia only, D=extirpation.

TABLE 2. Projected future likelihoods of habitat outcomes[a] for Coho Salmon under selected Forest Management Options

Option	Outcome[a]			
	A	B	C	D
1 (very little harvest)	80	15	5	0
7 (greatest harvest)	10	20	50	20
9 (intermediate harvest)	65	20	15	0

[a]Habitat outcomes: A=well distributed, B=gaps, C=refugia only, D=extirpation.

Habitat for northern spotted owls was projected to have a greater than 80% likelihood of Outcome A under Options 1 (*very little harvest*) and 9 (*intermediate harvest*), with about a 70% likelihood of this outcome for Option 7 (*greatest harvest*). Panelists did not anticipate likelihood's of extirpation for any of these alternatives (Table 1). Coho salmon (*Oncorhynchus kisutch*) were projected to have 80 and 65% likelihood's of Outcome A under Options 1 and 9 respectively. However, Option 7 was judged to be unlikely (< 10%) to result in Outcome A for Coho salmon, and the risk of Outcome D (*extirpation*) reached 20% (Table 2). Apparently, the expert panels felt that Coho salmon are even more sensitive to forest management impacts than the threatened northern spotted owl. Coho are presently being considered for listing as a threatened species under the Endangered Species Act.

A similar pattern was observed with most other taxa, with decreasing likelihood's of Outcome A under Options 1, 9, and 7, respectively (Figure 1), which probably reflects a correlation between reserved forest area and number of taxa for which Outcome A is likely (Figure 2). However, habitat for invertebrate animals and nonvascular plants was significantly less likely to achieve Outcome A than that for vertebrate animals and vascular plants. Even under the restrictive Option 1, habitat for only 25% of invertebrate animal and nonvascular plant species assessed would have greater than an 80% likelihood of Outcome A. Apparently, viability of these species is significantly affected by factors other than those resulting from federal forest management policies.

Based on the FEMAT analysis (Forest Ecosystem Management Team 1993), in July, 1993, President Clinton chose a preferred alternative ("Option 9"). Although Option 9 provided at least an 80% likelihood of achieving Outcome A for most vertebrate animal and vascular plant taxa, only 14% of invertebrate animal and nonvascular plant taxa would be similarly protected (Table 1). Additional mitigating measures were developed to increase protection for these groups. However, only 25% of these taxa would achieve an 80% likelihood of Outcome A under Option 1, which would allow very little timber

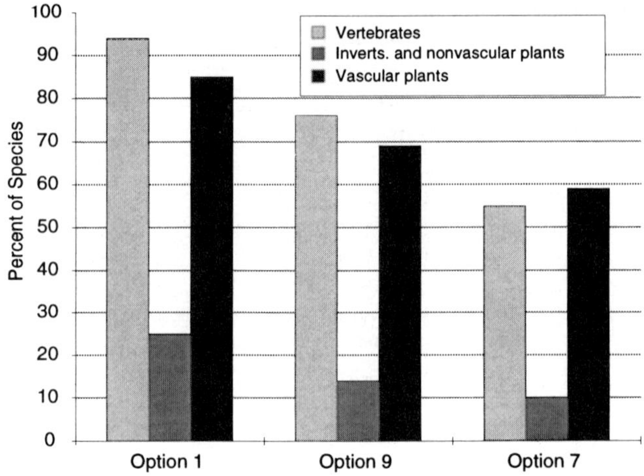

Figure 1. Percent of vertebrate, nonvascular plant and invertebrate, and vascular plant species with >80% likelihood of maintaining populations well distributed on federal lands in next 100 years (Outcome A) under three management options.

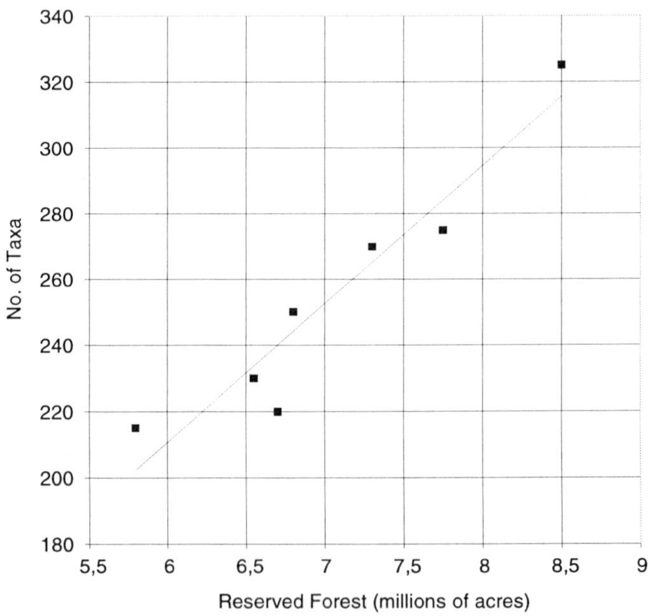

Figure 2. Relationship of forest area reserved by seven management options and number of taxa with >60% likelihood of habitat adequate to support well distributed populations (Outcome A) on federal lands for the next 100 years (FEMAT 1993, p. II-29)

harvest Therefore, concerns about viability of these species cannot be resolved by applying mitigating measures exclusively to federal lands.

External factors such as global warming, pollution, and hydroelectric dams are also significant influences on the viability of many taxa. For example, populations of anadromous fish such as Coho Salmon are greatly affected by ocean harvest, water quality, and changes in ocean currents. Conservation of these species will require broad cooperation among governments and land owners.

Following further public, legal and technical review (U. S. Department of Agriculture and U. S. Department of Interior 1994a), in April, 1994 the federal government began implementation of the Northwest Forest Plan (U. S. Department of Agriculture and U. S. Department of Interior 1994b). Under this plan, land allocations include approximately 10 million acres of Late-Successional and Riparian Reserves and 3.9 million acres which will continue to be available for timber harvest.

6. Lessons Learned

Assessment panels were an effective and efficient method for obtaining professional judgments from experts. Application of results was limited to comparison of various alternatives, and FEMAT made no attempt to actually predict future outcomes or viabilities. The distinction between comparison and prediction is important, so that policy makers and the general public do not develop false expectations for future environments. Professional judgment cannot accurately predict the future, nor can mathematical modeling.

Scientific credibility of panelists is obviously critical, and leading experts usually welcomed the opportunity to participate on panels. Discussion among participants was generally congenial and positive, and in many cases consensus was reached. However, a few panelists expressed doubt that harvest restrictions and biodiversity conservation measures would be implemented as described and may have underestimated the conservation value of some options. Panelists were clearly instructed to assume full implementation of each option, yet experts have biases like many other people. Thus, in future assessment efforts, establishment of trust among scientists, managers and policy makers will be critical.

Another potential problem resulted from some panelists' lack of understanding of silvicultural and forest management practices. Foresters and biologists sometimes "speak different languages". For example, some panelists perceived thinning as a practice applied to regenerating forests to produce even aged stands. Foresters considered thinning to be a silvicultural tool which could be used to shape the structure of forests to achieve a variety of objectives. Such fundamental communication problems can result in widely varying visions of future managed forests. Thus, foresters should work closely with panels to make sure participants clearly understand proposed actions. Multi-discipline field trips are especially valuable tools to enhance communication and understanding.

Biological risk assessment as conducted by the FEMAT was a valuable tool which contributed greatly to the development of the Northwest Forest Plan. However, the following key social and policy questions remain unanswered:

Which of the assessment Outcomes are acceptable? Outcome A *(well distributed habitat)* was considered to meet the legal requirements of the National Forest

Management Act, but is it desirable or realistic to require that habitats be well distributed across federal lands. Some species may have never been well distributed, or have already suffered serious losses of habitat which cannot be restored.

What levels of risk are acceptable? The FEMAT assessment process assumed that an 80% likelihood of achieving Outcome A represented a relatively secure level of habitat protection. Is an 80% likelihood adequate? Some would argue that species protection requires a much greater likelihood, others would accept more uncertainty. Is any likelihood of extirpation acceptable?

Various combinations of uncertainty levels and Outcomes may interact in determining social acceptability. Would a 95% likelihood of Outcome B be more acceptable than an 80% likelihood of Outcome A?

These questions are not scientific, they are social. They will not be answered by scientific panels, but through public discussion and political debate.

References

Booth, D.E. 1991. Estimating prelogging old-growth in the Pacific Northwest. J. Forestry. 89: 25-29.

Forsman, E.D., Meslow, E.C. and Strub, M.J. 1977. Spotted owl abundance in young versus old-growth forests. Oregon Wildlife Society Bulletin. 5: 43-47.

Forsman, E.D., Meslow, E.C. and Wight, H.M. 1984. Distribution and ecology of the spotted owl in Oregon. Wildlife Monographs. 87: 1-64.

Forest Ecosystem Management Assessment Team 1993. Forest ecosystem management: An ecological, economic and social assessment. USDA Forest Service and USDI Bureau of Land Management. Portland, OR. 833 p. + appendices.

Meslow, E.C., Holthausen, R.S. and Cleaves, D.A. 1994. Assessment of terrestrial species and ecosystems. J. Forestry. 92: 24-27.

Thomas, J.W. 1994. Forest ecosystem management assessment team: Objectives, process and options. J. Forestry. 92: 12-19.

U. S. Department of Agriculture and U. S. Department of Interior 1994a. Final supplemental environmental impact statement on management of habitat for late-successional and old-growth forest related species within the range of the northern spotted owl. USDA Forest Service and USDI Bureau of Land Management. Portland, OR. 501 p. + appendices.

U. S. Department of Agriculture and U. S. Department of Interior. 1994b. Record of decision for Amendments to Forest Service and Bureau of Land Management planning documents within the range of the northern spotted owl. USDA Forest Service and USDI Bureau of Land Management. Portland, OR. 74 p.

BIODIVERSITY ISSUES IN ROMANIAN FOREST MANAGEMENT PLANNING

S. TAMAS
Transylvania University of Brasov
B-dul Eroilor 29
2200 Brasov, Romania

Abstract

Loss of biodiversity is a global issue affecting on an increasing scale European forests too. The forests of Romania should be managed taking into account the necessity of preserving their high level of biodiversity. The paper outlines the ways biodiversity conservation is achieved in protected areas set up within the forest estate of Romania. Based on the fact that an exhaustive research project for solving the basic issues concerning the theoretical and practical aspects of biodiversity is time-consuming and requires an important financial effort, the paper focuses on some straightforward, practical means of biodiversity conservation within the commercially utilised forests of Romania.

1. Biodiversity and the Temperate Forests

Biological diversity has always been a primordial characteristic of all the ecosystems representing basic structural and functional components of the biosphere. It is only relatively recently however that a major scientific interest was focused on biodiversity and its implications in the management of natural resources. In principle, the term of biodiversity refers to the variety, variability and population size of living organisms to be found in a certain environment (Pearce and Moran 1994). The importance of biodiversity and the theoretical and practical consequences of its overwhelming presence in the most diversified ecosystems came under scrutiny in the circumstances determined by its recent continuous decline.

Although biodiversity is a feature present in practically all populations of living organisms, economic considerations determined the analysis of its influences mainly on the management of biological resources. Biological resources are viewed as those components of the biosphere which are of primary interest in terms of direct human uses.

Forests represent one of the main biological resources of the world and it is within the forests that the loss of biodiversity came into the spotlight (Wilson 1988). Indeed, forests are increasingly regarded as critical areas for managing and protecting biodiversity. The long-time management of a large proportion of the world's forests that fulfil socio-economic requirements has contributed to the accumulation of a substantial amount of data on most of the populations integrated into the forest ecosystem. A complete

P. Bachmann et al. (eds.), Assessment of Biodiversity for Improved Forest Planning, 397–404.
European Forest Institute Proceedings no. 18. © 1998 *Kluwer Academic Publishers. Printed in the Netherlands.*

characterisation of forest ecosystems is at the moment, and for a long time to come, impossible due to the extremely intricate multilevel structure and functioning of the biological components of the forest. A prevailing trend is however emphasised more and more resolutely: the decline in the biological diversity of the forests. If biodiversity refers to the presence of variability for the most diverse components of a given system comprising living organisms, loss of biodiversity occurs when variability has limited conditions to express itself, resulting in an increased homogeneity of the system. The presence of a high level of structural and functional heterogeneity is a fundamental condition of existence for systems which support living organisms. That is why a reduced heterogeneity is viewed as an expression of poor system performance and constitutes a warning for taking as urgently as possible the required management decision for reversing this trend. The variability of biological resources is strongly influenced both by latitude and altitude: as a rule, biodiversity decreases with increasing latitude and increasing altitude. Species richness is the highest in tropical forests. It is estimated that one square kilometre of tropical forest in Central or south America can contain many thousands of species of butterflies, beetles and other insects and several hundred species of trees and birds. At such large numbers of species and with the recent strong human-generated alterations in the tropical environment, it is normal that the decline in biodiversity expressed as loss in the number of species is highest in the tropical rain forest. In percentage terms however, this loss may be comparable with that of temperate forests, with a much lower biodiversity. Without minimising biodiversity decline in tropical forests, it is a top theoretical and practical priority to focus on the biodiversity issues of temperate forests, which both as area and productivity are of an outstanding importance to mankind. The long-term management of temperate forests has been an important source of scientific information which together with renewed research work in the field could contribute to the solution of the problems related to declining biodiversity in the temperate region.

Biodiversity loss becomes thus a global issue, its importance being practically the same throughout the world, regardless of the fact that in certain areas it affects a megadiversity and in others a minidiversity.

2. Circumstances of Biodiversity Presence in Romania

The forests of Romania, covering an area of 6.25 million ha, are situated in the temperate zone and are of high biodiversity in terms of ecosystems, species composition and genetic variability induced by the high proportion of unaltered ecosystems. Natural and semi-natural ecosystems represent 47% of the territory of Romania. The latitude and altitude variability place Romania, a central-European country, under various influences of arctic, alpine, western, balcanic, sub-Mediterranean and eastern, Caucasian origin. That is why 217 forest sites and over 250 forest types have been identified in Romania. The flora in general and the forests in particular are characterised by the presence of a wide diversity of species – 3,100 species, 540 subspecies and varieties, 60 species of forest trees and shrubs. A number of 228 plant species and over 1,000 species of the fauna are endemic to Romania and some of them are perfectly adapted to the local conditions, forest ecosystems playing a major role in their presence and conservation (Tudor et al. 1996). Endangered endemic species represent a major issue in biodiversity loss and should constitute an essential objective for practical forest management interventions aiming at their conservation.

By geographic areas, the distribution of forests is 66% mountain conifers and beech forests, 24% hill oak and beech forests and 10% lowland mixed broad-leaved forests. The area of the forests managed for timber production and protection functions covers 56% of the overall forest area, while special protection forests represent 44% of this area.

3. Forest Management and Biodiversity Conservation in Protected Areas

Aware of the high value of biodiversity of Romania's forests, the national forest administration has always aimed at a differentiated management of the forests taking into account the achievement of a high economic efficiency and the maintenance of the forests in their best bioecological condition. These apparently conflicting interests could not be met simultaneously throughout the forests, over-exploitation resulting in both loss of biodiversity and site quality deterioration, on the one hand, and non-intervention on a large scale being totally unacceptable for economic reasons, on the other hand. It has been and still it is relatively easy to find the right balance between these two extreme situations as until 1990 all the forests were state-owned and at present only about 6% of them are privately owned forests of up to 1 ha in size for each owner. Unfortunately, due to the present economic circumstances, to the inadequate agro-forestry education, to the lack of interest in the sustainable management of privately-owned forests and to the relatively timid enforcement of the law, a sizeable area of the privately-owned forests has been clear-felled. Due to the generally patchy distribution of private forests, these clear-fellings did not have a major negative influence on biodiversity, although it is questionable whether these felled areas would soon regain their forest status.

The problem of biodiversity conservation has been dealt with differently within the two major functional groups set up for management purposes: timber production and general protection forests, and forests with special protection functions. The latter functional group comprises forests which perform such as

a) watershed protection,
b) soil protection for the prevention of erosion,
c) protection against harmful climatic factors or represent
d) forests of social interest, and
e) natural monuments and reserves.

If within the above-mentioned categories a) - c) forest management planning can often find the suitable alternatives for balancing economic and protective effects, forests representing natural monuments and reserves are exempt from commercial management practices and play a major role in the conservation of biodiversity.

The setting up of forest reserves has been a long-standing tradition in Romanian forest management practice. The first project aiming at setting up a forest area as a national park dates back to 1924. In 1973 a major conceptual project for establishing a system of 14 national parks has been worked out, but its accomplishment was met by the resistance of the higher decision bodies. However, at various administrative levels an important number of natural reserves of forestry, botanical, geological, paleonthological, speological, or zoological, and of mixed interest have been set up. They comprise 424 natural reserves, 8 landscape reserves and 31 natural monuments. The whole area of these reserves is 96,800 ha, out of which 83% are set up within the forest ecosystems, either as mixed or typical forestry-related interest reserves. Out of the 463 natural reserves, only 41 have a valid legal status, the rest having been constituted by local authorities and being regarded as such by the forest management working plans (Oltean et al. 1992). After 1990 the conditions for setting up national parks have improved and 13 national parks have been established by Order 7/1990 of the Ministry of Waters, Forests and Environment. The present area of the national parks and reserves covers about 8% of the total area of the forests. A set of technical regulations for the management of national parks, natural reserves, natural monuments and recreation forests has been worked out by the ROMSILVA National Forest Administration (Tudor and Patrascoiu 1990) specifying that the protected natural reserves are structured into two zones: a scientific or integral reserve zone and a buffer protection zone. Each zone is managed specifically, i.e. non-intervention in the integral reserve, conservation cuts in the buffer protection zone and delicate silvicultural interventions enabling commercial timber production in the peripheral zone. Due to the relatively large area of national parks and natural reserves with a sensible effect on the economic efficiency of forestry and to the very broad criteria taken into account when setting up these protected areas, under the "Research program for biodiversity conservation and ecological reconstruction" of the Ministry of Waters, Forests and Environment, the network of national parks has been restructured. The basic idea in this process has been a clear specification of the objectives to be protected and a scientific substantiation of the protective capabilities of the ecosystems in the specified areas. Only 6 of the 13 national parks set up in 1990 have retained their status, some parks being included in the biosphere reserves category or being reassigned their previous functions. The conservation of biodiversity within protected areas is achieved by a specific forest management planning document which sets the aims and objectives for the protected area and specifies the necessary technical planning means and interventions for the plenary manifestation of the protection functions. Based on the national conservation strategy, a general management plan is worked out and detailed on an annual basis with the necessary operations and specific projects to be worked out. An important topic in the management plan deals with the organisation of tourist access and economic activities specifying the restrictions imposed in each of the three zones of a national reserve. Properly managed through the forest management working plans, the national forest parks and natural forest reserves represent excellent areas for the conservation and monitoring of biodiversity.

4. The Role of Forest Management Planning in Biodiversity Conservation within Commercially Utilised Forests

The whole forested area of Romania is managed in accordance with management plans worked out by the Forest Research and Management Institute (ICAS) and by several private forest management planning companies, in accordance with the provisions of the technical regulations worked out by the ROMSILVA National Forest Administration Board. For each management unit plans are worked out for a period of 10 years at stand level based on the detailed data capture of both site and stand characteristics at subcompartment level. Subcompartments, compartments or even whole management units which are assigned special protection functions are, according to their specific status, either exempted from commercial utilisation, or subjected to silvicultural interventions which enable the full achievement of their protection functions. If in the field phase of the forest management planning process special protection objectives referring to the presence of endemic or rare species, exceptionally high quality stands etc. are singled out, such subcompartments are constituted individually or in groups as special management sub-units within the management unit set up mainly on orographical criteria or based on permanent artificial limits. The silvicultural interventions in these management sub-units are in accordance with the individual requirements of each stand. Under these circumstances, even if specific stands are not legally designated as protection areas of national or local interest, they are managed in accordance with their status specified by the forest management working plans, all the necessary conditions for biodiversity conservation being met.

If areas within the economically managed forests needing special protection are dealt with accordingly by the forest management plans, the majority of forests are, however, assigned the main function to fulfil economic requirements in terms of timber production and in the subsidiary, protection functions. Even in these forests the conservation of biodiversity is, to a very reasonable extent, achieved by setting realistic production targets, structural stand and/or forest management unit models and devising the necessary silvicultural interventions for achieving these models based on a comprehensive decision process.

The main provisions of the forest management working plans which contribute to the conservation of biodiversity are:

The setting up of subcompartments with a minimum area of 0.5 ha. This provision is important in identifying forest sites and stands which have different characteristics compared to their surrounding area and which require specific silvicultural interventions. Although, in principle, 0.5 ha subcompartments incur an economic sacrifice by preserving a patchy, biodiversity conservation friendly structure rather than having homogenised stands of up to 35 ha, this provision of the forest management planning regulations is strictly observed throughout the country. On the other hand, even if site conditions and stands are homogeneous on large areas, a maximum allowed subcompartment area of 35 ha prevents radical silvicultural interventions on large areas, enabling thus a balanced stand development in accordance with, and not seriously affecting site conditions.

The reliance exclusively on natural regeneration whenever it is possible. Natural regeneration is an important means of promoting and conserving biodiversity, through the achievement of both inter- and intraspecific variability. The permanent improvement of species suitability to site conditions even under variable site and climatic circumstances determines a versatile and favourable stand development, with a proper representation of all specific ecosystem components.

The acceptance of a natural regeneration period of 10 to 20 years for each stand in which final felling is planned. The suggested natural regeneration period in the final years preceding the 1989 political changes was up to 60 years, but it proved to be economically and biologically impracticable because of the exaggerated logging costs, the sometimes heavy damage of the recruitment and the possibility of an arbitrary overcut for a supposedly better natural regeneration. The present silvicultural practice, in accordance with the forest management working plans, applies a 10 to 20 years long natural regeneration period which allows for at least three cuts both in the typical and in the irregular shelterwood or group felling.

The strict limitation of clear-felled areas. Forest management planning regulations provide for clear-fellings to be performed mainly in strips with coupes not exceeding 3 ha. Thus the clear-felled areas are to a large extent regenerated naturally by seed from the residual stand. Such situations occur typically in pure spruce stands where afforestation is a common practice on a variable proportion of the maximum 3 ha coupes. Seed sources for the nursery production of transplants are carefully selected in order to achieve a new stand capable of fully benefiting from the available site conditions as soon as possible.

The specification of differentiated species mixture targets at different stand development stages. In principle, forest management planning designates an optimal species mixture for each stage in the development of a stand. The species mixture at regeneration results from the natural regeneration and it is usually much richer in species than the felled stand. If necessary – for special soil protection functions or game management –, extra tree and shrub species can be added by planting, the idea being to have a starting composition as varied as possible, in accordance with the peculiarities of the site and the economic requirements. The role of tending fellings – cleaning and thinning – is to perform the necessary positive and negative selection in order to attain the intermediate target compositions, which combine economic interests with species biological peculiarities as well as with biodiversity conservation interests.

Setting selection forest structures as targets in commercially managed forests. The selection forests, typically made up of spruce, fir and beech are present in Romanian upland forestry and although their value in terms of timber quality, yield and biodiversity potential is widely recognised, the further development of this structural high forest type is limited by economic considerations. Selection forest structures are however present in many forests of social interest and computer-aided simulation strategies (Florescu et al. 1990) are available for converting even-aged high forest structures to selection forest structures.

The use of rotation as an instrument in controlling growing stock size and quality. In general, rotation is based on stands maturity age, reflecting the moment the stand still fulfils optimally its designated role. Rotation is usually a multiple of 10 or 20 years and a higher rotation, recommended in many forest management working plans helps to increase both growing stock volume and biodiversity.

The ecological restoration of certain forest areas. Forest management planning has a major influence in the ecological restoration of stands which, due to bad management practices, have been substantially altered by the extension of spruce beyond its natural limits, both into the beech zone and into lowland sites where it could not properly adapt to the specific sites. The present trend is to re-establish on all these sites the previous natural species mixtures by eliminating in a stepwise process pure spruce stands and replacing them with broad-leaved species stands capable of an optimal use of local site potentials and to re-establish the much higher biodiversity characteristic to the former stands.

5. Further Research Priorities and Practical Steps in Enhancing Biodiversity in the Forests of Romania

Although the forests of Romania have been managed for a long time on a sound scientific basis, in accordance with the economic requirements and site capabilities, enabling an adequate representation of biodiversity for the whole forest domain of the country, the conservation and further development of biodiversity should be achieved by taking into account the following possibilities:

- the wider application of the systems approach at forest management planning level by a better harmonisation of silvicultural and logging interests taking into account the economic effects of forestry and the social effects of the forests in a market-oriented economy;
- the use of the systems approach for the more detailed characterisation through scientific research of all biological components of the forest, paying special attention not only to tree species of primordial economic interest, but to all the other plant and fauna species which may influence the bioproductive and bioprotective performances of the forest;
- a more detailed characterisation and monitoring through periodical data collection within the field stage of the forest management planning process of the various species present in the forest ecosystem;
- an interdisciplinary approach to forest management planning with a more active participation of all interested organisations and of specialists from the most diversified biological scientific research areas in evaluating forest management planning decision alternatives and their possible effects on biodiversity conservation;
- an evaluation, based on a major research program, of the current practical status of biodiversity in central-European forests and an economic assessment of the influence of biodiversity-conservation-generated silvicultural interventions on the general economic performance of forestry. This economic effect should be combined with an evaluation of specific research costs and the problem of funding these researches should be tackled too;
- an increased role of forest management planning institutes in identifying areas of outstanding interest in terms of biodiversity and in imposing through the forest management working plans specific interventions for restoring and conserving biodiversity. A legal basis should be created for compensating from internal or external sources the losses incurred to the owner of the forest by pursuing interests other than the highest economic efficiency of the forest company;
- the provision in the forest management working plans of the necessity of creating in specified areas non-intervention strips for monitoring potential biodiversity and of leaving behind, after final felling, a certain amount of coarse wood debris. These provisions should be however substantiated through specific research projects.

References

Florescu, I., Tamas, S. and Chitea, G. 1990.Cercetari privind elaborarea unor modele provizorii de structura pentru arborete in curs de transformare spre gradinarit. In: Metode si tehnologii moderne in cultura si exploatarea padurilor. University of Brasov. Pp. 107-114.

Oltean, M., Romanca, G. and Zamfir, M. 1992. List of protected areas in Romania (1932-1991). Ocrot. nat. med. inconj. 36: 23-33.

Pearce, D. and Moran, D. 1994. The Economic Value of Biodiversity. Earthscan Publications Ltd. London.

Tudor, T., Galinescu, V. and Corduneanu, F. 1996. Problematica parcurilor nationale din fondul forestier al Romaniei. ROMSILVA. R.A. Report. Bucharest.

Tudor, T. and Patrascoiu, N. 1990. Indrumari tehnice privind gospodarirea parcurilor nationale, rezervatiilor naturale, monumentelor naturii si a padurilor cu functii de recreere din fondul forestier. ROMSILVA. Bucharest.

Wilson, E.O. 1988. The current state of biological diversity. In: Wilson, E.O. (ed.). Biodiversity. National Academy Press. Washington D.C.

THE BIODIVERSITY AND FOREST MANAGEMENT IN THE CZECH REPUBLIC

V. BURIÁNEK
Forestry and Game Management Research Institute
Jíloviští - Strnady; Czech Republic

Abstract

This paper addresses the biodiversity of forest ecosystems as an important criterion of sustainable forest management. Possibilities of biodiversity classification and evaluation are discussed and the situation in the Czech Republic described, including information on measures proposed for maintaining, preserving and increasing biodiversity.

The political tools on biodiversity protection, including the Helsinki resolution H2 are discussed. On the basis of numbers of forest plants and trees, it is shown how small part of the global biodiversity is represented by the forestry in Central Europe.

Some most important aspects of the current unsatisfactory state of biodiversity of the forest ecosystems in the Czech Republic are described. The data have been collected and elaborated from the Report on forestry of the Czech Republic, Red List of plant communities and protected and threatened species of the Czech flora. Today, unsuitable species composition is considered to be one of the main causes of decline and threat to many plant and animal species and whole forest communities, resulting in a lower level of biodiversity.

Alternatives of measuring and evaluating biodiversity are mentioned. In conclusion, a survey of the most important measures for sustaining and increasing the biodiversity of forest ecosystems, as proposed and implemented in the framework of the ecological transformation of forest management in Czech Republic, is presented. It is for instance suggested changes in tree species composition, improvement of the forest typology, realisation of measures on preservation and reproduction of gene resources of forest tree species, deflection of clear-cut management system, increased share of the natural regeneration, increased care for the specific biotopes, creation and use of the "generals" of the regional system of ecological stability and reduction of damages by hoofed game. The attention is pointed to gene bases, their characteristic, purpose and management as a basic measure of forest management for sustaining and improving the biodiversity of forest ecosystems.

1. Introduction

In 1992 in Helsinki, the Resolution H2 "General guidelines for the conservation and management of the biodiversity of European forests" was adopted practically by all the European states. In general it is an application of the General Agreement on Biodiversity,

P. Bachmann et al. (eds.), Assessment of Biodiversity for Improved Forest Planning, 405–409.
European Forest Institute Proceedings no. 18. © 1998 *Kluwer Academic Publishers. Printed in the Netherlands.*

as agreed in UNCED Conference in Rio de Janeiro in 1992, in conditions of Europeans forest ecosystems. The signatory states are obliged to prepare and realise practical measures to stop the current unsatisfactory trends.

The ideas of the Resolution H2 are taken into account in the "Principles of the state forest policy" agreed by the Czech Government in May 1994. Also the New Forest Act, adopted by the Czech Parliament in November 1995, is based on these international commitments.

2. Current situation in the Czech Republic

Considering the biodiversity in forest ecosystems, it is obvious that the share of forest associated plants and especially woody species in Central Europe is very small. From about 100,000 organism species and 3,000 vascular plant species determined in the Czech Republic there are only 800 considered as forest associated, from which 145 are woody species and 52 trees. However, wood species are the most important from the forest ecosystem point of view, influencing the variability and living conditions of other forest organisms. A decrease in wood species diversity can thus impoverish other groups of organisms.

The current state of biodiversity is mostly unsatisfactory. Nowadays the attention in Czech forestry is given mainly to the problem of reconstruction of the natural tree species composition in forest stands. The reconstructed natural species composition, its current state and changes proposed are presented in Table 1, according to the 'Report on forestry of the Czech Republic' (Anonymous 1995). Today, the unsuitable species composition is considered to be one of the main causes of decline and a threat to many plant and animal species and whole forest communities, resulting in a lower level of biodiversity. The main impacts of unsuitable tree species composition, resulting in decreasing biodiversity, are

1) Dislodging of natural, ecologically very specialised species
2) Expansion of ecologically adaptable, often aggressive and expansive uninvited species (clear-cut, synanthropic species) – e.g. *Calamagrostis epigeios, Rubus fruticosus agg., Sambucus nigra.*

This fact is documented by the inventory of forest communities and vascular plants connected with the forest ecosystems and level of their threat. It is assumed that practically all natural forest communities can be characterised as decreasing due to anthropogenic activity, thirteen associations being directly endangered by instant extinction (Table 2). They are mainly natural communities of fir. In the "Red List" of protected and threatened vascular plant species of Czech Republic there are about 150 forest plant species listed (Míchal 1992). About 40 % of woody species is considered as endangered (Table 3). Except for *Rosa*, critical groups such as *Rubus, Taraxacum, Hieracium* taxa are omitted because of insufficient knowledge.

3. Measurement and assessment of forest biodiversity

Essential biodiversity characteristics and key elements which need to be measured or assessed at forest stand level to monitor forest biodiversity and for the use of improved

TABLE 1. Species composition in %

Tree species	reconstructed natural	current 1994	supposed target composition
Spruce	11,2	54,3	30
Pine	3,4	17,5	19,6
Larch	0	3,4	5
Fir	19,8	0,9	4
Douglas fir	0	0,2	0,5
Dwarf pine	0,3	0,3	0,3
Other conifers	0	0,6	0,6
Oak	19,3	6,1	8
Beech	40,1	5,6	18
Hornbeam	1,5	1,2	1
Maple	0,7	0,7	4
Ash	0,6	1,0	1
Lime	0,8	0,9	4
Birch	0,8	3,0	2
Alder	0,6	1,3	1
Elm	0,3	0	0
Poplar	0,3	0,5	0,4
Willow	0,1	0,1	0,1
Robinia	0	0,5	0,2
Other broadleaves	0,2	0,3	0,3
Conifers	34,7	77,2	60
Broadleaves	65,3	21,3	40

TABLE 2. Forest plant communities in the Czech Republic and how endangered they are (after Moravec 1995)

Forest communities (Associations)	rare	sufficiently abundant	total	%
Assoc. immediately endangered by human activities and threatened by extiction	5	8	13	17,8
Assoc. with decline caused by human activities	31	24	55	75,3
Without endangerment	2	3	5	6,9
Total	38	35	73	100,0

forest management should not be difficult to define (Jeffers 1995). However, practical investigation is usually much more complicated. Simple archiving of lists of species is not sufficient.

The following variables are regularly assessed and measured on the chosen monitoring plots e.g. in the framework of the ICP Forests Programme or other projects in the Czech Republic:

- topography and surface structure of the site, slope, exposition,
- geological basis, characterisation of the soils, its microbiology and chemical analysis

TABLE 3. Forest plants in the Czech Republic and how endangered they are – after Red List (Procházka et al. 1983)

Forest species	woody species	herbs	total	%
Total number of vascular plants species associated in forest ecosystems	145	cca 655	800	100
Threatened forest species				
Extinct or missing	3	3	6	0,8
Critically threatened	13	12	25	3,1
Strongly threatened	12	14	26	3,3
Threatened	13	29	42	5,3
Rare or dispersed occurrence	17	28	45	5,6
Total threatened forest species	58	86	144	18,0

- measurement of the diameters, heights and canopy characteristics of all forest trees above a certain diameter, stand structure, layering
- description of the ground vegetation of all layers including rare species, assessment of degree of coverage in tree, shrub, herb and moss layer (method of phytocenological relevés)
- assessment of the presence, numbers and quality of tree regeneration.

These long term data show the dynamic changes in time and can then be applied for improved forest planning. As further characteristics of the biodiversity are considered, also the presence and location of any surface water, depth and nature of the humus layers, record of presence of visible fungi, birds, game and other animals, assessment of the amounts and nature of any dead or decaying wood in the stumps, standing or fallen trees and logs and records of any human influence in the past history.

4. Measures for biodiversity preservation and improvement

Concrete measures on the protection, conservation and increase of biodiversity have been proposed and are gradually being implemented in frame of the ecological transformation of the Czech forest management. Some of them have been carried out for more than for twenty years. The most important principals of the improvement of forest ecosystem biodiversity are the following:

- changes in the species composition in the frame of the newly prepared economical units, deflection from the conifer monoculture
- improvement of silvicultural measures on the basis of a typological survey
- realisation of measures on preservation and reproduction of gene resources of forest tree species, including rare, commercially less important species
- deflection of clear-cut management system
- increased care for the specific biotopes on forest land (forest edges, forest-steppes, detritus slopes, peat-bogs, avalanche slopes etc.)
- creation and use of the "generals" of the regional system of ecological stability on the forest land

- increased share of the natural regeneration
- reduction of damages caused by game by radical lowering of the state of hoofed game.

For the sake of forest management promotion, the state offers funds which may be provided i.a. to increase the share of the stabilising and soil-improving tree species — mainly broadleaves ones. There are measures for gene-pool conservation in general. In situ: gene bases, approved stands for seed collection and procurement, protected areas of various levels (national parks, landscape protected areas, nature reserves. Ex situ: seed orchards, clonal archives, seed banks trial-testing plots and reproductive plantations – so called seed stands.

Gene bases are considered to be the basic measure for the preservation of gene sources and genetic variability of forest tree populations "in situ". They are defined as a complex of, hopefully, autochtonous forest stands and are declared in forestry management plans as special purpose forests. Their area varies from 100 to 2,000 ha. In comparison with current silvicultural measures they have special management regimes (e.g. longer rotation and regeneration period, reproduction from autochtonous material, preferably natural). The network of these forest areas has been established in the 80s. Until present, 163 gene bases (about 82,000 ha) have been decreed. The next 108 ones (45,000 ha) are proposed.

References

Anonymous 1995. Zpráva o stavu lesního hospodárství Ceské republiky (Report on forestry of the Czech Republic). Ministry of Agriculture of the Czech Republic. Prague.

Jeffers, J.N.R. 1995. Measurement and characterisation of biodiversity in forest ecosystems – New methods and models. In: Bachmann, P., Kuusela, K. and Uuttera, J. (eds.). Assessment of biodiversity for improved forest management, Proceedings of the International Workshop. 12-17 June 1995. Koli. Finland. EFI Proceedings No. 6. Pp. 59-67.

Míchal, I. 1992. Obnova ekologické stability lesu (Recovery of forest ecological stability). Academia. Prague.

Moravec, J. (ed.) 1995. Rostlinná spolecenstva Ceské republiky a jejich ohrození (Red list of plant communities of the Czech Republic and their endangerment). Severoceskou prírodou. 1: 1-206.

Procházka, F., Cerovský, J., and Holub, J. (1983) Chránené a ohrozené druhy kveteny CSR (Protected and threatened species of the flora of the Czech Republic). SZN. Prague.

Appendices

APPENDIX 1: LIST OF PARTICIPANTS

Dr. Per Angelstam
Swedish Univ. of Agricultural Sciences
Dept of Wildlife Ecology
Grimsö Wildlife Research Center
730 91 Riddarhyttan
Sweden
Tel. +46 581 660433
Fax +46 581 92019
Email per.angelstam@vilt.slu.se

Prof. Dr. Peter Bachmann
Swiss Federal Institute of Technology
ETH Zentrum
8092 Zürich
Switzerland
Tel. +41 1 632 1111
Fax +41 1 632 1127

Dr. Lucio Barabesi
University of Siena
Dip. di Metodi Quantitativi
P.ta S. Francesco 17
53100 Siena
Italy
Tel. +39 557 29 86 30

Mr. Davide Bettelini
Servicio Forestale Ticino
Ufficio Pianificazione Forestale
Via Guasta 6
6504 Bellinzona
Switzerland
Tel. +41 91 804 36 53
Fax +41 91 804 44 38

Dr. Markus Bollinger
Swiss Forest Agency
Federal office of Environment, Forests and Landscape
BUWAL/Eidg. Forstdirektion
3003 Bern
Switzerland
Tel. +41 31 3247787
Fax +41 31 3247789

Dr. Vaclav Burianek
Forestry and Game Management Research Institute
Biology and Forest Trees Breeding
Jiloviata-Strnady
15604 Praha 5 Zbraslav
Czech Republic
Tel. +42 02 591612
Fax +42 02 591413

Prof. Maria Giulia Cantiani
Universita di Trento
Mesiano di Povo
38050 Trento
Italy
Tel. +39 461 88 26 78
Fax +39 461 88 26 72

Mr. Moses Chakanga
National Forest Inventory of Namibia
P.O. Box 23930
7000 Windhoek
Namibia
Tel. +264 61 221478
Fax +264 61 222830

Dr. N. Mark Collins
World Conservation Monitoring Centre
219, Huntingdon Road
Cambridge CB3 0DL
United Kingdom
Tel. +44 1223 277 314
Fax +44 1223 277 316
Email collins@wcmc.co.uk

Dr. Francisco Dallmeier
Smithsonian Institution
SI/MAB
S. Dillon Ripley Center, Suite 3123, MRC 705
Washington D.C 20560
USA
Tel. +1 202 357 4248
Fax +1 202 786 2557
Email SIWP01.IC.OHM@ic.si.edu

Mr. Michael Dietrich
WSL
Zürcherstrasse 111
8903 Birmensdorf
Switzerland
Tel. +41 1 739 21 11
Fax +41 1 739 22 15
Email dietrich@wsl.ch

Mr. C.Paul Djogbenou
DFRN-Benin
Projet plantations de bois de feu dans le sud-benin
c/o DFS Wittelbacherstrasse 11
85622 Feldkirchen
Germany
Tel. +49 89 9038844
Fax +49 89 9038870

Mr. Glen B. Dunsworth
Macmillan Bloedel Ltd.
65 Front Street
Nanaimo, B.C. V9R 5H9
Canada
Tel. +1 604 7553425
Fax +1 604 7553550

Ms. Margorzata Falencka-Jablonska
Forest Research Institute
Department of Ecology
Bitwy Warszawskiej 1920 No 3
00973 Warsaw
Poland
Tel. +48 22 462016
Fax +48 22 462078

Prof. Lorenzo Fattorini
University of Siena
Dip.di Metodi Quantitativi
P.ta S. Francesco 17
53100 Siena
Italy
Tel. +39 557 29 86 30

Dr. Richard Ferris-Kaan
Forestry Commission Research Division
Woodland Ecology Branch
Alice Holt Lodge, Wrecclesham, Farnhams
Surrey GU10 4LH
United Kingdom
Tel. +44 1420 22255
Fax +44 1420 23653

Dr. Silvia Fineschi
WSL ENGREF/MAI
Zürcherstrasse 111
8903 Birmensdorf
Switzerland
Tel. +41 1 739 21 11
Fax +41 1 739 22 15
Email Fineschi@wsl.ch

Dr. Alain Franc
19 Avenue de Maine
75732 Paris Cedex
France
Tel. +33 1 45 49 89 05
Fax +33 1 45 49 88 27
Email franc@engref.fr

Prof. George Z. Gertner
University of Illinois
Department of Forestry
W503 Turner Hall,
1102 South Goodwin Drive
Urbana, IL 61801
USA
Tel. +1 217 333 9346
Fax +1 217 244 3219
Email gertner@uxh.cso.uiuc.edu

Dr. Andrew N. Gillison
CIFOR
P.O. Box 6596
JKPWB Jakarta 10065
Indonesia
Tel. +62 251 34 36 52
Fax +62 251 32 64 33
Email a.gillison@cgnet.com

Ms. Elena Gordillo Alonso
Universidad Politecnica de Madrid
Escuela Tecnica Superior de Ingenieros de Montes
Opto de Economia y Gestion
28040 Madrid
Spain
Tel. +34 1 336 71 31
Fax +34 1 543 95 57

Dr. Jeffrey H. Gove
USDA Forest Service
Northeastern Forest Experiment Station
P.O. Box 640
Durham NH 03824

USA
Tel. +1 603 868 76 67
Fax +1 603 868 76 04
Email jhgove@christa.unh.edu

Mr. Maurice Heusèrr
Wageningen Agricultural University
P.O. Box 342
Netherlands
Tel. +31 317 484 906
Fax +31 317 483 542
Email maurice.heuserr@bgs.bosb.wau.nl

Mr. Simon Hodge
Forestry Commission Research Division
Northern Research Station, Woodland
Ecology Branch
6700 AH Wageningen
Roslin
Midlothian EH25 9SY
United Kingdom
Tel. +44 131 445 2176
Fax +44 131 445 5124

Dr. Howard Hoganson
University of Minnesota
115 Green Hall, 1530 N. Cleveland Av.
St.Paul MN 55108-1027
USA
Tel. +1 218 327 4490
Fax +1 218 327 4126
Email hogan001@maroon.tc.umn.edu

Mr. Markus Holopainen
University of Helsinki
Dept. of Forest Resource Management
PL 24
00014 Helsingin Yliopisto
Finland
Tel. +358 9 1917675
Fax +358 9 191 7755
Email markus.holopainen@helsinki.fi

Dr. Jean-Paul Jeanrenaud
WWF
Avenue du Mont Blanc
1196 Gland
Switzerland
Tel. +41 22 3649528
Fax +41 22 3645829
Email claude.martin@wwf-int.ch

Mr. Martin Jenkins
World Conservation Monitoring Centre
219 Huntingdon Road
Cambridge CB3 0DL
United Kingdom
Tel. +44 1223 27 73 14
Fax +44 1223 27 71 36

Ms. Irina Jusupova
Swiss-Kyrguz Support Programme
Les-IC
Glavpochtampt p/o 2011
72000 Bishkek
Kyrgyzstan
Tel. +7 3312 27 96 34
Fax +7 3312 27 96 34

Ms. Michèle Kaennel
WSL
Zürcherstrasse 111
8903 Birmensdorf
Switzerland
Tel. +41 1 739 25 96
Fax +41 1 739 22 15
Email kaennel@wsl.ch

Dr. Markus Keller
WSL
Zürcherstr. 111
8903 Birmensdorf
Switzerland
Tel. +41 1 739 2218
Fax +41 1 739 2215
Email: keller@wsl.ch

Prof. Dr. Michael Köhl
Dresden University of Technology
Chair of Forest Biometrics and
Computer Sciences
Wilsdruffe Str. 18
01737 Thanrandt
Germany
Tel. +49 35203 381613
Fax +49 35203 381628
Email koehl@forst.tu-dresden.de

Ms. Marja Kolström
University of Joensuu
Faculty of Forestry
P.O.Box 111
80101 Joensuu
Finland
Tel. +358 13 251 4404
Fax +358 13 251 4444
Email Marja.Kolstrom@forest.joensuu.fi

Mr Kari Korhonen
University of Joensuu
P.O. 111
80101 Joensuu
Finland
Tel. +358 13 251 3631
Fax +358 13 251 4444
Email korhonen@forest.joensuu.fi

Prof. Kullervo Kuusela
European Forest Institute
Torikatu 34
80100 Joensuu
Finland
Tel. +358 9 485 927
Fax +358 9 625 308

Mr. Tomas Lämås
Swedish University of Agricultural Sciences
Dept. of Forest Resource Management and Geomatics
901 83 Umeå
Sweden
Tel. +46 90 165800
Fax +46 90 778116
Email tomas.lamas@resgeom.slu.se

Prof. Majid F. Makhdoum
University of Tehran
Faculty of Natural Resources
Karaj
Iran
Tel. +98 261 46 28 15
Fax +98 21 65 81 11

Dr. Marco Marchetti
Ministry of Environment
Rome
Italy
Tel. +39 6 782 3504
Fax +39 6 780 3377
Email rdm@cdc.it

Mr. Jean-David Moore
Laval University
6 Jardins Mérici, App.3
G1S AN7 Québec
Canada
Tel. +1 41 852 75474

Dr. Ueli Müller
Forestry Support Programme Kyrgyzistan
Orobekova 116, Appt.2
72000 Bishkek
Kyrgyzstan

Mr. Mike Müller
National Forest Inventory of Namibia
P.O. Box 23930
7000 Windhoek
Namibia
Tel. +264 61 221478
Fax +264 61 222830

Dr. Turatbek Musuzaliev
State Forest Agency
Isanov 94-7
72000 Bishkek
Kyrgyzstan
Tel. +7 3312 21 36 79
Fax +7 3312 26 10 78

Mr. Niall McCormick
FIRS/EMAP/SAI/JRC
CCR-Euratom
TP 442
21020 Ispra
Italy
Tel. +39 332 78 50 09
Fax +39 332 78 94 69

Mr. Steven Northway
Macmillan Bloedel Ltd.
Resource Analysis
65 Front Street
Nanaimo, B.C. V9R 5H9
Canada
Tel. +1 604 7553484
Fax +1 604 7553550

Mr. Stig Ohlsson
Swedish Environmental Protection Agency
Blekholmsterrassen 36
10648 Stockholm
Sweden
Tel. +46 8 698 13 35
Fax+46 8 698 16 62
Email sto@environ.se

Mr. Andreas Ottitsch
University of Natural Resources
Institute of Forest Socio-Economics
Gregor-Mendel Str. 33
1180 Wien
Austria
Tel. +43 1 47 654 44 07
Fax +43 1 47 654 44 04
Email ottitsch@hp02.boku.ac.at

Dr. Risto Päivinen
European Forest Institute
Torikatu 34
80100 Joensuu
Finland
Tel. +358 13 252 020
Fax +358 13 124 393
Email risto.paivinen@efi.joensuu.fi

Dr. Piotr Paschalis
Faculty of Forestry
Wawelska 52/54
0922 Warsaw
Poland
Tel. +48 22 490 774
Fax +48 22 420 192
Email paschalis@delta.sggw.waw.pl

Prof. Dr. G.P Patil
Pennsylvania State University
Center for Statistical Ecology and
Environmental Statistics
University Park PA 16802
USA
Tel. +1 814-865 59 42
Fax +1 814-863 71 14
Email gpp@psuvm.psu.edu

Mr. Andrew Peace
Forestry Commission Research Division
Statistics & Computing Branch
Alice Holt Lodge, Wrecclesham, Farnham
Surrey GU10 4LH
United Kingdom
Tel. +44 1420 22255
Fax +44 1420 23653

Dr. Ajith Perera
Ontario Forest Research Institute
Landscape ecology
1235 Queen Street East
Sault Ste Marie, Ontario P6A 5N5
Canada
Tel. +1 705 9462981
Fax +1 705 9462030
Email Pereraa@gov.on.ca

Mr. Carlo Ricotta
Universita di Roma
Dipartemento di Biologia Vegetale
Via Tripoli 2
00199 Roma
Italy
Tel. +39 6 4991 37 23
Fax +39 6 445 75 40
Email avena@caspur.it

Ms. Leena Roihuvuo
European Forest Institute
Torikatu 34
80100 Joensuu
Finland
Tel. +358 13 252 020
Fax +358 13 124 393
Email leena@efi.joensuu.fi

Prof. Dietmar Rose
University of Minnesota
College of Natural Resources
1530 N. Cleveland Avenue
St. Paul, MN 55108
USA
Tel. +1 612 6249711
Fax +1 612 6255212

Dr. Karl Schieler
Forest Research Institute
Seckendorf Gudent Weg 8
1131 Wien
Austria
Tel. +43 1 878 380
Fax +43 1 877 5907

Dr. A. Schuler
Swiss Federal Institute of Technology
Rämistr. 101
8092 Zürich
Switzerland
Tel. +41 1 632 3196
Fax +41 1 632 1127
Email schuler@waho.ethz.ch

Mr. Charles G. Shaw
USDA Forest Service
Pacific Northwest Research Station
2770 Sherwood Lane, Suite 2 A
99801-8545 Juneau AK
USA
Tel. +1 303 4981251

Ms. Eugenya Sherbinina
National Academy of Sciences
Institute of Forest and Walnut farming
Karagachovaya Roska 15, Institut Lesa
720015 Bishkek
Kyrgyszia
Tel. +7 3312 27 90 62
Fax +7 3312 24 36 07

Mr. Ulf Söderberg
SLU
Dept. of Forest Res. Management and Geomatics
901 83 Umeå
Sweden
Tel. +46 90 16 58 20
Fax +46 90 77 81 16
Email Ulf.Soderberg@resgeom.slu.se

Dr. Birger Solberg
ACCN/NISK
P.O.Box 5044
1432 Ås
Norway
Tel. + 47 64 948 890
Fax + 47 64 948 928
Email Birger.Solberg@nisk.no

Dr Ian F. Spellerberg
Lincoln University
Dept. of Resource Management
P.O. Box 56
New Zealand
Tel. +64 33252811
Fax +64 33253841
Email Spelleri@lincoln.ac.nz

Mrs Sabine Stab
Beni Biological Station and Biosphere Reserve
Heimstättenstrasse 5
32758 Detmold
Germany
Tel. +49 523122052

Dr. Göran Ståhl
Swedish University of Agricultural Sciences
Dept of Forest Resource management and Geomatics
901 83 Umeå
Sweden
Tel. +46 90 165837
Fax +46 90 778116

Mr. Tibor Standovar
L. Eötvös University
Dept. Plant Taxonomy and Ecology
Ludovika ter 2
Budapest H-1083
Hungary
Tel. +36 1 210 1084
Fax +36 1 1338 764
Email standy@cudens.elte.hu

Dr. Edward E. Starkey
U.S. Department of Interior
National Biological Service
3200 SW Jefferson Way
91331 Corvallis, Oregon
USA
Tel. +1 541 131 2056
Fax +1 541 131 3049
Email starkeye@ccmail.orst.edu

Mr. Franz Starlinger
Forest Research Institute
Seckendorff-Gudent-weg 8
1130 Vienna
Austria
Tel. +43 1 87838317
Fax +43 1 8775907

Mr. Kanybek Sultankulov
Forest Inventory Unit
Kolmo uch N17
72000 Bishkek
Kyrgyzstan
Tel. +7 3312 21 36 79

Mr. Stefan Tamas
Faculty of Forestry and Forest Exploitation
Transylvania University
Maior Cranta 45
Brasov
Romania
Tel. +40 68 11 86 30
Fax +40 68 11 86 30

Ms. Tiina Tonteri
Finnish Forest Research Institute
P.O. Box 18
01301 Vantaa
Finland
Tel. +358 9 85 70 55 56
Fax +358 9 85 70 55 69
Email Tiina.Tonteri@Metla.fi

Prof. Vittorio Tosi
ISAFA
Forest and Game Management Research Institute
Piazza G. Nicolini 6
38050 Vlazzario di Trento
Italy
Tel. +39 461 924248
Fax +39 461 914551
Email isafa@tgs.it

Mr. Berthold Traub
WSL
Zürcherstr. 111
8903 Birmensdorf
Switzerland
+41 1 739 2210
+41 1 739 2215
Email traub@wsl.ch

Ms. Eva Ujvári Jármay
Forest Research Institute
Forest Tree Breeding
Hegyalja ut 10
3232 Mátrafüred
Hungary
Tel. +36 37 320177
Fax +36 37 320406

Mr. Janne Uuttera
European Forest Institute
Torikatu 34
80100 Joensuu
Finland
Tel. +358 13 2520219
Fax +358 13 124393

Mr. Michel Vallance
ONF
Département des Recherches Techniques
Boulevard de Constance
77300 Fontainebleau
France
Tel. +33 1 60 74 92 10
Fax +33 1 64 22 49 73

Prof. Dr. Jerome K. Vanclay
CIFOR
P.O. Box 6596 JKPWB
Jakarta 10065
Indonesia
Tel. +62 251 34 3652
Fax +62 251 32 6433
Email j.vanclayacgnet.com

Mr. Kris Vandekerkhove
Institute for forestry and game management
Gaverstraat 4
9500 Geraardsbergen
Belgium
Tel. +32 2 6570386
Fax +32 2 6579682
Email Vandekerkhove@ibw.be

Dr. Bronislav Venglovsky
National Academy of Sciences
Institute of Forest and Walnut farming
Karagachovaya Roska 15, Institut Lesa
720015 Bishkek
Kyrgyzstan
Tel. +7 3312 27 90 62

Dr. Paolo Viskanic
FAO
Uganda Forest Department
P.O. Box 40184
Kampala
Uganda
Tel. +256 41 22 16 72
Fax +256 41 22 16 72
Email viskanic@imul.com

Mr. Johan Wester
Skogsvårdsstyrelsen
Västerbottens
P.O. Box 284
90183 Umeå
Sweden
Email Johan.Wester.SVSAC@SVO.SE

Ms. Emilia Wisniewska
Forest Research Institute
Dept. of Forest Fires
ul. Bitwy Warsawskiej 3
00973 Warsaw
Poland
Tel. +48 22 462 018
Fax +48 22 462 018

Dr. Ronghua Ye
Swiss Federal Institute of Technology
Forest Inventory and Planning
ETH-Zentrum
8092 Zurich
Switzerland
Tel. +41 1 632 31 94
Fax +41 1 632 11 27
Email ye@pfe.waho.ethz.ch

Prof. Mahmoud Zobeiry
University of Karaj
Faculty of Natural Resources
Karaj
Iran
Tel. +98 261 42 55 3
Fax +98 21 800 79 88

APPENDIX 2: LIST OF REVIEWERS

Each paper published in this volume has been reviewed independently by at least two reviewers. We thankfully acknowledge the reviewers for their valuable comments and suggestions which helped to ensure the scientific quality of this publication.

P. Angelstam	Sweden
L. Barabesi	Italy
A. Bitter	Germany
P. Brassel	Switzerland
H.C. Bodmer	Switzerland
F. Dallmeier	USA
G. Dunsworth	Canada
C. Elliott	Switzerland
B. Eriksson	Sweden
L. Fattorini	Italy
S. Folving	Italy
A. Franc	France
G. Z. Gertner	USA
J. H. Gove	USA
G. Gertner	USA
A.N. Gillison	Indonesia
T. Gregoire	USA
M. Heusèrr	The Netherlands
M. Holopainen	Finland
S. Hodge	UK
H. Hoganson	USA
E. Huessendorfer	Germany
I. Hunter	Finland
J. Innes	Switzerland
S. Jaakola	Switzerland
A. Kangas	Finland
J. Kangas	Finland
M. Kaennel	Switzerland
M. Köhl	Germany
K. Korhonen	Finland
I. Kupka	Czech Republic
K. Kuusela	Finland
E. Landis	USA
T. Lämas	Sweden
J. Lappi	Finland
G. Lund	USA
K. Kuusela	Finland
D. Mandallaz	Switzerland
M. Marchetti	Italy
P. Niemelä	Finland
S. Northway	Canada
T. Nuutinen	Finland
G. Oesten	Germany
S. Ohlsson	Sweden
A. Ottitsch	Austria
R. Päivinen	Finland
J. Parviainen	Germany
P. Paschalis	Poland
G. P. Patil	USA
S. Poso	Finland
B. Ranneby	Sweden
C. Ricotta	Italy
D. Rose	USA
H. Saarenmaa	Denmark
O. Sallnäs	Sweden
A. Schuler	Switzerland
C.T. Scott	USA
K. Seeland	Switzerland
M. Siitonen	Finland
A. Singh	New Zealand
J. P. Skovsgaard	Denmark
U. Söderberg	Sweden
B. Solberg	Norway
I. Spellerberg	USA
H. Spiecker	Germany
E. Starkey	USA
T. Standovar	Hungary
W. Suter	Switzerland
B. Traub	Switzerland
S. Tamas	Romania
T. Tonteri	Finland
J. Uuttera	Finland
J. Vanclay	Indonesia
P. Wallenius	Finland
A. Zingg	Switzerland
M. Zobeiry	Iran

FORESTRY SCIENCES

37. J.J.A. Janssen: *Mechanical Properties of Bamboo*. 1991 ISBN 0-7923-1260-0
38. J.M. Bonga and P. Von Aderkas: *In Vitro Culture of Trees*. 1992 ISBN 0-7923-1540-5
39. L. Fins, S.T. Friedman and J.V. Brotschol (eds.): *Handbook of Quantitative Forest Genetics*. 1992 ISBN 0-7923-1568-5
40. M.J. Kelty, B.C. Larson and C.D. Oliver (eds.): *The Ecology and Silviculture of Mixed-Species Forests*. A Festschrift for David M. Smith. 1992 ISBN 0-7923-1643-6
41. M.R. Ahuja (ed.): *Micropropagation of Woody Plants*. 1992 ISBN 0-7923-1807-2
42. W.T. Adams, S.H. Strauss, D.L. Copes and A.R. Griffin (eds.): *Population Genetics of Forest Trees*. Proceedings of an International Symposium (Corvallis, Oregon, USA, 1990). 1992 ISBN 0-7923-1857-9
43. R.T. Prinsley (ed.): *The Role of Trees in Sustainable Agriculture*. 1993 ISBN 0-7923-2030-1
44. S.M. Jain, P.K. Gupta and R.J. Newton (eds.): *Somatic Embryogenesis in Woody Plants*, Vol. 3: Gymnosperms. 1995 ISBN 0-7923-2938-4
45. S.M. Jain, P.K. Gupta and R.J. Newton (eds.): *Somatic Embryogenesis in Woody Plants*, Vol. 1: History, Molecular and Biochemical Aspects, and Applications. 1995 ISBN 0-7923-3035-8
46. S.M. Jain, P.K Gupta and R.J. Newton (eds.): *Somatic Embryogenesis in Woody Plants*, Vol. 2: Angiosperms. 1995 ISBN 0-7923-3070-6
    Set ISBN (Volumes 44-46) 0-7923-2939-2
47. F.L. Sinclair (ed.): *Agroforestry: Science, Policy and Practice*. Selected Papers from the Agroforestry Sessions of the IUFRO 20th World Congress (Tampere, Finland, 6–12 August 1995). 1995 ISBN 0-7923-3696-8
48. J.H. Goldammer and V.V. Furyaev (eds.): *Fire in Ecosystems of Boreal Eurasia*. 1996 ISBN 0-7923-4137-6
49. M.R. Ahuja, W. Boerjan and D.B. Neale (eds.): *Somatic Cell Genetics and Molecular Genetics of Trees*. 1996 ISBN 0-7923-4179-1
50. H.L. Gholz, K. Nakane and H. Shimoda (eds.): *The Use of Remote Sensing in the Modeling of Forest Productivity*. 1996 ISBN 0-7923-4278-X
51. P. Bachmann, M. Köhl and R. Päivinen (eds.): *Assessment of Biodiversity for Improved Forest Planning*. Proceedings of the Conference on Assessment of Biodiversity for Improved Planning (Monte Verità, Switzerland, 1996). 1998 ISBN 0-7923-4872-9

KLUWER ACADEMIC PUBLISHERS – DORDRECHT / BOSTON / LONDON